Neuroendocrine Regulation and Altered Behaviour

EDITED BY PAVEL D. HRDINA AND RADHEY L. SINGHAL

PLENUM PRESS — NEW YORK AND LONDON

CONTENTS

Preface

Part One: Hormones and Regulation of Central Nervous System Function

1. Neuropeptide Influences on the Central Nervous System:
   A Psychobiological Perspective *C.A. Sandman, A.J. Kastin
   and A.V. Schally*   3
2. Behavioural Effects of Pituitary Hormones
   V.M. Wiegant and D. de Wied   29
3. Dopamine Agonist and Antagonist Drugs and Hypothalamic-
   Pituitary Dysfunction *E.E. Müller, F. Camanni, A.R. Genazzani,
   D. Cocchi, F. Casanueva, F. Massara, V. Locatelli,
   A. Martinez-Campos and P. Mantegazza*   51
4. Some Aspects of the Neuroendocrine Regulation of Mammalian
   Sexual Behaviour *K. Larsson and C. Beyer*   95

**Part Two: Brain Endorphins, Psychotropic Drugs and
Neuroendocrine Regulation**

5. Opiates and Neuroendocrine Regulation
   A. Dupont, N. Barden, F. Labrie, L. Ferland and L. Cusan   121
6. Behavioural Modulation by Systemic Administration of
   Enkephalins and Endorphins *R.D. Olson, A.J. Kastin,
   G.A. Olson and D.H. Coy*   141
7. Effect of Cannabinoids and Narcotics on Gonadal Functions
   A. Jakubovic, E.G. McGeer and P.L. McGeer   151
8. Effect of Neuroleptics on Pituitary Function in Man
   S. Lal and R.B. Rastogi   169

**Part Three: Neuroendocrine Regulation and Pathogenesis
of Mental Illness**

9. Neuroendocrine Studies on the Pathogenesis of Depression
   N. Hatotani, J. Nomura and I. Kitayama   187
10. Thyroid Hormone in the Regulation of Neurotransmitter
    Function and Behaviour *R.L. Singhal and R.B. Rastogi*   205
11. Behavioural Studies with Lithium in Rats: Implications for
    Animal Models of Mania and Depression *P.E. Harrison-Read*   223
12. Experimental Models of Mental Illness: Separation-induced
    Depression in Primates *P.D. Hrdina and M.E. Henry*   263
13. The Role of Neuroendocrine Function in Anorexia Nervosa
    P.E. Garfinkel, G.M. Brown and P.L. Darby   277

Contents

Part Four: Biological Markers of Altered Mental Function

14. Alteration in Brain Receptors in Affective Disorders
    M.S. Briley   297
15. Psychophysiological Correlates of Mental Disorders
    Y.D. Lapierre and V.J. Knott   315
16. Hormonal Markers in Schizophrenia and Depression
    G.M. Brown, J.M. Cleghorn, P.G. Ettigi and P. Brown   339
17. Contemporary Neuroendocrine Research Strategies and
    Methodologies in Psychiatry *R.E. Poland and R.T. Rubin*   363
18. Relevance of Plasma and Urine Amine Metabolites as
    Diagnostic and Pharmacodynamic Indicators in Psychiatric
    Disorders *J.J. Warsh, H.C. Stancer and P.P. Li*   381

Notes on Contributors   401
Index   405

PREFACE

In recent years, it has become clearly recognized that many behavioural disturbances and psychiatric illnesses are intimately associated with alterations in neuroendocrine function. This volume is designed to provide a thorough, up-to-date review of our current knowledge of the neuroendocrine correlates of altered behaviour in man and experimental animals. Particular emphasis has been focused on the mechanisms which may underlie the coupling of mental functions with endocrine changes and the possible common links in the central regulation of both endocrine and psychic activities. One of the main objectives of this book is to consider both the experimental and clinical approaches in studying the interrelationship between neuroendocrine regulation and altered behaviour, and to assess its importance in the pathogenesis, diagnosis and treatment of mental disorders.

The book has been organized into four major sections which focus on: (i) the role of pituitary hormones in the regulation of CNS function; (ii) the relationship between brain endorphin systems, psychotropic drugs and neuroendocrine regulation; (iii) the importance of neuroendocrine regulation in the pathogenesis of mental disorders; and (iv) biological markers of altered mental function. Each chapter is organized for ease of comprehension as well as for rapid retrieval of progress and essential information concerning the neuroendocrine basis of altered behaviour and psychiatric illnesses. The contributing authors were selected because of their widely recognized expertise in the field. In addition, an attempt was made to cover some aspects of the subject which, in our opinion, have not received adequate exposure in recent years. The editors have imposed as few restrictions as possible, thus enabling the authors to encompass a wide range of material and to express their ideas freely. The responsibility for precision and accuracy of data and references, allocation of priorities, expression of evaluation, etc., therefore lies with the individual authors.

We hope that this book will not only reflect the significant growth of the body of knowledge in the field and serve as an important reference source for biologists, psychopharmacologists, psychiatrists, endocrinologists, behavioural and neuroscientists, psychologists, biochemists and all those interested in the biological basis of altered behaviour, but will also stimulate more research and suggest new directions for future investigations in the field.

The editors would like to express their gratitude to Mr Tim Hardwick of Croom Helm Ltd Publishers for his valuable advice in the initial phases of organization of this volume, and to Mrs Diane McNeil for her devoted assistance in administrative and editorial work.

Pavel D. Hrdina and Radhey L. Singhal
Ottawa, Canada

PART ONE

HORMONES AND REGULATION OF CENTRAL NERVOUS SYSTEM FUNCTION

1 NEUROPEPTIDE INFLUENCES ON THE CENTRAL NERVOUS SYSTEM: A PSYCHOBIOLOGICAL PERSPECTIVE

Curt A. Sandman, Abba J. Kastin and Andrew V. Schally

TABLE OF CONTENTS

I	CLASSICAL VIEW	5
	A. Structure	6
II	ALTERNATIVE VIEW	6
III	ANIMAL STUDIES	8
	A. MSH, ACTH and Their Fragments	8
	B. Developmental Studies	9
	C. Endorphins	11
	D. Developmental Findings	12
IV	STUDIES IN NORMAL HUMAN VOLUNTEERS	15
	A. Physiological Effects	16
	B. Behavioural Effects	17
	C. Sex Differences	18
	D. Personality Influences	19
V	STUDIES OF PATIENT GROUPS	19
	A. MSH/ACTH Fragments	19
	B. Endorphins	21
VI	SUMMARY AND CONCLUSION	22
	REFERENCES	25

I. CLASSICAL VIEW

The pituitary gland is a small organ (weighing about 60 mg in a human adult male and slightly more in an adult female) which is located at the base of the brain immediately below the hypothalamus. It is divided into three basic regions: the anterior lobe, the intermediate lobe and the posterior lobe. Each region secretes a different group of polypeptide hormones, which has specific physiological significance for the biological integrity of essential organismic functions. Many of the effects occur at a distance from the pituitary and appear to energize other organs of the endocrine system, hence the name tropic hormones. This regulatory capacity of the pituitary hormones led early investigators to refer to the pituitary as the master gland (Lewin, 1972).

However, the pituitary is far from being an autonomous organ. Regulation of the anterior pituitary is actually controlled by neurohumoral products of the hypothalamus, *releasing factors,* which are polypeptides synthesized within the hypothalamus (Schally *et al.,* 1973). For example, although there is some controversy over its identity and production, corticotropic-releasing factor (CRF) is widely accepted as the factor which is secreted by the hypothalamus to stimulate the release of ACTH (Turner and Bagnera, 1971). Relatedly, MSH is controlled, in part, by an inhibitory releasing factor, MSH-inhibiting factor (MIF) (Kastin and Ross, 1964; Kastin and Schally, 1967), and perhaps in part by MSH releasing factor (MSH-RF) (Siegal and Eisenman, 1972).

Hypothalamic control of the pituitary appears to be modulated by the limbic system. The limbic system includes portions of the cerebral hemispheres and the diencephalon, which have many interconnections with the hypothalamus (Grossman, 1967). That these interconnections may modulate the pituitary's release of ACTH is suggested by the fact that lesions of the septum (Usher and Lamble, 1969) and hippocampus (Antelman and Brown, 1972) increased basal levels of ACTH, whereas lesions of the amygdala (Bush *et al.,* 1973) decreased levels of ACTH.

This hierarchical regulatory system also interacts with the pineal gland. Release of melatonin by the pineal increased MSH content of the pituitary, whereas suppression of melatonin resulted in decreased pituitary content of MSH (Kastin and Schally, 1967). In addition to the central regulation of poly-peptide-hormone release, peripheral organs participate in a complex feedback loop to refine further the influence of these molecules (Table 1.1). Thus, control of pituitary functions is achieved at several levels in the brain and the body.

Table 1.1: Classically Defined Influence of Pituitary Peptide Hormones on Organisms and Functions of the Body

Anterior lobe	Target action
Adrenocorticotropin	Adrenal cortex
Gonadotropins	Ovaries, testes, corpus luteum
Lipotropin	Fat
Prolactin	Mammary, corpus luteum
Somatotropin	Metabolic processes
Thyrotropin	Thyroid
Intermediate lobe	
Melanocyte-stimulating hormone	Pigment cell
Posterior lobe	
Vasopressin	Kidney, arterial vessels
Oxytocin	Uterus, kidney

A. Structure

The pituitary hormones are classified as polypeptides. A polypeptide is a series of amino acids linked by peptide bonds resulting from the elimination of water between an amino group in one molecule and a carboxyl group in an adjacent molecule. The exact sequence of amino acids in a group is very important, for it determines the properties of the molecule. Indeed, recent evidence suggests that the fit of a molecule with its putative receptor may be the most significant mechanism determining its function.

Several peptides which have been subjected to intensive scientific scrutiny (MSH, ACTH and endorphins) can be derived from a larger neuronal glycoprotein of about 31,000 daltons (Mains *et al.*, 1977). Further, MSH (which shares a behaviourally active segment with ACTH) and the endorphins are contained within the structure of the 90-amino-acid-chain lipotropin (LPH). It is conjectured that these larger prohormones may be reduced by enzymatic action to smaller chains. The smaller chains in turn may mediate specific behavioural functions. Thus, there is developing a new view of the function of the endocrine system, which may complement the classical view.

II. ALTERNATIVE VIEW

In the past decade, dramatic changes have occurred in the way the structure of brain is viewed. It was once thought that the structure of the central nervous system was fixed during a period of development, never to change except by traumatic events. The proposal that the brain and its receptors may continue to develop in response to stimulation has permitted a much more fluid and dynamic understanding of the workings of the brain (Browning *et al.,* 1978; Lynch *et al.,* 1978).

Consistent with this dynamic model were the conclusions drawn from neuro-chemical events in the brain. For more than 50 years it has been known that neural information can be transmitted chemically. Further, it is well recognized that some of the chemicals important for neural transmission, the biogenic amines, are subject to flux and replacement. Thus, the function and even the structure of the brain may be partially controlled by a very dynamic system.

Since the 1960s, parallel developments regarding the functioning of the endocrine system and its relationship to the central nervous system also have been accumulating. Controversial studies of the influence of amino acids on behaviour catalyzed studies of biochemistry and behaviour. The early report of Thompson and McConnell (1955), and then the plethora of replications and extensions, indicated that experience may be coded in amino-acid structure. This line of development culminated with the discovery by Ungar (1973) of a poly-peptide with ten amino acids (scotophobin), which appeared to mediate the learned aversive response to darkness in rats.

The convergence of these findings and approaches has suggested clearly that the relationship of the organism to the environment should be considered. It is apparent that changes in the environment are reflected by chemical and structural changes in the nervous system. Thus, experience may be chemically and subse-quently structurally coded. The initial but prevalent evaluation of this alternative view is illustrated in Table 1.2. The majority of data collected have been from studies in which peptide fragments were administered and the behavioural consequences observed. Although this approach is primarily pharmacological, sufficient evidence has accumulated to warrant the conclusion that peptide fragments are behaviourally active.

Table 1.2: Proposed Effects of Several Peptides on the Brain and Behaviour

Peptide	Proposed Functions
Scotophobin	Fear of dark
ACTH (4-10 fragment)	Improve memory, improve attention, increased fear
MSH	Improve attention
Vasopressin	Enhance memory
Luteinizing hormone	Increase libido
Endorphins	Analgesic properties, sexual arousal, decreased sensitivity
Enkephalins	Improve learning, increase rewarded behaviour, transient analgesic properties
DSIP	Induce sleep
MIF	Decrease Parkinsonian tremor, reverse affective disorders

III. ANIMAL STUDIES

A. MSH, ACTH and Their Fragments

The effects of neuropeptides on the central nervous system have been described most completely with ACTH and MSH. The earliest explanation of the effects of these peptides was that they influenced the emotional state of the organism and made them more fearful. This explanation relied heavily on the assumption that avoidance situations elicited a fear or anxiety response. The ACTH secretion in response to the stress of shock was considered a manifestation of the drive of fear. The view that injections of ACTH produced the fear dominated scientific thinking for about 15 years.

A departure from the popular fear hypothesis was proposed by de Wied and Bohus (1966). They interpreted the prolonged extinction after treatment with MSH/ACTH as evidence that these peptides enhanced short-term memory. Although it was argued that trial-to-trial memory was primarily influenced, refinements of this position suggested that this effect was secondary to the increased general motivational state of the organism. Nevertheless, several experiments have implicated MSH/ACTH fragments in the primary retrieval processes. Rigter and his colleagues (Rigter and Van Riezen, 1975; Rigter *et al.*, 1976) trained animals to inhibit a response to avoid shock and then induced amnesia by applying electroconvulsive shock or partial asphyxiation by CO_2. Treatment of rats with MSH/ACTH 4-10 before the retention test restored the memory of the experience. These data were interpreted as supporting the proposal that short-term memory, especially those processes involved in information retrieval from long-term memory, was enhanced by treatment with MSH/ACTH fragments.

Another early departure from classical thinking involved the examination of the effects of peptides in an appetitive paradigm (Sandman *et al.*, 1969). Until this initial study, all of the effects of ACTH fragments had been tested in aversive, stressful situations. MSH/ACTH was found to prolong extinction even in this appetitive task, thereby casting serious doubt on the fear hypothesis. In a series of studies, it was concluded that the perceptual/attentional functioning of the organism was influenced primarily by MSH/ACTH. The major support for this conclusion was developed with the visual-discrimination and reversal-shift problem in rats.

In several studies (Sandman *et al.*, 1972, 1973, 1974, 1980b), rats were trained with a two-choice visual-discrimination problem to avoid shock by running to a white door. After the animals acquired the response and successfully avoided shock, the task was reversed so that the simultaneously available black door was the correct response. The initial stage of the experiment measured the animals' ability to learn a new response. The reversal stage measured the animals' selective attention (Mackintosh, 1965, 1969). An attentive animal solved the reversal problem faster than an unattentive animal because it learned about the dimension of brightness during the original problem and not only that white was correct. Thus, when the problem was changed, the attentive animal tested values on the dimension of brightness (black-white) rather than irrelevant dimensions (e.g., in this case spatial localization).

In the initial studies, treatment of rats with MSH had no significant effect on original learning. However, rats treated with MSH required approximately 50% *fewer* trials to solve the reversal learning problem. It was concluded from these data that the MSH/ACTH peptide enhanced attentional processes.

In a recent, refined analysis (Sandman *et al.*, 1980a) the influence of MSH/ACTH 4-10, α-MSH (1-13), β_p-MSH (1-18), β_h-MSH (1-22) and ACTH 1-24 was determined with tests of learning and attention. This study permitted the investigation of the putative effects of the redundant chemical information stored in these related peptide chains. Although the prevailing view among investigators studying structure-activity relationships was that behavioural information in these molecules was redundant, Greven and de Wied (1977) recently indicated that the proposed redundancy was specific to extinction of the pole-jumping avoidance response.

The results of the study indicated that the speed of learning in the original problem diminished with administration of the same dose of compounds of increasing molecular weight. Thus, the initial stage was enhanced significantly with administration of MSH/ACTH 4-10. Except for ACTH 1-24, all of the other peptides also improved learning, though without achieving acceptable levels of statistical significance. The structure-activity relationships were much different for reversal learning and extinction. Maximal enhancement of reversal learning (an index of attention) was achieved with administration of α-, β- and (human) β-MSH but not with MSH/ACTH 4-10 or ACTH 1-24.

The results of the early phases of the learning process (original learning) are in agreement with the conclusions of de Wied and Bohus (1966). If behavioural information was coded redundantly in these related molecules, a monotonic relationship would be predicted between performance and molecular weight. The relationships observed in this study support such a speculation and suggested that trial-to-trial memory may be influenced by the 4-10 fragment.

However, the results of the reversal learning problem (i.e., attention) indicated that only compounds with MSH-like configurations improved performance. These findings suggested that attentional function may be specific to a particular peptide sequence. Thus, the fit of a molecule with its putative receptor may influence discrete behavioural patterns.

B. Developmental Studies

The influence of early treatment with MSH-like peptides on later behaviour has been investigated in several studies. The rationale of this approach is that the brain and endocrine system are not fully developed in immature organisms, and thus may be extremely pliable. Administration of peptides at early stages of development may result in structural changes which are reflected in permanent alterations in behaviour. Relatively little developmental work has been done with MSH, ACTH and their fragments.

In a series of studies summarized in Figure 1.1, it was reported that MSH treatments postnatally after between two and seven days resulted in significant improvement in learning of adult rats (Beckwith *et al.,* 1977b; Sandman *et al.,* 1977b). Increased efficiency in receiving reinforcement with a difficult operant schedule (DRL-20) characterized adult rats given MSH as infants. Similarly, rats given MSH as infants acquired and extinguished an active avoidance response more proficiently than controls when tested as adults.

Figure 1.1: Schematic design of phases of experiment showing developmental studies of MSH. Three different tasks were used to test the influence of early administration of MSH to rats. Early treatment of rats with MSH improved performance on each of these tasks.

Physiology and Behavior, 1977

Subjects: 105 albino rats

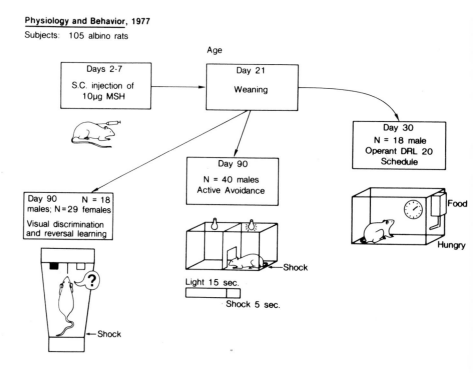

Similar findings were also generated with the visual-discrimination procedure. Rats treated as infants with MSH performed the discrimination and reversal problem more accurately than rats treated with the vehicle solution. However, this result pertained only to the male animals since there was no effect of MSH on the learning performance of female rats (Beckwith *et al.,* 1977b). A second study (Champney *et al.,* 1976) with a behaviourally active analog of ACTH 4-9 — met $(0_2)^4$, D-lys^8, phe^9 — confirmed the positive findings of male animals and also confirmed that the influence of the peptide on the learning of female rats was different than for male rats.

In a study of social behaviour (Beckwith *et al.,* 1977a), infant rats were again treated with MSH between the ages of two and seven days and then observed in the open field when they were 45 and 120 days of age. Pairs of rats of the same sex were placed in the open field for five minutes, and the time in contact with one another was measured. The findings indicated that females treated with MSH as infants spent the greatest amount of time in contact with each other. This effect was apparent at 45 but not at 120 days. Treatment of infant male rats with MSH also increased contact time compared to control animals, and the effect persisted for at least 120 days. These complex findings indicated that early treatment with MSH influenced behaviour and that the effect was sexually dimorphic.

C. Endorphins

Embedded in the C-terminal of the LPH molecule are enkephalin (LPH 61-65) and the endorphins (α 61-76; γ 61-77; β 61-91), although the physiologically active enkephalin may not be derived from the LPH chain. The discovery of these molecules catalyzed enormous interest in the study of peptides.

Even though it is now considered somewhat of an epiphenomenon, injections of the endorphins into the ventricles or into discrete areas of the brain can produce profound analgesia. The analgesic effects of methionine and leucine enkephalin have been small and short-lived. However, by substituting d-alanine in the second position, an analog was produced with very powerful analgesic properties (Walker *et al.*, 1977a).

Similarly, injections into the brain of α-, γ- and β-endorphin produced analgesia (Walker *et al.,* 1977b). Among these related molecules, β-endorphin is the most potent in producing analgesia. As with the enkephalins, substitution of d-alanine in the second position resulted in profound analgesia persisting for up to six hours. However, peripheral administration of enkephalin, endorphins or their potent d-ala-2 analogs even in larger doses (1-10 mg) did not result in analgesia. Thus, other effects of these peptides have been explored.

The extra-opiate effects of the enkephalins and endorphins have not been studied extensively; however, the studies which have been conducted clearly indicated that these peptides are behaviourally active. For instance, intracerebral injections of opioid peptides increased grooming behaviour (Gispen *et al.,* 1977), decreased electrically induced self-stimulation (Stein and Belluzzi, 1978), depressed levels of sexual activity of male rats in the presence of estrous females (Meyerson and Terenius, 1977), stimulated penile erection and spontaneous seminal emission (Walker *et al.,* 1977a) and increased food intake in rats (Belluzzi and Stein, 1978). All of these effects are reversible by naloxone.

Peripheral administration of opioid peptides also results in a number of behavioural responses. In one of the earliest studies of the behavioural effects of enkephalin, Kastin *et al.* (1976) reported that small doses (80 μg/kg) enhanced maze learning of hungry rats. It was unlikely that this effect was due to appetite, general motor activity or arousal. In a series of studies, de Wied *et al.* (1978) reported that met-enkephalin was as potent as MSH/ACTH 4-10 in delaying extinction of the avoidance response.

Rigter (1978) has reported that met- and leu-enkephalin can attenuate amnesia by CO_2. (This study was identical in design to earlier ones testing MSH/ACTH 4-10.) The results presented a puzzling profile, suggesting that met- and leu-enkephalin influenced memory in different ways. Met-enkephalin reduced amnesia when given before acquisition, before retrieval or at both times. Leu-enkephalin reversed amnesia only when given before retrieval or if given both before acquisition and retrieval. It could be argued that the endogenous opiods influenced both consolidation and retrieval processes of memory, whereas the MSH/ACTH peptides influenced only retrieval.

More recently, Le Moal *et al.* (1979) reported opposite effects of α- and γ-endorphins on extinction of the pole jumping task, finding that α-endorphins inhibited (or delayed) extinction and γ-endorphins facilitated extinction. However, the influence of these similar peptides was identical in an appetitive task.

Among the most striking studies of the multiple independent actions of peripheral endorphin administration is that of Veith *et al.* (1978b), in which the different effects of α-, γ- and β-endorphins on the behaviour of rats were examined in the open field. They reported that each of these related peptides exerted specific, nonoverlapping effects. Treatment with α-endorphin typically produced penile erection and spontaneous seminal discharge—an effect interpreted as relating to pleasure. The effect of γ-endorphin was to increase defecation and diminish exploratory behaviour. Peripheral administration of β-endorphin resulted in increased grooming, an effect reported by other investigations as well (Gispen *et al.*, 1977). These data suggested that peptides which shared an amino-acid core nevertheless exerted discrete and specific influences on behaviour.

D. Developmental Findings

Several factors converged to suggest that the effects of exposure to β-endorphin early in life would have a profound influence on the organism. First, earlier studies with the MSH/ACTH fragments suggested that neonatal exposure to their related sequence of amino acids exerted a persisting influence. Second, Gautray *et al.* (1977) reported that β-endorphin was elevated in the amniotic fluid during fetal distress, a condition which can result in profound physical and cognitive impairment. Third, it is known (Simon and Hiller, 1978) that the rat is not born with a fully developed complement of opiate receptors. The most rapid proliferation occurs during the first 21 days postnatally and matures by about 140 days of age. Thus, exposure to endorphin during periods of plasticity may influence ultimate development. These factors prompted our initial experiments.

In the first experiment (Sandman *et al.*, 1979) rats were injected with 50 μg (β-endorphin or a vehicle control), s.c. each day from the second to the seventh day of life (Figure 1.2). The rats were then left undisturbed until the 90th day, when tail-flick tests for analgesia were performed. These results indicated that

Figure 1.2: Diagram of study involving administration of β-endorphin to new-born rats and subsequent testing for analgesia. Early administration of β-endorphin to rats resulted in 'chronic' elevation in pain threshold.

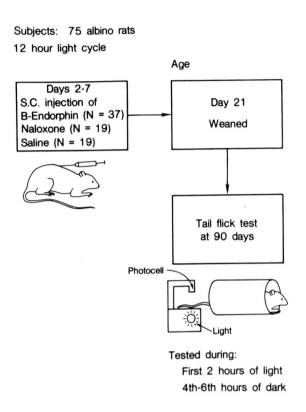

Subjects: 75 albino rats
12 hour light cycle

Age

Days 2-7
S.C. injection of
B-Endorphin (N = 37)
Naloxone (N = 19)
Saline (N = 19)

Day 21
Weaned

Tail flick test
at 90 days

Photocell

Light

Tested during:
First 2 hours of light
4th-6th hours of dark

early exposure to β-endorphin permanently increased, by over 70%, the threshold of thermal pain. At this time, the mechanism of action is unclear; however, alteration of opiate receptors or of levels of β-endorphins are possible factors.

In the second study, the consequences of injecting pregnant rats with either β-endorphin or a vehicle solution was examined (McGivern *et al.*, 1979). Pregnant rats were injected s.c. every other day, from the seventh to the 21st day of pregnancy. The male offspring were crossfostered and studied longitudinally from birth through to the 180th day (see Figure 1.3). Preliminary findings after injection of radioactive β-endorphin into separate rats suggested that approximately 5% of the radioactivity may pass the placental barrier.

The results (summarized in Table 1.3) indicated that early *in utero* exposure to β-endorphin retarded aspects of development. The rats given the peptide were delayed significantly in eye opening, were less active during early life and failed to exhibit the normal pattern of activity later. From the 40th day through

Figure 1.3: Design of study in which pregnant mothers were injected s.c. with β-endorphin and the offspring tested from birth. A variety of significant changes were observed, and characterized as depression of perceptual/attentional functioning.

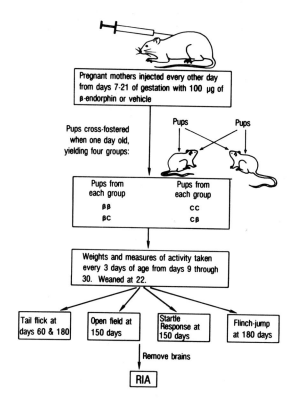

Table 1.3: Summary of Effects of *in utero* **Exposure to β-endorphin on Development and Behavioural Parameters in the Rat**

Delayed eye opening (by two days)
Depressed activity
Significant weight gain by fourth day
Attenuation of startle response to auditory stimulus by 350%
Less responsive to placement in center of field (100%)
Less activity in open field (200%)
Disturbed social behaviour
Improved passive avoidance learning (inhibiting response)
Hypersensitivity of morphine
Impaired performance of reversal learning of visual-discrimination problem

to maturity, the rats exposed to β-endorphin *in utero* evidenced a significant weight gain. At sacrifice (the 180th day) their wet brain weights were greater and their tails were longer than matched controls.

Behavioural tests indicated that rats exposed to β-endorphin *in utero* appeared less responsive to environmental stimulation than the animals in the control group. The startle response, measured by assessing the animal's somatomuscular reaction to a very loud (60-db) tone, was attenuated by more than 300% in animals given β-endorphin as compared with animals given the control solution.

Typically, when placed in the center of an open field, rats initially moved to a wall and subsequently explored the apparatus. Both of these tendencies were markedly reduced in the animals exposed to β-endorphin. In a second condition, pairs of animals were observed in the open field. The rats exposed to β-endorphin spent less time in physical contact and less time in close approximation to each other than the control group. Clearly, *in utero* exposure disrupted the normal open-field and social behaviour of rats.

Two tests of learning also distinguished the rats exposed *in utero* to β-endorphin from those exposed only to the vehicle solution. In a test of passive avoidance, rats exposed to β-endorphin took significantly longer than control animals to re-enter a chamber in which they had received shock 24 hours earlier. In a test of visual discrimination, there were no differences between groups during initial learning; however, the rats treated with β-endorphin required significantly more trials than controls to learn the reversal shift. This pattern of results suggested impaired attention without a concomitant impairment of memory.

In view of the report of permanent elevation in threshold for a thermal stimulus in rats treated as neonates with β-endorphins, it was of extreme interest to test the animals exposed *in utero* to an identical test. There were no differences between the rats exposed *in utero* to β-endorphin or vehicle on the tail-flick test. However, rats exposed to β-endorphin were significantly more sensitive to analgesic doses of morphine than the control animals. The increased sensitivity to morphine after *in utero* exposure to β-endorphin is in direct opposition to the effects reported for *in utero* exposure to morphine (Paul *et al.*, 1978). Prenatal exposure to morphine produces tolerance to subsequent treatment with morphine. It is conceivable that the effects observed after *in utero* exposure to β-endorphin may reflect, at least partially, its extra-opiate influences. The dose or type of opiate peptide may influence later behaviour when injected during the first week of life. The differences we have described in rats after perinatal and postnatal treatment with β-endorphin may agree with the differences reported by Bloom *et al.* (1979) in the concentration of β-endorphin in the brains during early development.

IV. STUDIES IN NORMAL HUMAN VOLUNTEERS

A growing number of studies indicate that neuropeptides, especially the MSH/

ACTH fragments, influence the physiology and behaviour of normal human subjects.

A. Physiological Effects

In the earliest study of the influence of MSH/ACTH fragments, male subjects received either MSH/ACTH 4-10 or ACTH 1-24 and were monitored for basal changes in physiological functions as well as during specific tasks (Miller *et al.*, 1974). No effect on the EEG was observed in subjects receiving ACTH 1-24. However, spectral analysis of the EEG $(0_1 0_2)$ indicated a decrease in the power output of the 3-7-Hz frequency but an increase in the 8-12-Hz and 12-Hz frequencies in subjects injected with MSH/ACTH 4-10 (see Table 1.4). However, the most striking finding in this study was the delay in the α-blocking EEG response to repetitive stimulation. Typically, during the first few trials there is a characteristic increase in EEG frequency to external stimulation. After several trials, supposedly after the subject has habituated to the stimulus, the EEG response diminishes and the predominant frequency is in the α range (8-12-Hz). Treatment with MSH/AXTH 4-10 attenuated the habituation.

Table 1.4: Summary of the Physiological Effects of MSH and ACTH Fragments in Normal Volunteers

Central nervous system
  Increase in 8-12- and above-12-Hz activity
  Decrease in 3-7-Hz activity
  Delay of α-blocking response of the EEG to repetitive stimulation.
  Augmented components of the cortical evoked potential.
Autonomic nervous system
  No effect on several basal autonomic measures
  Heart-rate deceleration to novel stimulus (orienting response)
Endocrine system
  No effect on measures of cortisol, FSH, LH, estrogen or progesterone

Two studies have been conducted to assess the influence of MSH/ACTH fragments on the AEP. In the first study (Miller *et al.*, 1976), the visual AEP was measured while subjects performed a complex task. The subjects given MSH/ACTH 4-10 evidenced an augmentation of the negative peak about 350 ms after stimulation. Further, MSH/ACTH 4-10 resulted in increased latency and decreased amplitude of the P200 complex.

In a second study, with a simple task, the influence of an orally administered analog of MSH/ACTH 4-9 was examined. Five men and five women were given 0, 5, 10 and 20 mg of the analog or d-amphetamine (10 mg) as a positive control in a double-blind procedure. Immediately after ingesting the coded capsule, brief, bright flashes of light were projected, while EEGs were recorded from the right and left hemisphere of the occipital cortex. The results indicated that dosage, time after ingestion, hemisphere of the brain and sex of the subject all were influential factors determining the effects of the peptide. The P200 complex was enhanced in both hemispheres of women, and P100 was augmented in the right hemisphere of men. Area-under-the-curve measures were taken to supplement interpretation of the findings. Several striking results were obtained. The most dramatic effect was the interaction among dosage, hemisphere of the brain and sex of the subject. In men, the major effect of the peptide was seen in the right hemisphere of the brain and peaked around 60 minutes. Both hemispheres in women appeared to be influenced by the peptide, but the most obvious effect was in the left hemisphere. In the right hemisphere of both men and women, the influence of d-amphetamine and of 20 mg of the peptide appeared to be similar.

In two studies (Miller *et al.*, 1974; Sandman *et al.*, 1977a) no effect of the peptide on basal autonomic responses has been observed. However, significant heart-rate deceleration (a reliable index of the orienting responses) to novel stimulation has been reported after treatment with MSH/ACTH 4-10 (Sandman *et al.*, 1977a). This finding is consistent with the α-blocking response discussed above, since both responses reflect increased awareness of the environment.

B. Behavioural Effects

In a series of early studies, the effects of MSH/ACTH 4-10 were examined on a number of behavioural parameters in normal volunteers (see Table 1.5). Among the most reliable findings were increased visual retention, decreased anxiety and enhanced visual discrimination (Miller *et al.*, 1974; Sandman *et al.*, 1975). Several parameters were not affected by the peptide, including short-term memory for digits, measures of emotionality, reaction time and verbal memory. However, interpretation of the findings has not been uniform. Several studies were initiated in our laboratory to explore the primary processes affected by the peptide.

In the first study, the influence of MSH/ACTH 4-10 on perceptual threshold was examined (Sandman *et al.*, 1977a). Tests of detection and discrimination were employed. Infusion of MSH/ACTH 4-10 raised the threshold for detection and impaired the subjects' ability to accurately report the presence of a stimulus. However, when subjects were administered the peptide, their ability to discriminate the two stimuli was improved. These results suggested that MSH/ACTH 4-10 facilitated stimulus processing or selective attention, whereas simple intake or detection of threshold stimuli was impaired. Conceivably the peptide raised the absolute threshold for stimuli and thus functioned as a filtering mechanism to protect the organism from distraction 'perceptual noise'. However, when stimuli were above the threshold, the processing of information was facilitated.

A recent study in our laboratory (Ward *et al.*, 1979) was designed specifically to test the influence of MSH/ACTH 4-10 on attention and memory with the

Table 1.5: Summary of the Behavioural Effects of MSH/ACTH Fragments in Human Beings

---

MSH/ACTH fragments

Increased visual retention on Bention visual-retention test

Decreased anxiety

Improved concept-formation learning (men only)

Elevated threshold for detection

Improved ability to discriminate tachistoscopic stimulus

Increased reaction time for all set sizes in item-recognition test (only intercept affected)

Improved verbal memory (women only)

Delay of fatigue in continuous-performance test

Increased work productivity (mentally retarded)

Increased social contact (mentally retarded)

---

item-recognition test. In this test, subjects were presented with a memory set consisting of one, two, three or four items. After they had memorized the set, probe stimuli were presented. Half of the stimuli were in the set and half were not. The subject depressed one key if the probe was a member of the memory set, and a second key if it was not.

Treatment of subjects with MSH/ACTH 4-10 decreased reaction time for all four memory sets. In conjunction with other data, the most parsimonious interpretation of these findings is that MSH/ACTH 4-10 facilitated selective 'encoding' or attention to environmental stimuli.

C. Sex Differences

Several studies have suggested that MSH/ACTH 4-10 influenced men and women differently. As reviewed earlier, data gathered with rats indicated that MSH enhanced visual attention only in males. Conversely, spatial abilities appeared to be augmented by MSH in females. Electrophysiological data have also indicated that men and women were influenced differently by MSH/ACTH fragments. Evoked potentials from the right hemisphere were enhanced in men but potentials from both hemispheres were affected in women. A study by Veith *et al.* (1978a) was designed to measure in women many of the variables affected by MSH in men.

Women were injected with MSH/ACTH 4-10 either during their menstrual phase (endogenous ACTH is low) or during midcycle (ACTH is high). In addition to indexes of emotion and cognitive state, radioimmunoassays were performed on plasma for FSH, LH, 17β-estradiol and cortisol.

Although there were differences in hormonal levels during the menstrual phase compared with midcycle phase, there was no indication that MSH/ACTH 4-10 influenced any of the hormones. The test of visual memory found to be sensitive to peptide effects in men was not influenced by the peptide in this study.

However, tests of verbal memory (not influenced by the peptide in men) did show significant improvement in women after treatment with MSH/ACTH 4-10. Further, a measure of intradimensional shift (an index of visual attention) was impaired in women receiving the peptide. This finding is also different from previous reports with male subjects.

These results are in accordance with behavioural studies of rats and electrophysiological studies of human subjects. The results of this and earlier studies suggested that MSH/ACTH 4-10 augmented verbal abilities in women and visual processes in men. While the precise mechanisms of these effects are not known, the results are also consistent with the hemispheric differences discussed earlier.

D. Personality Influences

Although the possibility that personality may be determined by neurochemicals has been a dominant theme at various times in the history of psychology, few recent studies have examined the interactions between peptides and personality. Only Miller *et al.* (1976) controlled for possible dispositional differences among subjects. When they selected subjects based upon low anxiety and field-dependence scores, they found minimal effect of MSH/ACTH 4-10 on behaviour.

In a recent study, Breier *et al.* (1979) reported the interactions between the introversion-extroversion dimension and response to MSH/ACTH 4-10. They found that the dimension of personality determined, to some degree, response to the peptide. The extroverted subjects performed a series of 'mental performance' tasks better after injections of MSH/ACTH 4-10 than during placebo injections. Introverted subjects showed no improvement in performance after receiving the peptide. Further, MSH/ACTH 4-10 decreased forearm blood flow in extroverted subjects but increased flow among introverted subjects. The authors suggested that MSH/ACTH 4-10 may act as a mild central stimulant, and thus is more effective in subjects characterized as cortically inhibited (extroverts).

V. STUDIES OF PATIENT GROUPS

A. MSH/ACTH Fragments

One of the earliest reports of the behavioural actions of MSH and the first clinical study was conducted in amenorrheic women (Kastin *et al.*, 1968). Slowing of the EEG, increased heart rate, menstrual bleeding and increased feelings of nervousness and anxiety appeared to be related to infusion of MSH. The second clinical study of the effects of MSH was done with endocrine patients and controls (Kastin *et al.*, 1971). The major finding was enhanced somatosensory evoked potential after infusion of MSH. Recently, there has been an intense interest in the ameliorative effects on behaviour of the MSH/ACTH fragments and the endorphins.

Among the most dramatic effects of MSH/ACTH fragments have been on the behaviour of mentally retarded individuals. To date, three studies have been completed. In the first two studies (Sandman *et al.*, 1976; Walker and Sandman, 1979), mentally retarded men were injected with MSH/ACTH 4-10 or given an oral analog of MSH/ACTH 4-9 and then given tests similar to those administered to normal volunteers. Treatment with the peptide resulted in significant deceleration of heart rate (as already noted, an index of the orienting response) to novel stimulation. In addition, treatment with the peptide improved learning of intra-dimensional and extradimensional shifts, visual retention, spatial localization and matching auditory patterns.

In the third study, four doses (0, 5, 10 and 20 mg) of the ACTH 4-9 analog were examined in retarded clients while they performed their day-to-day activities (Sandman *et al.*, 1980b). The clients were paid a wage to bend electrical leads to fit a mold. There were four steps in the process which varied in difficulty from the bending of resistors to quality-control inspection. During the course of the study the clients performed the same task each day. The peptides were administered in the morning every day for two weeks. Placebo weeks preceded and succeeded the treatment weeks. Observations of productivity and of social behaviour were made at regular intervals during the morning hours.

The doses of the peptide interacted with the difficulty of the task to produce distinctive curves. The high dose, 20 mg, interfered with productivity in each step, while 5 mg had mixed effects, enhancing performance only for the more complex tasks, and 10 mg improved productivity in all but the first step.

These data provided indirect support for the hypothesis that the MSH/ACTH 4-9 analog exerted an influence on perceptual/attentional mechanisms. Since performance was disrupted by extraneous movement, lack of coordination or inattention to detail, these data may be construed as ecological validation that MSH/ACTH 4-9 influenced attention.

Other data collected in this study indicated that the peptide might also influence social behaviour in a dose-dependent way. The evidence suggested that patient-patient and patient-supervisor contact increased during treatment, especially with 10 and 20 mg. Self-stimulation also increased during treatment with the peptide. These data are in agreement with the reports of Beckwith *et al.* (1977a), in which rats increased contact time after injections with MSH.

An unusual pattern emerged from this study. Increased productivity coupled with greater interpersonal awareness and self-stimulation was evident. Apparently, when the clients worked after treatment with the peptide, they did so with greater concentration and intensity. These data can be reconciled with earlier findings that normal and retarded subjects evidenced enhanced orienting responses to novel stimuli and also retained the ability to discriminate relevant from irrelevant information after treatment with MSH/ACTH fragments.

There is suggestive evidence that MSH/ACTH fragments ameliorated the behaviour of elderly subjects (Ferris *et al.*, 1976; Will *et al.*, 1978; Branconnier *et al.*, 1979; Miller *et al.*, 1980). Among these studies, the work of Branconnier and his colleagues is especially significant. Eighteen mildly senile, organically impaired subjects displayed reduced depression and confusion and increased vigour after treatment with MSH/ACTH 4-10. In addition, and consistent with the findings of Gaillard and Varney (1979), the peptide delayed fatigue associated with a reaction-time task. Further, the peptide produced a shift in the EEG to lower frequencies (3.5-4.5 and 7.5 Hz). More recently, Miller *et al.* (1980) reported improvement in visual retention after receiving MSH/ACTH 4-10 in the elderly. This effect was greater in men than in women.

B. Endorphins

The initial speculation that β-endorphin may underly schizophrenia was based upon two reports in rats (Bloom *et al.*, 1976; Jacquet and Marks, 1976). These studies described 'waxy flexibility' and rigidity in rats after large doses of centrally administered β-endorphin. Even though similar observations were made in both studies, the authors offered very different conclusions. Bloom and his coworkers suggested that the 'catatonic-like' states observed in rats and parallelled in psychiatric patients may result from excessive amounts of endorphin. Conversely, Jacquet and Marks suggested that β-endorphin itself may be an anti-psychotogen. The literature with clinical groups reflects this dilemma, as there are studies proposing either naloxone or endorphin to treat psychotic states.

The first issue to be resolved was whether or not endorphin levels were altered in schizophrenia. At least two studies (Lindstrom *et al.*, 1978; Domschke *et al.*, 1979) reported significantly elevated basal levels of β-endorphin in the CSF compared to control subjects. However, Domschke *et al.* (1979) found that the elevation in β-endorphin was significant only for acute schizophrenic states and that chronic states were associated with significantly lower levels. Conversely, Terenius *et al.* (1976) reported that CSF endorphins are elevated in chronic schizophrenia. It appears that both elevated and depressed levels of endorphin may reflect psychological disorders and careful diagnosis is a prerequisite for effective treatment.

Approaches testing the possibility that schizophrenia may involve an excess of endorphin have used the opiate antagonist, naloxone, as treatment. The results are equivocal: some (Gunne *et al.*, 1977) report improvement in schizophrenia, while others (Janowsky *et al.*, 1977; Volavka *et al.*, 1977) report no effect. Further controlled tests are required before definite conclusions can be reached.

The hypothesis that endorphin deficiency may underlie schizophrenia also presents a confusing picture. The initial study of Kline *et al.* (1977) suggested that symptoms of schizophrenia and depression were ameliorated by β-endorphin injections. More recently, Verhoeven *et al.* (1979) reported dramatic improvement in psychotic symptoms after treatment with an endorphin analog (des-tyr) -γ-endorphin.

Although the role of endorphins in the modulation of behaviour appears significant, there is an absence of controlled studies in human subjects. Acceptance of the initial and somewhat tenuous assumption that the endorphins form the biochemical basis of schizophrenia may have misguided research efforts. The results of studies in animals suggest that examination of the influence of endorphin on learning and perception will generate more definitive information concerning the physiological significance of these interesting peptides.

VI. SUMMARY AND CONCLUSIONS

Most of the behavioural influences described for MSH/ACTH and the endorphins can be viewed as a modulation of attentional/perceptual function. A proposed model for understanding the influence of LPH fragments on attention is presented in Figure 1.4. Although the model suggests that selective attention is improved by MSH/ACTH fragments, and impaired by endorphins, the model should be viewed as heuristic. The evidence reviewed above clearly implicated MSH/ACTH fragments as modulators of the component of selective attention described in the model. To date, the results indicate that MSH/ACTH molecules raise perceptual threshold to act as a filter and appear to increase the probability of detecting stimuli which are interesting and important. From the model it could be argued that enhanced short-term memory is a secondary result of improved selective attention. Thus, only elements which pass the scrutiny of the attentional filter can enter short- or long-term memory. However, it could also be argued that an initial perceptual apparatus is essential for activation of the attentional mechanism.

As reviewed, the majority of the research on the effects of the endorphins has pursued the possibility that these molecules underlie pathological states. Distressingly few studies have examined the behavioural significance of the endorphins. Some studies suggest that the effects of endorphins are identical to those of the MSH/ACTH fragments, while others suggest these molecules have different effects. The developmental study of the behavioural effects of the endorphins, which measured a large number of variables, indicated an overall depression of the organism's response to the environment (McGivern *et al.*, 1979). Further, the pervasive finding that endorphins administered centrally induce analgesia is quite consistent with this possibility. Thus, for heuristic purposes, it may be useful to consider the possibility that the endorphins and MSH fragments may have opposite effects on perceptual/attentional modulation.

Although there is evidence that the opiates and MSH/ACTH fragments enhance learning, a compelling case can be made that a functional reciprocity exists between these peptides. Of central interest is the finding that MSH and the endorphins are stored and released by the intermediate lobe of the pituitary. Further, the provocative findings of O'Donohue *et al.* (1979) and of Watson and Akil (1979) that both MSH and β-endorphin are present in the same cells of the arcuate nucleus underscore the probability of a functional relationship between these molecules.

Among the most direct evidence of reciprocity is the attenuation of opioid-induced analgesia by MSH/ACTH fragments (Krivoy *et al.*, 1977). Szekely *et al.* (1979) reported that α-MSH administered concomitantly with morphine attenuates tolerance and dependence on morphine. The related peptide MIF also attenuates the effects of morphine (Walter *et al.*, 1979). The analgesic potency of opioids have been related to decreased levels of cyclic AMP in the brain.

Figure 1.4: Model of the influence of MSH/ACTH fragments and endorphin on perceptual modulation. The model suggests that the major influence of MSH/ACTH is on selective attention. Heuristically, the model implies complementary influences of MSH/ACTH and the endorphins.

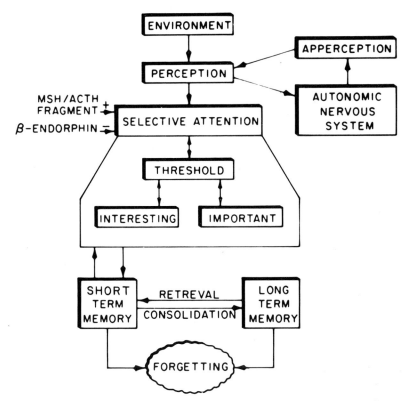

Thus, Gispen *et al.* (1977) reasoned that the effects of MSH/ACTH fragments on opioid-induced analgesia may be mediated by adenylate-cyclase activity. Indeed, increases in cyclic AMP in the diencephalon and mesencephalon (Gispen *et al.*, 1977) and in the cortex (Christensen *et al.*, 1976) have been reported after treatment with MSH/ACTH fragments. Therefore, it is conceivable that the attenuation of analgesia by MSH/ACTH may relate partially to the reciprocal influence of MSH/ACTH and endorphin on cyclic-AMP activity.

Recent evidence from our laboratory indicates further reciprocity between MSH and endorphin. Injection of 10 μg of β-endorphin into the lateral ventricle of rats produces a marked attenuation of the early components of the auditory evoked potential. Subsequent injection of 50 μg of MSH/ACTH 4-10 not only reverses the attenuation, but may even enhance this component.

In a related study (preliminary findings), β-endorphin (10 μg) injected into the periaqueductal gray (PAG) of rats produced reliable and long-lasting (2-h)

Table 1.6: Suggestive Evidence of Relationship between MSH/ACTH Fragments and Endorphin/Enkephalin Molecules

MSH/ACTH fragments	*Endorphin/enkephalin*
Behavioural effects	
1. Improve acquisition of maze	Improve acquisition of maze
2. Delay extinction of avoidance learning	Delay extinction of avoidance learning
3. Facilitates reversal learning	Delays reversal learning
4. Reverses amnestic treatment with prior administration	Attenuates amnestic treatment before or after ECT
5. Increases exploratory activity	Decreases exploratory activity
6. Increases gregariousness	Decreases gregariousness
7. Increases sexual activity	Decreases sexual activity
8. 4-10 has no effect on grooming	Increased grooming
9. Increased self-stimulation	Decreased self-stimulation
10. Accelerates eye opening in newborn	Retards eye opening in newborn
Opiate-related effects	
11. Hyperalgesia	Analgesia
12. Attenuates morphine-induced analgesia	Addiction and tolerance
Electrophysiological effects	
13. Increases early components of AEP	General suppression of AEP
14. Abolishes elliptogenic spiking in nucleus gigantocellularis produced by β-endorphin	Initiates spiking in nucleus gigantocellularis
15. Increases spinal-neuron excitability	Decreases spinal-neuron excitability
Biochemical effects	
16. Increases dopamine turnover	Decreased dopamine turnover
17. Increases membrane permeability to Ca^{2+}	Decreases membrane permeability to Ca^{2+}
18. Increase in cyclic AMP	Decrease in cyclic AMP
Neuroanatomy	
19. Released by intermediate and anterior lobe of pituitary	Released by intermediate and anterior lobe of pituitary
20. Found in cells of arcuate nucleus	Found in cells of arcuate nucleus

'elliptogenic' spiking in the area of the nucleus gigantocellularis. Subsequent injections of MSH/ACTH 4-10 (20 μg) or of naloxone into the PAG immediately blocked the spiking. The ameliorative effect of MSH/ACTH 4-10 persisted for up to 30 minutes before the spiking returned.

Inferences drawn from the behavioural evidence also support the hypothesis of a reciprocal relationship between the two parts of LPH. For instance, MSH/ACTH 4-10 antagonizes the morphine-induced behavioural arousal of the mouse (Katz, 1979). Increases in social behaviour have been reported for the MSH/ACTH fragments (Beckwith *et al.*, 1977a; Sandman *et al.*, 1980a) but treatment with opioids results in decreased social contact (Panksepp *et al.*, 1978; McGivern *et al.*, 1979). The MSH/ACTH fragments have reliably resulted in enhanced reversal learning (Sandman *et al.*, 1972, 1973; Beckwith *et al.*, 1977b) but rats treated *in utero* with β-endorphin evidence a significant deficit in reversal learning (McGivern *et al.*, 1979).

Thus, although there are contraindications in the literature, an emerging array of evidence supports the possibility that the opioid and MSH/ACTH peptides share a reciprocal and modulating influence on the brain and behaviour (see Table 1.6). Further, as illustrated in Figure 1.4, the nature of the reciprocity may be viewed as a modulation of attentional/perceptual processes.

REFERENCES

Antelman, S.M. and Brown, T.W. (1972). *Physiol. Behav., 9*, 15
Beckwith, B.E., O'Quin, R.R., Petro, M.S., Kastin, A.J. and Sandman, C.A. (1977a). *Physiol. Psychol., 5*, 295
Beckwith, B.E., Sandman, C.A., Hoterall, D. and Kastin, A.J. (1977b). *Physiol. Behav., 18*, 63
Belluzzi, J.D. and Stein, L. (1978) *Neuroscience Abs.*, p. 405
Bloom, F., Bayon, A., French, E., Henrickson, S., Koob, G., LeMoal, M., Rossier, J. (1979). In *Peptides: Structure and Biological Functions*, ed. E. Gross (Pierce Chemical Co., Rockford, IL) p. 811
Bloom, F., Segal, D., Ling, N. and Guillimen, R. (1976). *Science 194*, 630
Branconnier, R.J., Cole, J.O. and Gardos, G. (1979). *Psychopharmacology 61*, 161
Breier, C. Kain, H. and Konzett, H. (1979). *Psychopharmacology 65*, 239
Browning, M., Dunwiddie, T., Bennett, W., Gispen, W. and Lynch, G. (1978). *Science 203*, 60
Bush, D.F., Lovely, R.H. and Pagano, R.R. (1973). *Physiol. Psychol., 83*, 168
Champney, T.F., Sahley, T.C. and Sandman, C.A. (1976). *Pharmacol. Biochem. Behav., 5*, 3
Christensen, C.W., Harston, C.T., Kastin, A.J., Kostrzewa, R.M. and Spirtes, M.A. (1976). *Pharmacol. Biochem. Behav., 5*, 117
Domschke, W., Dickschas, A. and Mitznegg, P. (1979). *Lancet, I*, 1024
Ferris, S.H., Sathananthan, G., Gershon, S., Clark, C. and Moshinsky, J. (1976). *Pharmacol. Biochem. Behav., 5*, 73
Gaillard, A.W.K. and Varney, C.A. (1979). *Physiol. Behav., 23*, 79
Gautray, J.P. Jolivet, A., Bielh, J.P. and Guillemin, R. (1977). *Am. J. Obstet. Gynecol., 123*, 211
Gispen, W.H., Reith, M.D., Schotman, P., Weigant, V.M., Zwiers, H. and de Wied, D. (1977). In *Neuropeptide Influences on the Brain and Behavior*, ed. L.H. Miller, C.A. Sandman and A.J. Kastin (Raven Press, New York) p. 61
Greven, H.M. and de Wied, D. (1977). *Frontiers of Hormone Research 4*, 140

Grossman, S.P. (1967). *A Textbook of Physiological Psychology* (Wiley, New York)
Gunne, L.M., Lindstrom, L. and Terenius, L. (1977). *J. Neural Trans., 40,* 3
Jacquet, Y.F. and Marks, N. (1976). *Science 194,* 632
Janowsky, D.S., Bloom, F., Abrams, A. and Guillemin, R. (1977). *Am. J. Psychiatry 134,* 926
Kastin, A.J. and Ross, G.T. (1964). *Endocrinology 75,* 187
Kastin, A.J. and Schally, A.V. (1967). *Nature 213,* 1238
Kastin, A.J., Kullander, S., Borglin, N.E., Dahlberg, B., Dyster-Aas, K., Kraukau, C.E.T., Ingvar, D.H., Miller, M.D., Bowers, C.Y. and Schally, A.V. (1968). *Lancet, 1,* 1007
Kastin, A.J., Miller, L.H., Gonzalez-Barcena, W.D., Hawley, W.D., Dyster-Aas, K., Schally, A.V., De Parra, M.L.V., and Velasco, M. (1971). *Physiol. Behav. 7,* 893
Kastin, A.J., Scollan, E.L., King, M.G., Schally, A.V. and Coy, D.H. (1976). *Pharmacol. Biochem. Behav., 5,* 691
Katz, R.J. (1979). *Eur. J. Parmac., 53,* 383
Kline, N.S., Li, C.H., Lehman, H.E., *et al.* (1977). *Arch. Gen. Psychiatry 34,* 111
Krivoy, W.A., Kroeger, D.C. and Zimmerman, E. (1977). *Psychoneuroendocrinology 21,* 43
LeMoal, M., Koob, G.F. and Bloom, F.E. (1979). *Life Sci., 24,* 1631
Lewin, R. (1972). *Chemical Communicators* (Anchor, Garden City)
Lindstrom, L.H., Widerlov, E., Gunne, L.M., Wahlstrom, A. and Terenius, L. (1978). *Acta. Psychiat. Scand., 57,* 153
Lynch, G., Gall, C. and Dunwiddie, T.V. (1978). In *Maturation of the Nervous System, Progress in Brain Research,* Vol. 48, ed. M.A. Conner *et al.* (Elsevier Press) p. 113
Mackintosh, N.J. (1965) *Psychol. Bul., 64,* 124
Mackintosh, N.J. (1969). *J. Comp. Physiol. Psychol., 67,* 1
Mains, R.E., Eipper, B.A. and Ling, N. (1977). *Proc. Nat. Acad. Sci., 74,* 3014
McGivern, R.F., Sandman, C.A., Kastin, A.J. and Coy, D.H. (1979). Paper presented at the International Congress of Psychoneuroendocrinology (Park City, Utah)
Meyerson, B.S. and Terenius, L. (1977). *Eur. J. Pharmacol. 42,* 191
Miller, L.H., Kastin, A.J., Sandman, C.A., Fink, M. and van Veen, W.J. (1974). *Pharmacol. Biochem. Behav., 2,* 663
Miller, L.H., Harris, L.C., Van Riezen, H. and Kastin, A.J. (1976). *Pharmacol. Biochem. Behav., 5,* 17
Miller, L.H., Groves, G.A., Bupp, M.J. and Kastin, A.J. (1980). *Peptides 1,* 55
Mirsky, A., Miller, R. and Stein, M. (1953). *Psychosom. Med., 15,* 574
O'Donohue, T.L., Miller, R.L. and Jacobowitz, D.M. (1979). *Brain Res., 175,* 1
Panksepp, J., Herman, H., Conner, R., Bishop, P. and Scott, J.P. (1978). *Biol. Psychiat.,* 13,607
Paul, L., Diaz, J. and Barley, B. (1978). *Neuropharmacology 17,* 655
Rigter, H. (1978). *Science, 200,* 83
Rigter, H. and Van Riezen, H. (1975). *Physiol. Behav., 14,* 563
Rigter, H., Jamssens-Elbertse, R. and Van Riezen, H. (1976). *Pharmacol. Biochem. Behav., 5,* 53
Sandman, C.A., Kastin, A.J. and Schally, A.V. (1969). *Experentia 25,* 1001
Sandman, C.A., Miller, L.H., Kastin, A.J. and Schally, A.V. (1972). *J. Comp. Physiol. Psychol., 80,* 54
Sandman, C.A., Alexander, W.D. and Kastin, A.J. (1973). *Physiol. Behav., 11,* 613
Sandman, C.A., Beckwith, B.E., Gittis, M.M. and Kastin, A.J. (1974). *Physiol. Behav.,* 13,163
Sandman, C.A., George, J. Nolon, J.D., Van Reizen, H. and Kastin, A.J. (1975). *Physiol. Behav., 15,* 427
Sandman, C.A., George, J., Walker, B.B., Nolan, J.D. and Kastin, A.J. (1976). *Pharmacol. Biochem. Behav., 5,* 23
Sandman, C.A., George, J., McCanne, T.R., Nolan, J.D., Kaswan, J. and Kastin, A.J. (1977a). *J. Clin. Endocrin. Metab., 44,* 884
Sandman, C.A., Kastin, A.J. and Miller, L.H. (1977b). In *Clinical neuroendocrinology,* ed. L. Martini and G.M. Besser (Academic Press, New York)
Sandman, C.A., McGivern, R.F., Berka, C., Walker, J.M., Coy, D.H. and Kastin, A.J. (1979). *Life Sci., 25,* 1755
Sandman, C.A., Beckwith, B.E. and Kastin, A.J. (1980a). *Peptides, 1,* 277
Sandman, C.A., Walker, B.B. and Lawton, C.A. (1980b). *Peptides 1,* 109
Schally, A.V., Arimura, A.V. and Kastin, A.J. (1973). *Science 197,* 341

Selye, H. (1956). *The Stress of Life* (McGraw-Hill, New York)

Siegel, G.J. and Eisenman, J.S. (1972). In *Basic Neurochemistry*, ed. R.W. Albers, G.J. Siegel, R. Katzman and B.W. Agranoff (Little Brown, New York)

Simon, E.J. and Hiller, J.M. (1978). *Fed. Proc., 37,* 141

Stein, L. and Belluzzi, J.D. (1978). *Adv. Biochem. Psychopharmacol., 18,* 299

Szekely, J., Miglecz, E., Dunai-Kovacs, Z., Tarnawa, I., Ronai, A.Z., Graf, L. and Bajusz, S. (1979). *Life Sci., 24,* 1931

Terenius, L., Wahlstrom, A., Lindstrom, L. and Widerlov, E. (1976). *Neurosci. Le Hem., 3,* 157

Thompson, R. and McConnell, J.V. (1955). *J. Comp. Physiol. Psychol., 48,* 65

Turner, C.D. and Bagnera, J.T. (1971). *General Endocrinology* (Saunders, Philadelphia)

Ungar, G. (1973). *Naturwissenschaften 60,* 307

Usher, D.R. and Lamble, R.V. (1969). *Physiol. Behav., 4,* 923

Veith, J.L., Sandman, C.A., George, J. and Stevens, V.C. (1978a). *Physiol. Behav., 20,* 43

Veith, J.L. Sandman, C.A., Walker, J.M., Coy, D.H., Schally, A.V. and Kastin, A.J. (1978b). *Physiol. Behav., 20,* 539

Verhoeven, W.M.A., van Praag, H.M., van Ree, J.M. and de Wied, D. (1979). *Arch. Gen. Psychiatry 36,* 294

Volavka, J., Mallya, A., Baig, S. and Perez-Cruet, J. (1977). *Science 196,* 1227

Walker, B.B. and Sandman, C.A. (1979). *Amer. J. Ment. Def., 83,* 346

Walker, J.M. Berntson, G.B., Sandman, C.A., Coy, D., Schally, A.V. and Kastin, A.J. (1977a). *Science 196,* 85

Walker, J.M., Sandman, C.A., Berntson, G.B., McGivern, R.F., Coy, D.H. and Kastin, A.J. (1977b). *Pharmacol. Biochem. Behav., 7,* 543

Walter, R., Ritzman, R.F., Bhargava, H.N. and Flexner, L.B. (1979). *Proc. Nat. Acad. Sci., 76,* 518

Ward, M.M., Sandman, C.A., George, J. and Shulman, H. (1979). *Physiol. Behav., 22,* 669

Watson, S.J. and Akil, H. (1979). Eur. J. Pharmacol., *58,* 101

de Wied, D. and Bohus, B. (1966). *Nature 212,* 1484

de Wied, D., Bohus, B., Van Ree, J.M. and Urban, I. (1978). *Pharmacol. Exp. Therap., 204,* 570

Will, J.C., Abuzzahab, F.S. and Zimmermann, R.F. (1978). *Psychopharmacol. Bull., 14,* 25

2

BEHAVIOURAL EFFECTS OF PITUITARY HORMONES

Victor M. Wiegant and David de Wied

TABLE OF CONTENTS

I	INTRODUCTION	31
II	SHORT-TERM INFLUENCES OF ACTH ON MOTIVATIONAL PROCESSES	32
	A. Experiments with Hypophysectomized Rats	32
	B. Experiments with Intact Rats	32
	C. Structure-activity Studies with ACTH-like Peptides	34
III	PEPTIDES RELATED TO β-LPH AFFECT ADAPTIVE BEHAVIOUR IRRESPECTIVE OF OPIATE-RECEPTOR INTERACTION	35
IV	NEUROLEPTIC-LIKE EFFECT OF [DES-TYR^1]-γ-ENDORPHIN (DTγE; β-LPH$_{62-77}$)	37
V	THE INDUCTION OF EXCESSIVE GROOMING BY PEPTIDES RELATED TO ACTH AND β-LPH: A PHENOMENON INVOLVING OPIATE RECEPTORS	38
VI	MECHANISM OF ACTION OF ACTH-LIKE NEUROPEPTIDES	40
	A. ACTH: Neurotransmitter, Neuromodulator, Neurohormone	40
	B. Is cAMP a Second Messenger for ACTH-like Peptides in the Brain?	42
	C. Does ACTH Modulate Brain Processes through an Effect on Membrane Phosphorylation?	43
	D. A Model for ACTH Action on the Brain	44
VII	SUMMARY AND CONCLUSIONS	45
	REFERENCES	46

I. INTRODUCTION

Hormones secreted by the pituitary play an important role in maintaining home-ostasis for the organism. Observations by Selye (1950) on the 'general adaptation syndrome' implicated pituitary-adrenal-system hormones as functional principles in adaptational processes. Little attention has been paid to the relation of hormonal effects on behaviour and the functioning of central nervous structures. For, from a classic-endocrinological point of view, the brain was not recognized as a target organ for these hormones. The study of impaired learning behaviour of animals after removal of the pituitary (de Wied, 1969), however, clearly indicated that hypophyseal principles are involved in a number of brain functions and that they are important for the maintenance of normal behavioural patterns. Impaired behaviour, as mentioned above, was readily restored by the substitution of ACTH or α-MSH, but also by treatment with fragments of these peptide hor-mones that lack the classic endocrine activity of the parent hormone. Also, in intact animals it was found that learning and memory processes can be modulated by peptides related to ACTH or β-LPH (de Wied, 1969; de Wied et al., 1978a). Based on such observations, it was postulated that the pituitary manufactures peptides that are released upon adequate stimulation and influence processes of learning, memory and motivation by direct action on the central nervous system (de Wied, 1969). Indeed, ACTH, α-MSH, β-LPH and fragments of this hormone, the endorphins, have been found not only in the pituitary, but in many brain structures as well (Krieger et al., 1977; Orwall et al., 1979; Rossier et al., 1977). Recently, it has been suggested that specific enzyme systems present in pituitary and brain generate bioactive peptides from inactive precursor molecules (Walter et al., 1973; Austen et al., 1977; Burbach et al., 1979). In this way, β-endorphin (β-LPH$_{61-91}$), a peptide with opiate-like properties, can be generated from the nonopiate-like hormone β-LPH (Gráf et al., 1976; Bradbury et al., 1976b). β-Endorphin, in turn, may serve as precursor for a series of shorter sequences with a variety of behavioural activities. For, in the presence of brain membranes, β-endorphin can be metabolized to α-endorphin, [des-tyr[1]] -α-endorphin, γ-endorphin and [des-tyr[1]] -γ-endorphin (Burbach et al., 1979, 1980). Likewise, ACTH may function as a precursor molecule for smaller peptides with differential activities (de Wied, 1974). Interestingly, it has been demonstrated that ACTH, β-LPH and possibly still other peptides are derived from the same large precursor molecule (Mains et al., 1977; Peng Loh, 1979). In addition, immunohistochemical studies revealed a widespread and diffuse neuronal system in the central nervous system, containing β-LPH, β-endorphin and ACTH- immunoreactivity (see Watson et al., 1978a; Watson and Akil, 1980). Thus, on environmental stimula-

31

tion, peptides with different intrinsic activities may be released from pituitary or central cells and modulate the activity of neuronal systems in the brain. This altered activity finally results in behavioural adaptation of the organism to the environmental stimulus.

II. SHORT-TERM INFLUENCES OF ACTH ON MOTIVATIONAL PROCESSES

A. *Experiments with Hypophysectomized Rats*

The idea that hypophyseal hormones are involved in learning processes originates from studies on the rate of acquisition of an active avoidance response in hypophysectomized rats (Applezweig and Baudry, 1955; Applezweig and Moeller, 1959). These studies showed that complete removal of the pituitary gland in rats severely impairs acquisition of a shuttle-box avoidance response. Extirpation of the adenohypophysis resulted in a similar behavioural deficit (de Wied, 1964), indicating that the deficiency in avoidance learning is caused by absence of anterior pituitary hormones. Indeed, the removal of the posterior lobe of the pituitary, which includes the intermediate lobe, does not interfere with avoidance acquisition (de Wied, 1965).

Administration of natural ACTH in amounts that maintain adrenal activity in hypophysectomized rats restored avoidance learning almost completely. This effect is not the result of the ACTH-mediated adrenal activity, since dexamethasone failed to stimulate shuttle-box avoidance behaviour in hypophysectomized rats (de Wied, 1971). Thus, the behavioural effect of ACTH must be due to an extra target effect of this polypeptide hormone. That indeed it is was demonstrated in experiments using structurally related peptides such as α-MSH, $ACTH_{1-10}$ and $ACTH_{4-10}$ (for the structure of the ACTH-like peptides, see Table 2.1). These peptides lack corticotropic activity, yet they are as effective as $ACTH_{1-24}$ in restoring the rate of acquisition of the avoidance response in hypophysectomized rats (de Wied, 1969). In view of these findings it was postulated that peptides like $ACTH_{4-10}$, or closely resembling this heptapeptide, normally operate in the acquisition of new behavioural patterns. Such peptides may be synthesized by the pituitary gland or generated from precursor molecules and released on adequate stimulation to affect brain structures involved in learning processes.

B. *Experiments with Intact Rats*

The effect of ACTH and related peptides on the acquisition of conditioned avoidance behaviour, which can be readily demonstrated in hypothysectomized rats, is rather difficult to observe in intact rats (Murphy and Miller, 1955; Beatty *et al.,* 1970; Kelsey, 1975). Extinction or retention of conditioned avoidance behaviour, which can be readily demonstrated in hypophysectomized peptides. Murphy and Miller (1955) first showed that ACTH, when injected during shuttle-box training, delays subsequent extinction of the behaviour. A more pronounced effect was found, however, when ACTH was given during the

Table 2.1: Amino-acid Sequences of a Number of ACTH Analogs

	1	2	3	4	5	6	7	8	9	10	11	12	13	14	15	16	17	18	19	20	21	22	23	24
ACTH$_{1-24}$	H-ser	-tyr	-ser	-met	-glu	-his	-phe	-arg	-trp	-gly	-lys	-pro	-val	-gly	-lys	-lys	-arg	-arg	-pro	-val	-lys	-val	-tyr	-pro -OH
α-MSH	Ac-ser	-tyr	-ser	-met	-glu	-his	-phe	-arg	-trp	-gly	-lys	-pro	-val -OH											
ACTH$_{4-10}$				H-met	-glu	-his	-phe	-arg	-trp	-gly -OH														
				0																				
				↑			D																	
Org 2766				H-met	-glu	-his	-phe	-lys	-phe -OH															
				0																				
				↑			D																	
Org 5041				H-met	-glu	-his	-phe	-lys	-phe	-glu	-lys	-pro	-val	-gly	-lys	-lys	-NH$_2$							

extinction period (de Wied, 1967). This behavioural influence occurs independently of the action of ACTH on the adrenal cortex, for ACTH is also active on the extinction of shuttle-box avoidance behaviour in adrenalectomized rats (Miller and Ogawa, 1962). Moreover, α-MSH, β-MSH, ACTH$_{1-10}$ and ACTH$_{4-10}$ are as active as natural ACTH in delaying extinction of the avoidance response (de Wied, 1966). In addition, ACTH and related peptides delay the extinction of one-way active pole-jumping avoidance behaviour (de Wied, 1966), improve maze performance (Flood *et al.,* 1976), facilitate passive avoidance behaviour (Levine and Jones, 1965; Lissák and Bohus, 1972; Kastin *et al.,* 1973; de Wied, 1974; Flood *et al.,* 1976), delay extinction of food-motivated behaviour in hungry rats (Leonard, 1969; Sandman *et al.,* 1969; Gray, 1971; Garrud *et al.,* 1974; Flood *et al.,* 1976), and delay extinction of conditioned taste aversion (Rigter and Popping, 1976) and sexually motivated behaviour (Bohus *et al.,* 1975).

The behavioural effects of peptides related to ACTH appear to be of a short-term nature. A single injection of ACTH$_{4-10}$ delays extinction of a pole-jumping avoidance response or facilitates passive avoidance retention in intact rats for a few hours only (de Wied, 1974). Similarly, cessation of the administration of ACTH$_{4-10}$ in hypophysectomized rats, which normalized the level of performance in these animals, leads to a progressive deterioration of avoidance behaviour despite shock punishment (Bohus *et al.,* 1973). These observations led to the hypothesis that ACTH and related peptides temporarily increase the motivational influences of environmental stimuli, which results in an increase in the probability of generating stimulus-specific responses.

However, these peptides affect learning and memory processes as well. They attenuate carbon-dioxide-induced amnesia for a passive avoidance response when administered prior to the retention test, but not when given prior to acquisition and the induction of amnesia (Rigter *et al.,* 1974). Furthermore, they alleviate amnesia produced by electroconvulsive shock (Rigter *et al.,* 1974; Rigter and van

Riezen, 1975) or by intracerebral administration of the protein-synthesis inhibitor puromycin (Flexner and Flexner, 1971) or anisomycin (Flood *et al.,* 1976). Rigter *et al.* (1974) interpreted the effect of $ACTH_{4\text{-}10}$ on amnesia as an influence on retrieval processes. Gold and van Buskirk (1976) found that post-trial administration of ACTH can enhance or impair later retention, depending upon the dose of the peptide. These authors suggest that ACTH modulates the memory-storage processing of recent information. Isaacson *et al.* (1976) demonstrated that $ACTH_{4\text{-}10}$ improves the use of information provided on the location of reward in a four-table choice situation. These findings are obviously not in conflict with a motivational hypothesis, because motivational effects cannot be excluded from most of the paradigms used in these studies.

C. Structure-activity Studies with ACTH-like Peptides

Structure-activity studies to determine the essential elements required for the behavioural effect of ACTH revealed that not more than four amino-acid residues are needed. Thus, $ACTH_{4\text{-}7}$ is as effective as the whole ACTH molecule in delaying extinction of pole-jumping avoidance behaviour (Greven and de Wied, 1973; de Wied *et al.,* 1975). The amino-acid residue phenylalanine in position 7 plays a key role in this behavioural effect of ACTH. Replacement of this amino acid by the D-enantiomer in $ACTH_{1\text{-}10}$ (Bohus and de Wied, 1966), $ACTH_{4\text{-}10}$ or in $ACTH_{4\text{-}7}$ (Greven and de Wied, 1973) causes an effect on the extinction of avoidance behaviour opposite to that found with nonsubstituted ACTH fragments. Such [D-phe[7]] ACTH analogs facilitate the extinction of active avoidance behaviour or approach behaviour motivated by food (Garrud *et al.,* 1974). However, [D-phe[7]] $ACTH_{4\text{-}10}$, like $ACTH_{4\text{-}10}$, facilitates passive avoidance behaviour when given prior to the retention test (de Wied, 1974) but, in relatively high doses, attenuates passive avoidance behaviour when administered immediately following the learning trial (Flood *et al.,* 1976) and delays extinction of conditioned taste aversion (Rigter and Popping, 1976). Replacement of other amino-acid residues in [lys[8]] $ACTH_{4\text{-}9}$ does not facilitate the extinction of avoidance behaviour (Greven and de Wied, 1973). The behavioural activity of ACTH fragments can be completely dissociated from inherent endocrine, metabolic and opiate-like activities by modification of the molecule. Substitution of met[4] by methionine sulfoxide, arg[8] by D-lys and trp[9] by phe yields a peptide (Org 2766) which is behaviourally a thousand times more active than $ACTH_{4\text{-}10}$ (Greven and de Wied, 1973). It possesses, however, a thousand times less MSH activity and its steroidogenic action is markedly reduced. It has no fat-mobilizing activity nor opiate-like effect, as assessed in the guinea-pig ileum preparation.

Although the sequence $ACTH_{4\text{-}7}$ is the smallest part of the molecule with full behavioural activity, as determined on extinction of pole-jumping avoidance behaviour (de Wied *et al.,* 1975), more activity sites are present in ACTH since $ACTH_{7\text{-}10}$ and $ACTH_{11\text{-}24}$ also contain some activity. Eberle and Schwyzer (1975) also found a second activity site (region 11-13) in α-MSH. The residual potency of $ACTH_{7\text{-}10}$ can be increased to the same level as that of $ACTH_{4\text{-}10}$ by extending the carboxyl terminus with $ACTH_{11-16}$ to $ACTH_{7-16}$. Thus, the essential elements for avoidance behaviour are not located exclusively in the region $ACTH_{4-7}$ but also in other areas of the molecule. In these areas the information may be present in a latent form, which is potentiated by chain elongation.

The tripeptide H-Phe-D-Lys—Phe-OH, which is the major breakdown product of the highly potent hexapeptide (Org 2766), has only minor behavioural effects (de Wied *et al.*, 1975). Again, chain elongation with $ACTH_{10-16}$ restores the potency. Substitution of Lys^{11} by the D-enantiomer further augments the effect on avoidance behaviour while extension of the NH_2-terminus with $Met(0)^4$-Glu^5-His^6 further potentiates the action, to yield a peptide (Org 5041) which is three hundred thousand times more active than $ACTH_{4-10}$ (Greven and de Wied, 1977).

The elongation of the peptide chain, as well as the modification of the structure by amino-acid substitutions, certainly contributes to the metabolic stability of the peptide (Witter *et al.*, 1975). However, the strong behavioural potentiation may also reflect increased affinity and/or intrinsic activity for the postulated receptor sites in the central nervous system.

III. PEPTIDES RELATED TO β-LPH AFFECT ADAPTIVE BEHAVIOUR IRRESPECTIVE OF OPIATE-RECEPTOR INTERACTION

In view of the potent behaviour effects of ACTH-like peptides, a program was started around 1970 to isolate neuropeptides with behavioural activity from hog pituitary material (Lande *et al.*, 1973). The biological activity of the fractions isolated was assayed on extinction of active and passive avoidance behaviour. One of the peptides obtained in pure form yielded three small peptides upon tryptic digestion. The amino-acid composition of two of these fragments showed striking similarity with that of $β$-LPH_{61-69} and $β$-LPH_{70-79}. The peptides possessed potent activity in the behavioural tests, but the amount available at the time was small and did not allow structure analysis. Several years later, it was discovered that peptides with high affinity for opiate binding sites and other opiate-like characteristics occur naturally in the brain (Hughes *et al.*, 1975; Bradbury *et al.*, 1976c; Guillemin *et al.*, 1976). These peptides, designated as enkephalins and endorphins, appeared to be structurally related to C-terminal sequences of $β$-LPH (for the structure of endorphins, see Table 2.2).

Extinction of pole-jumping avoidance behaviour was used to assay the behavioural effect of the endorphins. It was found that these peptides, like ACTH and neurohypophyseal hormones, profoundly affect active and passive avoidance behaviour. Following subcutaneous injection, α-endorphin ($β$-LPH_{61-76}) appeared to be the most potent peptide in delaying the extinction of pole-jumping avoidance behaviour (de Wied *et al.*, 1978a). On a molar basis, α-endorphin was over 30 times more active than $ACTH_{4-10}$. It also was more active than $β$-endorphin

Table 2.2: Amino-acid Sequences of Various Endorphins

	1 2 3 4 5 6 7 8 9 10 11 12 13 14 15 16 17 18 19 20 21 22 23 24
β-Endorphin (β-LPH$_{61-91}$)	H-tyr-gly-gly-phe-met-thr-ser-glu-lys-ser-gln-thr-pro-leu-val-thr-leu-phe-lys-asn-ala-ile-val -lys

asn 25
ala 26
his 27
lys 28
lys 29
gly 30
glu 31
OH

γ-Endorphin (β-LPH$_{61-77}$)	H-tyr-gly-gly-phe-met-thr-ser-glu-lys-ser-gln-thr-pro-leu-val-thr-leu-OH
[Des-tyr^1] γ-endorphin (β-LPH$_{62-77}$)	H-gly-gly-phe-met-thr-ser-glu-lys-ser-gln-thr-pro-leu-val-thr-leu-OH
α-Endorphin (β-LPH$_{61-76}$)	H-tyr-gly-gly-phe-met-thr-ser-glu-lys-ser-gln-thr-pro-leu-val-thr-OH
β-LPH$_{61-69}$	H-tyr-gly-gly-phe-met-thr-ser-gly-lys-lys-OH
Met-enkephalin (β-LPH$_{61-65}$)	H-tyr-gly-gly-phe-met-OH

(β-LPH$_{61-91}$). Thus, in contrast to the analgesic activity of the endorphins, for which much higher doses are needed (Bradbury *et al.*, 1976a), the behavioural action increased upon shortening of the peptide chain. The relatively weak effect of β-endorphin on avoidance behaviour was considered to result from the metabolic formation of fragments of β-endorphin with opposite behavioural activity. γ-Endorphin (β-LPH$_{61-77}$), which differs from α-endorphin by only one additional C-terminal amino acid, facilitates the extinction of pole-jumping avoidance behaviour (de Wied *et al.*, 1978a). Intraventricular administration of the respective peptides mimicked the effect of systemic administration, but much less peptide was needed to elicit equipotent behavioural effects.

The opiate-like character of endorphins is completely dependent on the presence of the N-terminal tyrosine residue. Removal of this amino acid from, for instance, β-endorphin caused complete loss of opiate-like activity, as determined on guinea-pig ileum, and destroyed the affinity for opiate binding sites (Guillemin *et al.*, 1976; Frederickson, 1977; de Wied *et al.*, 1978b). In contrast [des-tyr^1] γ-endorphin (DTγE; β-LPH$_{62-77}$) was even more potent than γ-endorphin on avoidance behaviour, and significant effects were obtained in doses as low as 30 ng given subcutaneously or 300 pg intracerebroventricularly (de Wied *et al.*, 1978b). In addition DTγE lacks the capacity to induce excessive grooming behaviour in rats (Gispen *et al.*, 1980), a behavioural model for opioid activity (see elsewhere in this chapter). In line with these observations is the finding that neither endorphin nor ACTH effects on the extinction of pole-jumping avoidance behaviour could be blocked by specific opiate antagonists (de Wied *et al.*, 1978a). Thus, the influence of endorphins and of ACTH and related peptides on avoidance behaviour takes place independently of opiate receptor sites in the brain.

IV. NEUROLEPTIC-LIKE EFFECT OF [DES-TYR^1] -γ-ENDORPHIN (DTγE; β-LPH$_{62-77}$)

Intraventricular administration of microgram doses of β-endorphin in rats produced a naloxone-reversible catatonia (Bloom *et al.*, 1976). Injection of β-endorphin directly into the periaquaeductal gray caused profound sedation and catalepsia, while fragments of this peptide caused attenuated forms of this behaviour (Jacquet and Marks, 1976). These effects could also be blocked by naloxone pretreatment. The authors suggested a role for β-endorphin as an endogenous neuroleptic. Segal *et al.* (1977) showed that in the dose range used, however, β-endorphin had the profile of effects observed for morphine, and not for the neuroleptic haloperidol.

Since the classical studies of Courvoisier *et al.* (1952) acquisition and extinction of an avoidance task have been considered sensitive substrates for neuroleptic activity. Thus, haloperidol facilitates pole-jumping avoidance behaviour and attenuated passive avoidance behaviour, as did γ-endorphin-type peptides (Kovács and de Wied, 1978). In contrast, the effects of α-endorphin on avoidance behaviour were in some respects comparable to those of amphetamine (Kovács and de Wied, 1978). In addition, DTγE was active in so-called grip tests. For example, animals treated subcutaneously with this peptide hung suspended above the floor of the cage with their front paws grasping a pencil for a significantly longer period of time than did saline- or α-endorphin-treated rats (de Wied *et al.*, 1978b). Such an effect is characteristic for drugs with neuroleptic properties like haloperidol, and is not observed in animals treated with morphine or β-endorphin. Moreover, like haloperidol, DTγE interfered with ACTH-induced excessive grooming behaviour in rats, when injected into the neostriatum or the nucleus accumbens (Wiegant *et al.*, 1977; Cools *et al.*, 1978; Gispen *et al.*, 1980). In a similar experimental setting, α-endorphin did not block the grooming response.

The psychopharmacological actions of DTγE, such as facilitation of the extinction of active avoidance behaviour, attenuation of passive avoidance behaviour, interference with ACTH-induced excessive grooming behaviour and the various positive grip tests, are characteristic effects of neuroleptic drugs. In view of this, it was postulated that a fault in endorphin metabolism, resulting in reduced availability of DTγE or of a closely related peptide, is an etiological factor in psychopathological states for which neuroleptics are effective (de Wied, 1979). To substantiate this hypothesis, the influence of DTγE was studied in chronic schizophrenic patients who were resistant to treatment with conventional drugs. Nearly all patients showed at least a transient improvement upon treatment with DTγE (van Ree *et al.*, 1978; Verhoeven *et al.*, 1978, 1979).

V. THE INDUCTION OF EXCESSIVE GROOMING BY PEPTIDES RELATED TO ACTH AND β-LPH: A PHENOMENON INVOLVING OPIATE RECEPTORS

Intraventricular but not systemic injection of ACTH or N-terminal fragments of this hormone in mammals produces a characteristic stretching and yawning syndrome (Ferrari *et al.*, 1963; Gessa *et al.*, 1967). At least in some species, this syndrome is preceded by an enhanced display of grooming behaviour (Ferrari *et al.*, 1963; Izumi *et al.*, 1973; Gispen *et al.*, 1975). In view of the short latency to the onset of the effect and its independence of endocrine activity, the induction of excessive grooming behaviour seems the result of a direct effect of ACTH on the central nervous system (Gispen *et al.*, 1975). This is further substantiated by the observation that systemic administration of the peptide does not result in excessive grooming behaviour.

Structure-activity studies revealed that $ACTH_{1-16}-NH_2$, α-MSH and β-MSH are as active as $ACTH_{1-24}$ (Gispen *et al.*, 1975). The shortest sequence possessing activity was $ACTH_{4-7}$, as was the case in studies on avoidance behaviour (Wiegant and Gispen, 1977; Greven and de Wied, 1973). $ACTH_{4-10}$ and $ACTH_{1-10}$, peptides containing full information for the effect on avoidance behaviour, were inactive in the excessive-grooming model, but [D-phe^7]-substitution rendered them active (Gispen *et al.*, 1975). D-substitution in peptides often improves their metabolic stability (Greven and de Wied, 1975). However, the enhanced activity after [D-phe^7]-substitution cannot simply be explained by such an improved stability, for the peptides [met(0)^4, D-lys^8, phe^9]ACTH$_{4-9}$ (Org 2766) — a highly potentiated peptide in avoidance behaviour — and [D-arg^8]ACTH$_{4-10}$ were inactive in the induction of excessive grooming (Gispen *et al.*, 1975). For both effects on avoidance behaviour and excessive grooming the essential information is comprised in the 4-7 sequence, and phe^7 appears to play a key role in the ACTH/central-nervous-system interactions underlying these behavioural actions. Yet, the data mentioned above suggest different mechanisms of action for the peptides in the two behavioural models. This is further substantiated by experiments underscoring the opiate charater of the grooming response, in contrast to the avoidance effect.

Zimmerman and Krivoy (1973, 1974) showed that ACTH reduced the morphine-induced inhibition of spinal reflex activity in cat and frog spinal-cord preparations, both *in vivo* and *in vitro*. ACTH counteracts the analgesic effect of morphine in rodents (Winter and Flataker, 1951). Also, N-terminal fragments of the hormone without appreciable endocrine activity reduce the analgesic response of rats to morphine as measured with the hot-plate test (Gispen *et al.*, 1976a). These studies suggested interaction of ACTH with opiate-sensitive sites, and indeed it was shown that ACTH and N-terminal fragments of ACTH have affinity for rat-brain opiate-binding sites *in vitro* (Terenius, 1975; Terenius *et al.*, 1975). Therefore, the relation of ACTH-induced excessive grooming behaviour to opiate-sensitive systems was investigated. Peripheral administration of specific opiate antagonists (naloxone, naltrexone) completely inhibited the display of excessive grooming induced by $ACTH_{1-24}$ (Gispen and Wiegant, 1976). Moreover, morphine induced moderate grooming behaviour if injected intraventricularly (i.c.v.) in low doses (0.05-0.5 μg), whereas higher amounts resulted in overall behavioural depression (Gispen and Wiegant, 1976). Injection of low doses of β-endorphin produced an even more marked increase in grooming activity than ACTH (Gispen *et al.*, 1976b). A dose as low as 10 ng already induced significant grooming, whereas microgram doses resulted in behavioural depression (Wiegant *et al.*, 1978a). The behavioural repertoire also differed from that of ACTH, in that stretching and yawning were less prominent with β-endorphin (Gispen *et al.*, 1976b). Shortening the peptide sequence from the C-terminus resulted in a rapid and progressive loss of activity: γ-endorphin and α-endorphin only possessed moderate potency, whereas β-LPH_{61-69} was the shortest peptide of the endorphin series which induced notable grooming behaviour. Met-enkephalin was inactive even if injected in high doses (Gispen *et al.*, 1976b). This may be due to rapid metabolic inactivation of the peptide upon injection, since [D-ala^2]-met-enkephalin, a peptide highly resistant to enzymatic breakdown (Pert *et al.*, 1976) did induce a display of excessive grooming (Wiegant and Gispen, unpublished results). It is found that DTγE, a peptide lacking the N-terminal tyrosine residue essential for the opiate character of endorphins (Guillemin *et al.*, 1976; Frederickson, 1977; de Wied *et al.*, 1978b), does not possess grooming-inducing properties (Gispen *et al.*, 1980). The 'opiate-like' character of peptide-induced grooming was further suggested by the observation that acute tolerance occurs for this phenomenon; that is, an i.c.v. injection of $ACTH_{1-24}$ was ineffective in grooming induction when given between one and eight hours after a previous i.c.v. treatment of the rat with $ACTH_{1-24}$ (Jolles *et al.*, 1978). A similar phenomenon occurred with β-endorphin (Wiegant *et al.*, 1978b). Systemic treatment with naltrexone at the time of the first i.c.v. injection not only suppressed the excessive grooming activity normally resulting from such an injection, but also prevented the development of acute tolerance (Jolles *et al.*, 1978; Wiegant *et al.*, 1978b). Interestingly, cross-tolerance was demonstrated between $ACTH_{1-24}$ (second injection) and β-endorphin, [D-Phe^7]$ACTH_{4-10}$ or morphine (first injection) (Jolles *et al.*, 1978). These observations point towards a close relationship between ACTH- and endorphin-sensitive structures in the central nervous system. A functional connection between ACTH and endorphin systems in the brain is further suggested by observations that β-LPH, β-endorphin and ACTH-like peptides can occur in the same neuronal cells (Watson *et al.*, 1978a; Pelletier, 1979; Sofroniew, 1979; Watson and Akil, 1980). As it has been shown that ACTH-like peptides have affinity for opiate binding sites (Terenius, 1975; Terenius *et al.*, 1975) but also for β-endorphin binding sites (Akil *et al.*, 1980) in brain tissue *in vitro*, it may well be that these peptides interact with one and the same receptor site or with different receptors located on the same neuron, thereby generating excessive grooming behaviour. However, it cannot be excluded, as yet, that sites sensitive to ACTH-like peptides or endorphins reside on different but interconnected neuronal cells.

VI. MECHANISM OF ACTION OF ACTH-LIKE NEUROPEPTIDES

A. ACTH: Neurotransmitter, Neuromodulator, Neurohormone

It has been postulated that the pituitary is an important source of peptides with neurotropic activity (de Wied, 1969). Such peptides may be released upon adequate stimulation, to affect central nervous structures involved in motivational, learning and memory processes. Via the systemic circulation they may penetrate the blood-brain barrier and reach their target cells in the central nervous system. That indeed the circulation can serve as a route to the brain for peptides may be inferred from studies by Verhoef *et al.* (1977) and Rapoport *et al.* (1980). These authors showed that radio-labeled peptides (ACTH, endorphins) penetrate into the central nervous system after systemic administration. Pituitary peptides could otherwise reach the brain by a direct route via the stalk. The application of a radio-labeled ACTH analog in the anterior pituitary showed a rapid retrograde transport of the peptide to the brain, particularly to the medial basal and periventricular hypothalamic regions (Mezey *et al.*, 1979).

Various approaches have been used to localize the site(s) of action of ACTH-like peptides in the central nervous system. Microinjection and micro-implantation of peptide material, as well as destruction of specific brain regions, indicated that a multiplicity of brain structures is involved in the behavioural effects of the peptides. An intact limbic-midbrain circuitry is essential for the effects of ACTH on avoidance behaviour (van Wimersma Greidanus *et al.*, 1979). Studies using a ^{3}H-ACTH$_{4-9}$ analog which has potent effects on avoidance learning showed specific uptake of this peptide in the dorsal septal nucleus after systemic administration (Verhoef *et al.*, 1977). In the induction of excessive grooming by ACTH, the dopaminergic nigrostriatal tract plays a crucial role (Cools *et al.*, 1978; Gispen *et al.*, 1980). Lesion studies indicated that the integrity of the hippocampus is also important for this effect, whereas the amygdala, the mammillary bodies and the hippocampus appeared to be involved in the induction of the stretching and yawning syndrome by ACTH (Colbern *et al.*, 1977). These data suggest that structures sensitive to ACTH and to neuropeptides related to this hormone are not confined to a single neuronal circuit, but indeed are widely distributed throughout the central nervous system.

The occurrence throughout the brain of ACTH-like immunoreactive material independent of the pituitary has been demonstrated by Krieger *et al.* (1977). Cytochemical studies visualized neuronal systems, showing ACTH-immunoreactivity originating from cell bodies located in the arcuate nucleus and with terminals largely distributed throughout limbic-midbrain areas (Pelletier and Leclerc, 1979; Watson *et al.*, 1978b). The existence of a central ACTH-containing system underscores the physiological importance of ACTH-like neuropeptides. Yet a function for ACTH as a classical neurotransmitter in the central nervous system lacks experimental support. Currently, the influence of ACTH-like neuropeptides on brain neuronal activity is best formulated as a neuromodulatory or a neurohormonal one. Neuromodulators are released in the intracellular space by intramural neurons, and alter the communications between neurons by affecting rapid processes at the level of the membrane. Their influence may extend over much larger distances than that of the classical neurotransmitters and their effect may or may not be localized in the synaptic region of the target cell. It might consist of altered neurotransmitter release from the presynaptic neuron, or a change in the effectiveness of neurotransmission by an effect on the postsynaptic element. Pearse (1969) suggested that there exists, in the organism, a diffuse neuroendocrine system constituted of a large variety of cells derived from the same ontogenic ancestors. Cells of this system all show so-called APUD-characteristics (APUD = amine content and/or amine precursor uptake, with presence of amino-acid decarboxylases). The other main property of these cells is that they secrete neuropeptides with intrinsic activities. It has been proposed (Barbeau, 1978) that the main function of peptidergic pathways is a trophic modulation of aminergic functions, pre- or postsynaptically. As to ACTH and related neuropeptides, indeed a multiplicity of influences on the metabolism of biogenic amines in the brain has been described (Dunn and Gispen, 1977; Versteeg, 1980). Neurohormones differ from neuromodulators in that they are released into the circulation by endocrine cells. They may affect central neuronal metabolism through a similar action on the (extra) synaptic membrane. Recently, it was recognized that peptide hormones may also enter their target cells and, once being 'internalized' influence cellular metabolism at the level of the cytosol. Whether or not ACTH-like peptides are internalized in neuronal cells is not clear as yet.

In general, peptide and amine hormones and neurotransmitters are thought to exert their actions on target cells through binding with stereospecific receptors located on the outer surface of the cell membrane. In the second-messenger model formulated by Sutherland (Sutherland and Robison, 1966), interaction of the hormone (the *first messenger*) with its receptor results in activation of the enzyme adenylatecyclase located on the inner side of the membrane. Thus, an increase in the intracellular concentration of 3', 5'-cyclic adenosine monophosphate (cAMP) is produced. Then cAMP (the *second messenger*) broadcasts the message to the various functional compartments of the cell, by activating protein kinases. By its phosphorylating action, a protein kinase may change the functional properties of its substrate proteins. For example, the attachment of a phosphate group to an enzyme may result in activation or inhibition of its catalytic properties (Rubin and Rosen, 1975). In this way, neuronal membranes may undergo changes in permeability resulting in altered neuronal activity (Greengard, 1976). Thus, phosphoproteins act as specifiers ensuring the specificity of the effector-cell response triggered by the hormone, as transduced and amplified by the adenylatecyclase-cAMP system. To date, it is well recognized that, although cAMP may play a cental role in may hormonal and neurotransmitter effects, other mechanisms may also serve as a channel for information transfer through the cell membrane. Such mechanisms may involve second messengers that govern the activity of specific intracellular protein kinases in a manner comparable to cAMP. A second cyclic nucleotide, cGMP, has been proposed as a second messenger in certain types of neurotransmission (Greengard, 1976). Likewise, influx of calcium ions through the cell membrane is thought to function as mediator for the effects of certain hormones and perhaps neurotransmitters (Rasmussen *et al.*, 1975). cGMP as well as the calcium ion are thought to exert their actions in the cell through a variety of protein kinases.

B. Is cAMP a Second Messenger for ACTH-like Peptides in the Brain?

In the light of the second-messenger concept it is expected that ACTH — as neuromodulator or neurohormone — interacts with receptors linked to adenylatecyclase. *In vitro* binding studies using a ^{3}H-ACTH$_{4-9}$ analog did not reveal binding sites in subcellular fractions of brain tissue. Yet behavioural and neuropharmacological evidence points to the existence of such ACTH receptors in the brain. The inability to detect high-affinity sites for ACTH in the brain may be the result of very high affinity and very low capacity of the receptor system (Witter, 1979). On the other hand, the ACTH-receptor complex may be too labile under the conditions used in the binding assay.

To date, the role of cAMP in ACTH/central-nervous-system interactions is not completely clear. Several authors were not able to detect an effect of ACTH on adenylatecyclase activity in cell-free preparations of rat brain tissue (Burkard and Gey, 1968; Von Hungen and Roberts, 1973) or rat cerebral-cortex slices (Forn and Krishna, 1971). On the other hand, indirect indications that ACTH-like peptides may affect brain cyclic nucleotide levels *in vivo* were presented by Rudman and coworkers (Rudman and Isaacs, 1975; Rudman, 1976). They showed that intrathecal injection of microgram quantities of ACTH or α-MSH in rabbits increased concentration of cAMP, but not of cGMP in cerebrospinal fluid. This increase in cAMP may originate from ACTH-sensitive adenylatecyclase located in circumventricular organs (Rudman, 1978). In a preliminary study, Christensen *et al.* (1976) showed that chronic treatment with α-MSH — that is, [AC-ser^1]ACTH$_{1-13}$-NH$_2$ — increased the level of cAMP in the occipital cortex of intact as well as hypox rats. Similar treatment left the level of cGMP unaltered (Spirtes *et al.*, 1978).

Recently, we investigated the influence of N-terminal fragments of ACTH on the accumulation of cAMP in rat brain, using three different approaches, that is, determination of adenylatecyclase activity in broken-cell preparations and in slices from posterior thalamus and neostriatum, and determination of *in vivo* brain cAMP levels (Wiegant and Gispen, 1975; Wiegant *et al.*, 1979). ACTH$_{1-24}$ had a bidirectional effect on the activity of adenylatecyclase in broken-cell preparations, in that it had a stimulatory effect in micromolar concentrations, whereas it inhibited adenylatecyclase in concentration of 0.01 mM and higher. Structure-activity studies revealed that in a concentration of 10^{-4} M ACTH$_{1-16}$-NH$_2$ and ACTH$_{4-7}$ also inhibited adenylatecyclase activity, whereas ACTH$_{11-24}$, ACTH$_{1-10}$, ACTH$_{4-10}$, [D-phe^7]ACTH$_{1-10}$ and [D-phe^7]ACTH$_{4-10}$ were inactive in this respect. These structure-activity relationships closely resemble the structural requirements for the induction of excessive grooming by ACTH (Gispen *et al.*, 1975; Wiegant and Gispen, 1977), except that the [D-phe[7]]- analogs of $ACTH_{4-10}$ and $ACTH_{1-10}$ are without effect in the adenylatecyclase system. Although these peptides exert marked effects in various behavioural paradigms, their mechanism of action seems to differ from that of the natural all-L peptides (Wiegant *et al.*, 1978b). In slices from posterior thalamus and neostriatum, $ACTH_{1-24}$ appeared to stimulate the activity of adenylatecyclase. The effect was rapid, and of short duration. Intraventricular injection of $ACTH_{1-16}$ $-NH_2$ in rats resulted in an increase in the septal cAMP concentration. No effect was observed in the other brain regions studied (Wiegant *et al.*, 1979). From these data it is concluded that ACTH-like peptides may serve as regulators of brain cAMP metabolism.

C. Does ACTH Modulate Brain Processes through an Effect on Membrane Phosphorylation?

Already in the early sixties, Heald (1962) proposed that changes in the state of phosphorylation of membrane proteins may govern the ion- permeability of the neuronal membrane, and thus play a key role in determining the functional activity of the neuron. Since then, much (circumstantial) evidence has been provided to lend support to this suggestion (Greengard, 1976; Williams and Rodnight, 1977), and it was deemed of interest to investigate the involvement of membrane phosphorylation in the effects of ACTH. *In vitro*, $ACTH_{1-24}$ dose-dependently inhibited the endogenous phosphorylation of at least five proteins in a preparation of synaptosomal plasma membrane (SPM) (Zwiers *et al.*, 1976), through direct action on the activity of membrane-bound protein kinase(s) (Zwiers *et al.*, 1978). Interestingly, this effect is not mediated by cAMP, for this nucleotide stimulated the phosphorylation of different membrane proteins. Structure-activity studies with ACTH-like peptides on endogenous phosphorylation of one particular SPM-protein (B-50) showed that $ACTH_{1-24}$, $ACTH_{1-16}$ and $ACTH_{1-13}$ are equally active, and also that $ACTH_{5-18}$ possesses activity. The sequences (11-24), (1-10) and 4-10) were inactive (Zwiers *et al.*, 1978). This structure-activity relation again is very similar to that found for the induction of excessive grooming (see elsewhere in this chapter; Gispen *et al.*, 1975). A relation between protein phosphorylation and the induction of excessive grooming was further suggested by experiments on the *in vivo* injection of ACTH and subsequent *in vitro* assay of endogenous phosphorylation of SPM proteins. In this experimental setting, i.c.v. injection of ACTH resulted in a dose-dependent increase in phosphorylation of SPM proteins. An increased *in vitro* phosphorylation under the experimental conditions used may be explained as the result of *in vivo* inhibition of phosphorylation activity (Zwiers *et al.*, 1977). The ACTH-sensitive protein kinase and its substrate protein B-50 have recently been purified and characterized (Zwiers *et al.*, 1979). B-50 phosphorylating activity in partially purified material showed strong dependence on magnesium and calcium. A fraction containing both the protein kinase and the

B-50 substrate as the only phosphoprotein appeared to have lipid kinase activity, in that it stimulated the formation of phosphatidylinositol-4,5-diphosphate (TPI) from phosphatidylinositol-4-phosphate (DPI) (Jolles *et al.*, 1980). Interestingly, the lipid-phosphorylating activity was dependent on the state of phosphorylation of B-50. These data imply that ACTH, by inhibiting the phosphorlylation of B-50 protein, stimulates the incorporation of phosphate in membrane lipids. Michell (1975) proposed that membrane phospholipids are intimately involved in the control of calcium influx into the cell and storage of the ion in the cell membrane. Thus, it may well be that ACTH modulates central neuronal activity by affecting calcium movements at the level of the membrane.

D. A Model for ACTH Action on the Brain

A large variety of effects of ACTH-like peptides in various behavioural paradigms (acquisition and extinction of conditioned avoidance behaviour, induction of excessive grooming behaviour, etc.; see elsewhere in this chapter) and actions of such peptides at the molecular level (on the metabolism of proteins, nucleic acids, catecholamines, etc.; see Dunn and Gispen, 1977) have been described. Although high-affinity binding sites for ACTH and related peptides have not been demonstrated in the brain (Witter, 1979), the multiplicity and the specificity of the ACTH/central-nervous-system interactions alone point toward the existence of specific ACTH receptors in the brain. Moreover, the studies cited above show the occurrence of membrane events upon ACTH action (modulation of the activity of adenylatecyclase and protein kinase) that are generally believed to take place as a consequence of the interaction of hormones or neurotransmitters with their receptors.

The information encoded in ACTH-like peptides is apparently not transferred to brain cells via a single receptor/second-messenger system, but clearly more than one channel is used, that is activation and inhibition of adenylatecyclase and of the phosphorylation of the variety of substrate proteins in the synaptic membrane (Figure 2.1). The function of one of these phosphoproteins was found to be related with the fate of calcium at the membrane, but other ACTH-regulated phosphoproteins may also play an important role in the modulation of neuronal activity. The structural requirements for the inhibitory effect of ACTH on adenylatecyclase and the induction of excessive grooming are alike. In addition, a strong correlation was observed between the structure-activity relations needed for the grooming response and for the inhibition of B-50 protein phosphorylation. Thus, it could be that two separate messenger systems in the brain, one involving cAMP and the other calcium as second messenger, underlie the induction of excessive grooming by ACTH.

Although the membranes used in our studies are mainly of presynaptic origin, it is not possible as yet to relate any of the effects described to events specifically localized either at the pre- or at the postsynaptic element. Likewise, it remains unclear whether the membrane effects observed occur in one and the same cell, or rather reflect responses of different neuronal elements. Certainly, more work

Figure 2.1: Schematic representation of the possible modes of action of ACTH and related neuropeptides in the central nervous system, where N denotes a neurotransmitter and R a receptor.

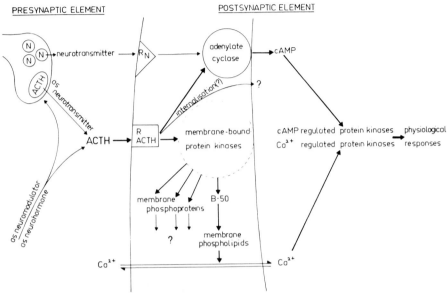

is needed to specify the exact nature of the central-nervous-system responses to ACTH-like peptides at the neurochemical level and their relation to behavioural effects of the peptides.

VII. SUMMARY AND CONCLUSIONS

The study of effects of pituitary hormones on behaviour has greatly enhanced our knowledge of brain functioning. Peptides derived from pituitary hormones profoundly affect behaviour in animals and man. Among neuropeptides, various classes may be distinguished by criteria based on their effects on consolidation and retrieval of information in conditioned avoidance situations. ACTH and related peptides exert a short-term effect, interpreted as an influence on motivational and attentional processes. Endorphins, like ACTH and related peptides, influence adaptive behaviour. α-Endorphin and γ-endorphin, peptides differing only in the terminal amino acid, affect avoidance behaviour in an opposite manner. The principle of closely related peptides with opposite action was also found for [L-phe[7]] ACTH$_{4-10}$ and [D-phe[7]] ACTH$_{4-10}$. A balance between such neuropeptides may thus be important for maintaining normal function of the brain. Disturbances in the control of pituitary hormones and related neuropeptides over brain functions may cause brain dysfunction, and may underlie psychopathological states.

ACTH-like peptides and endorphins induce excessive grooming behaviour in rats upon intraventricular administration. The generation of this behaviour is dependent on functional opiate receptors in the brain; this is in contrast to peptide effects on avoidance behaviour. The grooming response, though clearly pharmacological, has provided additional information on the intimate relationship of ACTH- and endorphin-sensitive central neuronal structures. The importance of such a relation is also apparent from the occurrence in the brain and pituitary of large molecules that serve as precursors for both classes of neuropeptides. Peptide sequences identical to ACTH, α-MSH and endorphins may be liberated enzymatically from the same precursor peptide upon adequate stimulation (Peng Loh, 1979). Subsequently, they may alter transsynaptic neuronal activities in their function as neuromodulators. The nature of the final response of target cells is not only determined by the information encoded in the peptide sequence, but also by the type of peptide-sensitive receptors present on the cell and the intracellular enzymatic machinery.

Indeed, the study of neurochemical actions of ACTH underscored the complexity of the interactions of ACTH with its target cells in the central nervous system (Dunn and Gispen, 1977). Neuropeptides may use various routes to transfer their intrinsic information to the neuronal substrate. It may well be that, for instance, in the induction of excessive grooming − a relatively simple behavioural response − at least two second-messenger systems in the brain are actively involved in decoding and transferring neuropeptide information. The type of intracellular messenger determines the nature of the neurochemical responses finally resulting in the specific electrophysiological output of the cell.

REFERENCES

Akil, H., Hewlet, W.A., Barchas, J.D. and Li, C.H. (1980). *Eur. J. Pharmacol., 60* (in press)
Applezweig, M.H. and Baudry, F.D. (1955). *Psychol. Rep., 1,* 417
Applezweig, M.H. and Moeller, G. (1959). *Acta Psychol., 15,* 602
Austen, B.M., Smyth, D.G. and Snell, C.R. (1977). *Nature 269,* 619
Barbeau, A. (1978). *Adv. Exp. Med. Biol., 113,* 101
Beatty, D.A., Beatty, W.A., Bowman, R.E. and Gilchrist, J.C. (1970). *Physiol. Behav., 5,* 939
Bloom, F., Segal, D., Ling, N. and Guillemin, R. (1976). *Science 194,* 630
Bohus, B. and de Wied, D. (1966). *Science 153,* 318
Bohus, B. and de Wied, D. (1980). In *General, Comparative and Clinical Endocrinology of the Adrenal Cortex,* vol. 3, ed. I. Chester Jones and I.W. Henderson (Academic Press, London) p. 256
Bohus, B., Gispen, W.H. and de Wied, D. (1973). *Neuroendocrinology II,* 137
Bohus, B., Hendrickx, H.H.L., van Kolfschoten, A.A. and Krediet, T.G. (1975). In *Sexual Behavior: Pharmacology and Biochemistry,* ed. M. Sandler and G.L. Gessa (Raven Press, New York) p. 269
Bradbury, A.F., Feldberg, W.F., Smyth, D.G. and Snell, C.R. (1976a). In *Opiates and Endogenous Opioid Peptides,* ed H.W. Kosterlitz (North Holland, Amsterdam) p. 9
Bradbury, A.F., Smyth, D.G. and Snell, C.R. (1976b). *Biochem. Biophys. Res. Comm., 69,* 950
Bradbury, A.F., Smyth, D.G., Snell, C.R., Birdsall, N.J.M. and Hulme, E.C. (1976c). *Nature 260,* 793

Burbach, J.P.H., Loeber, J.G., Verhoef, J., de Kloet, E.R. and de Wied, D. (1979). *Biochem. Biophys. Res. Comm., 86,* 1296

Burbach, J.P.H., Loeber, J.G., Verhoef, J., Wiegant, V.M., de Kloet, E.R. and de Wied, D. (1980). *Nature 283,* 96

Burkard, W.P. and Gey, K.F. (1968). *Helv. Physiol. Pharmacol. Acta 26,* 197

Christensen, C.W., Harston, C.T., Kastin, A.J., Kostrzewa, R.M. and Spirtes, M.A. (1976). *Pharmacol. Biochem. Behav., 5* (suppl. 1), 117

Colbern, D.L., Isaacson, R.L., Bohus, B. and Gispen, W.H. (1977). *Life Sci., 21,* 393

Cools, A.R., Wiegant, V.M. and Gispen, W.H. (1978). *Eur. J. Pharmacol., 50,* 265

Courvoisier, S., Fournel, J., Ducrot, R., Kolsky, M. and Koetschet, P. (1952). *Arch. Int. Pharmacodyn. Ther., 92,* 305

Dunn, A.J. and Gispen, W.H. (1977). *Biobehav. Rev., 1,* 15

Eberle, A. and Schwyzer, R. (1975). *Helv. Chim. Acta 58,* 1528

Ferrari, W., Gessa, G.L. and Vargiu, L. (1963). *Ann. N.Y. Acad. Sci., 104,* 330

Flexner, J.B. and Flexner, L.B. (1971). *Proc. Nat. Acad. Sci., 68,* 2519

Flood, J.F., Jarvik, M.E., Bennett, E.L. and Orme, A.E. (1976). *Pharmacol. Biochem. Behav. 5* (suppl. 1), 41

Forn, J. and Krishna, G. (1971). *Pharmacology 5,* 193

Frederickson, R.C.A. (1977). *Life Sci., 21,* 23

Garrud, P., Gray, J.A. and de Wied, D. (1974). *Physiol. Behav., 12,* 109

Gessa, G.L., Pisano, M., Vargiu, L., Crabai, F. and Ferrari, W. (1967). *Rev. Can. Biol., 26,* 229

Gispen, W.H. and Wiegant, V.M. (1976). *Neurosci. Lett., 2,* 159

Gispen, W.H., Wiegant, V.M., Greven, H.M. and de Wied, D. (1975). *Life Sci., 17,* 645

Gispen, W.H., Buitelaar, J., Wiegant, V.M., Terenius, L. and de Wied, D. (1976a). *Eur. J. Pharmacol., 39,* 393

Gispen, W.H., Wiegant, V.M., Bradbury, A.F., Hulme, E.C., Smyth, D.G., Snell, C.R. and de Wied, D. (1976b). *Nature 264,* 794

Gispen, W.H., Ormond, D., Tenhaaf, J. and de Wied, D. (1980). *Eur. J. Pharmacol., 65* (in press)

Gold, P.E. and van Buskirk, R. (1976). *Behav. Biol., 16,* 387

Gráf, L., Ronai, A.Z., Bajusz, S., Cseh, G. and Szekely, J.I. (1976). *FEBS Lett., 64,* 181

Gray, J.A. (1971). *Nature 229,* 52

Greengard, P. (1976). *Nature 260,* 101

Greven, H.M. and de Wied, D. (1973). *Progr. Brain Res., 39,* 429

Greven, H.M. and De Wied, D. (1977). In *Frontiers of Hormone Research,* Vol. 4, *Melanocyte Stimulating Hormone: Control, Chemistry and Effects,* ed. F.J.H. Tilders, *D.F. Swaab* and T j.B. van Wimersma Greidanus (S. Karger, Basel) p. 140

Guillemin, R., Ling, N. and Burgus, R. (1976). *C.R. Acad. Sci. Paris Ser. D 282,* 783

Heald, P.J. (1962). *Nature 193,* 451

Hughes, J., Smith, T.W., Kosterlitz, H.W., Fothergill, I.A., Morgan, B.A. and Morris, H.R. (1975). *Nature 258,* 577

Isaacson, R.L., Dunn, A.J., Rees, H.D. and Waldock, B. (1976). *Physiol. Psychol., 4,* 159

Izumi, K., Donaldson, J. and Barbeau, A. (1973). *Life Sci., 12,* 203

Jacquet, Y.F. and Marks, N. (1976). *Science 194,* 632

Jolles, J., Wiegant, V.M. and Gispen, W.H. (1978). *Neurosci. Lett., 9,* 261

Jolles, J., Zwiers, H., Schotman, P. and Gispen, W.H. (1980). In *Synaptic Constituents in Health and Disease,* ed. M. Brzin, D. Sket and M. Bachelard (Pergamon Press, London, in press)

Kastin, A.J., Miller, L.H., Nockton, R., Sandman, C.A., Schally, A.V. and Stratton, L.O. (1973). *Progr. Brain Res., 39,* 461

Kelsey, J.E. (1975). *J. Comp. Physiol. Psychol., 88,* 271

Kovács, G.L. and de Wied, D. (1978). *Eur. J. Pharmacol., 53,* 103

Krieger, D.T., Liotta, A. and Brownstein, M.J. (1977). *Proc. Nat. Acad. Sci. (Wash.) 74,* 648

Lande, S., de Wied, D. and Witter, A. (1973). *Progr. Brain Res., 39,* 421

Leonard, B.E. (1969). *Int. J. Neuropharmacol., 8,* 427

Levine, S. and Jones, L.E. (1965). *J. Comp. Physiol. Psychol., 59,* 357

Lissák, K. and Bohus, B. (1972). *Int. J. Psychobiol., 2,* 103

Mains, R.E., Eipper, B.A. and Ling, N. (1977). *Proc. Nat. Acad. Sci. USA 74,* 3014

Mezey, E.M., Kivovics, P. and Palkovits, M. (1979). *Trends Neurosci., 2,* 57
Michell, R.H. (1975). *Biochim. Biophys. Acta 415,* 81
Miller, R.E. and Ogawa, N. (1962). *J. Comp. Physiol. Psychol., 55,* 211
Murphy, J.V. and Miller, R.E. (1955). *J. Comp. Physiol. Psychol., 48,* 47
Orwall, E., Kendall, J.W., Lamorena, L. and McGievra, R. (1979). *Endocrinology 104,* 1845
Pearse, A.G.E. (1969). *J. Histochem. Cytochem., 17,* 303
Pelletier, G. (1979). *J. Histochem. Cytochem., 27,* 1046
Pelletier, G. and Leclerc, R. (1979). *Endocrinology 104,* 1426
Peng Loh, Y. (1979). *Proc. Nat. Acad. Sci. (USA) 76,* 796
Pert, C.B., Pert, A., Chang, J-k. and Fong, B.T.W. (1976). *Science 194,* 330
Rapoport, S.I., Klee, W.A., Pettigrew, K.D. and Ohno, K. (1980). *Science 207,* 84
Rasmussen, H., Jensen, P., Lake, W., Friedmann, N. and Goodman, D.B.P. (1975). *Adv. Cyclic Nucl. Res., 5,* 375
van Ree, J.M., Verhoeven, W.M.A., van Praag, H.M. and de Wied, D. (1978). In *Characteristics and Function of Opioids,* ed. J.M. van Ree and L. Terenius (Elsevier, Amsterdam) p. 181
Rigter, H. and van Riezen, H. (1975). *Physiol. Behav., 14,* 563
Rigter, H. and Popping, A. (1976). *Psychopharmacology 46,* 255
Rigter, H., van Riezen, H. and de Wied, D. (1974). *Physiol. Behav., 13,* 381
Rossier, J., Vargo, T.M., Minick, S., Ling, N., Bloom, F.E. and Guillemin, R. (1977). *Proc. Nat. Acad. Sci. (Wash.) 74,* 5162
Rubin, C.S. and Rosen, O.M. (1975). *Ann. Rev. Biochem., 44,* 831
Rudman, D. (1976). *Neuroendocrinology 20,* 235
Rudman, D. (1978). *Endocrinology 103,* 1556
Rudman, D. and Isaacs, J.W. (1975). *Endocrinology 97,* 1476
Sandman, C.A., Kastin, A.J. and Schally, A.V. (1969). *Experientia* (Basel) 25, 1001
Segal, D.S., Browne, R.G., Bloom, F., Ling, N. and Guillemin, R. (1977). *Science 198,* 411
Selye, H. (1950). *The Physiology and Pathology of Exposure to Stress* (Acta Med., Montreal)
Sofroniew, M.V. (1979). *Am. J. Anat., 154,* 283
Spirtes, M.A., Christensen, C.W., Hartson, C.T. and Kastin, A.J. (1978). *Brain Res., 144,* 189
Sutherland, E.W. and Robison, G.A. (1966). *Pharmacol. Rev., 18,* 145
Terenius, L. (1975). *J. Pharm. Pharmacol., 27,* 450
Terenius, L., Gispen, W.H. and de Wied, D. (1975). *Eur. J. Pharmacol., 33,* 395
Verhoef, J., Witter, A. and de Wied, D. (1977). *Brain Res., 131,* 117
Verhoeven, W.A., van Praag, H.M., Botter, P.A., Sunier, A., van Ree, J.M. and de Wied, D. (1978). *The Lancet i,* 1046
Verhoeven, W.M.A., van Praag, H.M., van Ree, J.M. and de Wied, D. (1979). *Arch. Gen. Psychiat., 36,* 294
Versteeg, D.H.G. (1980). *Pharmacol. Ther.,* (in press)
Von Hungen, K. and Roberts, S. (1973). *Eur. J. Biochem., 36,* 391
Walter, R., Griffiths, E.C. and Hooper, K.C. (1973). *Brain Res., 60,* 449
Watson, S.J. and Akil. H. (1980). *Brain Res., 82,* 217
Watson, S.J., Richard, C.W. III and Barchas, J.D. (1978a). *Science 200,* 1180
Watson, S.J., Akil, H., Richard, C.W., III and Barchas, J.D. (1978b). *Nature 275,* 226
de Wied, D. (1964). *Amer. J. Physiol., 207,* 255
de Wied, D. (1965). *Int. J. Neuropharmacol., 4,* 157
de Wied, D. (1966). *Proc. Soc. Exp. Biol. Med., 122,* 28
de Wied, D. (1967). *Excerpta Med. Int. Congr. Serv., 132,* 945
de Wied, D. (1969). In *Frontiers in Neuroendocrinology,* ed. W.F. Ganong and L. Martini (Oxford University Press, New York) p. 97
de Wied, D. (1971). In *Normal and Abnormal Development of Brain and Behavior,* ed. G.B.A. Stoelinga and J.J. van der Werff ten Bosch (Leiden University Press, Leiden) p. 315
de Wied, D. (1974). In *The Neurosciences: Third Study Program,* ed. F.O. Schmitt and F.G. Worden (MIT Press, Cambridge, Mass.) p. 653
de Wied, D. (1979). In *Central Regulation of the Endocrine System,* ed. K. Fuxe, T. Hökfelt and R. Luft (Plenum Press, New York) p. 297
de Wied, D., Witter, A. and Greven, H.M. (1975). *Biochem. Pharmacol., 24,* 1463

de Wied, D., Bohus, B., van Ree, J.M. and Urban, I. (1978a). *J. Pharmacol. Exp. Ther., 204,* 570

de Wied, D., Kovács, G.L., Bohus, B., van Ree, J.M. and Greven, H.M. (1978b). *Eur. J. Pharmacol., 49,* 427

Wiegant, V.M. and Gispen, W.H. (1975). *Exp. Brain Res., 23* (suppl) 219

Wiegant, V.M. and Gispen, W.H. (1977). *Behav. Biol., 19,* 554

Wiegant, V.M., Cools, A.R. and Gispen, W.H. (1977). *Eur. J. Pharmacol., 41,* 343

Wiegant, V.M., Jolles, J. and Gispen, W.H. (1978a). In *Characteristics and Function of Opioids,* ed. J.M. van Ree and L. Terenius (Elsevier, Amsterdam) p. 447

Wiegant, V.M., Colbern, D., van Wimersma Greidanus, Tj.B. and Gispen, W.H. (1978b). *Brain Res. Bull., 3,* 167

Wiegant, V.M., Dunn, A.J., Schotman, P. and Gispen, W.H. (1979). *Brain Res., 168,* 565

Williams, M. and Rodnight, R. (1977). *Progr. Neurobiol., 8,* 183

van Wimersma Greidanus, Tj. B., Croiset, G. and Schuiling, G.A. (1979). *Brain Res. Bull., 4,* 625

Winter, C.A. and Flataker, L. (1951). *J. Pharmacol. Exp. Ther., 101,* 93

Witter, A. (1979). In *Proceedings of the Colloquium on Receptors, Neurotransmitters and Peptide Hormones, Capri, May 1979,* ed. M. Kuhar, L. Enna and G.C. Pepeu (Raven Press, New York, in press)

Witter, A., Greven, H.M. and de Wied, D. (1975). *J. Pharmacol. Exp. Ther., 193,* 853

Zimmermann, E. and Krivoy, W.A. (1973). *Progr. Brain Res., 39,* 383

Zimmermann, E. and Krivoy, W.A. (1974). *Proc. Soc. Exp. Biol. Med., 146,* 575

Zwiers, H., Veldhuis, D., Schotman, P. and Gispen, W.H. (1976). *Neurochem. Res., 1,* 669

Zwiers, H., Wiegant, V.M., Schotman, P. and Gispen, W.H. (1977). In *Mechanism, Regulation and Special Functions of Protein Synthesis in the Brain,* ed. S. Roberts, A. Lajtha and W.H. Gispen (Elsevier, Amsterdam) p. 267

Zwiers, H., Wiegant, V.M., Schotman, P. and Gispen, W.H. (1978). *Neurochem. Res., 3,* 455

Zwiers, H., Tonnaer, J., Wiegant, V.M., Schotman, P. and Gispen, W.H. (1979). *J. Neurochem., 33,* 247

3 DOPAMINE AGONIST AND ANTAGONIST DRUGS AND HYPOTHALAMIC PITUITARY DYSFUNCTION

Eugenio E. Müller, Franco Camanni, Andrea R. Genazzani,
Daniela Cocchi, Felipe Casanueva, Ferdinando Massara,
Vittorio Locatelli, Abraham Martinez-Campos and
Paolo Mantegazza

TABLE OF CONTENTS

I	INTRODUCTION	53
II	HYPOTHALAMIC AND EXTRAHYPOTHALAMIC STRUCTURES RELATED TO PITUITARY CONTROL	54
	A The Median-eminence/Anterior-pituitary Complex	55
	(a) Neurobiology of the Median Eminence	55
	(b) Anterior Pituitary	57
	B Extrahypothalamic Structures for DA-neurohormonal Interaction	67
III	CHEMISTRY AND PHARMACOLOGY OF DA-RECEPTOR AGONISTS AND ANTAGONISTS	68
	A Directly and Indirectly Acting DA Agonists	68
	(a) Direct DA Agonists	68
	(b) Indirectly Acting DA Agonists	68
	B DA-antagonist Drugs	69
IV	INDIRECTLY ACTING DA-AGONIST AND DA-ANTAGONIST DRUGS IN THE DIAGNOSIS OF HYPERPROLACTINEMIC STATES	69
	A Dynamic Tests of PRL Secretion	71
	(a) Inhibitory Tests	71
	(b) L-DOPA plus Carbidopa and Nomifensine Tests	71
	(c) Stimulatory Tests	75
V	DA-AGONIST AND DA-ANTAGONIST DRUGS: TESTS OF GROWTH-HORMONE RESERVE IN CHILDREN	77
VI	DA-AGONIST DRUGS AND THE TREATMENT OF HYPERPROLACTINEMIC STATES	78
	A Puerperal and Pathological Lactation	79
	B Infertility	81
	C Prolactin-secreting Adenomas	82
	D Pregnancy	83
VII	DA-AGONIST DRUGS AND THE MEDICAL	

TREATMENT OF ACROMEGALY   85

VIII   BROMOCRYPTINE AND THE MEDICAL
        TREATMENT OF CUSHING'S DISEASE   88

IX   SUMMARY AND CONCLUSIONS   90

REFERENCES   91

I. INTRODUCTION

It is now well established that a family of neuropeptides, the hypothalamic releasing and inhibiting or regulatory hormones, produced in neural cell bodies, regulate the secretory activity of the anterior pituitary (AP) gland. They are secreted at the level of the median eminence (ME) into the primary plexus of the hypophysial-portal blood vessels and are then vehicled to the AP cells, where they stimulate or inhibit the synthesis and release of AP hormones.

The discovery of several of these hormones and their isolation, structural identification and synthesis provided evidence for the theory of neurohumoral control of the AP gland put forward by Green and Harris (1947). Along with the studies on the identification of specific neurohormonal influences, extensive investigation has focused on neurotransmitter control of secretion of the hypothalamic regulatory hormones (Müller *et al.,* 1978b).

The demonstration of rich innervation of the medial basal hypothalamus (MBH) by biogenic aminergic pathways, together with neuropharmacologic studies of the effects of blockade and stimulation of amine receptors, has led to the view that the secretion of hypothalamic regulatory hormones is controlled by aminergic neurons. Considerable progress in this field has been made possible by the development of a histochemical fluorescence technique that permits the detailed mapping of monoamine pathways, the introduction of methods to measure turnover of neurotransmitters and sensitive enzymatic-isotopic techniques for neurotransmitter determinations in microdissected hypothalamic nuclei or nuclear subdivisions (see Müller *et al.,* 1978b).

In addition to brain monoamines, notably catecholamines (CAs) such as dopamine (DA), norepinephrine (NE) and epinephrine (E) and indoleamines such as serotonin (5-HT) and melatonin, an expanding series of substances (i.e., acetylcholine, histamine, GABA, substance P, neurotensin, opioid peptides) whose neurotransmitter function is proven or postulated has to be considered.

Prominent among neurotransmitters, DA plays a key role in neuroendocrine regulation. The arcuate nucleus (ARC n) which forms a long body positioned laterally above the ME, contains primary cell bodies of the dopaminergic system which project nerve terminals to the ME. It thus represents an important synaptic area between hypothalamic-hypophysiotropic neurons and input aminergic pathways. Dopamine released directly into the hypophysial portal capillaries (Ben-Jonathan *et al.,* 1977) may act at the level of pituitary DA receptors to control AP function.

A number of organic brain diseases of neoplastic, inflammatory, degenerative or vascular nature or traumatic events, by causing damage to the hypothalamus, disrupt physiologic neurotransmitter-neurohormonal links and lead, ultimately, to abnormal AP-hormone secretion. In addition, the existence of a primary neurotransmitter dysfunction has been postulated in specific neuroendocrine disorders, i.e., growth-hormone (GH) deficiency states, acromegaly, Cushing's disease or tumorous or idiophatic hyperprolactinemia. However, since most of the above disturbances are always or very often associated with the presence of pituitary adenomas, it has been argued that the latter are primary causes for sustained hypersecretion of GH, corticotropin (ACTH) and prolactin (PRL) (Daughaday, 1977).

Whatever the physiopathology of these neuroendocrine disorders, direct DA-agonist drugs acting at the level of pituitary DA receptors may affect the exuberant secretion of AP hormones, while indirectly acting DA agonists, relying on the presence of an endogenous DA pool, may probe the functional state of DA neuronal function. In these endocrine abnormalities, the diagnostic use of DA-antagonist drugs can also be envisioned, since for their neuroendocrine effects to be manifested, they need receptors normally modulated by CNS dopaminergic inputs.

II. HYPOTHALAMIC AND EXTRAHYPOTHALAMIC STRUCTURES RELATED TO PITUITARY CONTROL

Description of the actions and uses of DA-agonist and DA-antagonist drugs in hypothalamo-pituitary dysfunction necessitates a comment on the nature and topographical localization of dopaminergic hypothalamic and extrahypothalamic structures related to AP control.

Dopamine neuronal systems in the brain are diverse in distribution and function. They comprise three main systems of long DA neurons i.e., the nigro-striatal, the mesolimbic and the mesocortical systems and also two systems composed of short intrahypothalamic axons, the tuberoinfundibular (TI) and incertohypothalamic systems, which project, respectively, from the ARC n or the periventricular n to the lateral parts of the external layer of the ME (TI neurons) and to the neurointermediate lobe (tuberohypophysial neurons) and from the rostral periventricular hypothalamus to the preoptic area.

It is now clear that CA projection to the external layer of the ME is almost entirely derived from the TIDA projection, although proof has also been given for the existence of a nigral hypothalamic-ME pathway, projecting from the pars compacta of the substantia nigra to the ventromedial nucleus and the ME (Kizer *et al.*, 1976).

At the level of the external layer of the ME, TIDA neurons interact function-ally with neurosecretory neurons manufacturing peptidergic regulatory hormones. In this way, DA neurons influence AP function directly, by regulating peptidergic neurons, or may contribute directly to neurohormonal control of PRL secretion by releasing DA directly into the portal capillary blood (Figure 3.1).

Figure 3.1: Hypothetical scheme of the interactions between peptidergic and neurotransmitter neurons for the control of prolactin secretion at the level of the median eminence. Key: (1) TIDA neurons responsible for direct release of DA into the hypophysial portal capillary system; (2) peptidergic neuron secreting the PRL-releasing factor (PRF); reciprocal connections between (1) and (2) are also shown; (3) median eminence; (4) interneurons modulating neural activity of TIDA and PRF neurons; the possibility that interneurons are arranged in series and may interact with each other through a recurrent inhibitory pathway is also depicted; (5) tanycite which extends processes from the floor of the third ventricle (III V) to the median eminence and to a hypophysiotropic neuron; (6) extrahypothalamic neuron projecting to the median eminence; (7) downward flow of blood in the portal capillary system vehicling DA and PRF to the anterior pituitary and retrograde flow vehicling peptides (x) from the pituitary to the brain; (8) lactotroph cell; (9) peripheral hypophysial vessel. Also, OC denotes optic chiasm; MB, mammillary body; AP, anterior pituitary; and PP, posterior pituitary.

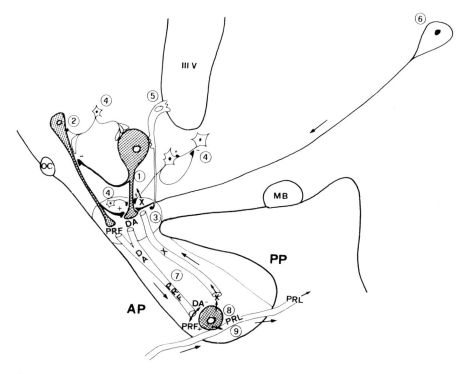

A. The Median-eminence/Anterior-pituitary Complex

(a) Neurobiology of the Median Eminence. The ME serves as the junctional zone between hypophysiotropic neurons and the portal circulation of the AP; in this

way it fulfils the function of a neuroendocrine transducer that converts neural into chemical information (Wurtman, 1973). However, although it provides a valuable neuroanatomical substrate for neurotransmitter-neurohormonal interactions within the CNS, it exhibits peculiar features, summarized below, which render it different from other CNS areas.

The DA cells of the ARC n are atypical neurons, in that no clear-cut morphologic evidence has been given for the existence of axoaxonic contacts at this level, whereas synapse-like structures have been observed between nerve terminals and glial profiles, suggesting that DA may influence the activity of these elements (Hökfelt, 1974).

The regulatory mechanisms governing the activity of TIDA neurons differ from those operative in the other DA systems. After DA is released into the synaptic cleft, it can activate presynaptic receptors (autoreceptors), with ensuing inhibition of DA synthesis and release. This 'feedback' mechanism, which is operative in the terminals of the mesocortical and nigrostriatal systems (Carlsson, 1975), seems not to be present in the TIDA system (Moore *et al.*, 1980). In addition, the activity of nigrostriatal DA nerves is regulated, in part, by a neuronal feedback loop whereby facilitation of DA transmission in the striatum causes a compensatory reduction of nigrostriatal nerve activity, while blockade of DA transmission leads to an increase in activity.

This is not true for TIDA neurons, where neither DA-receptor blockade by haloperidol nor stimulation by the DA agonist piribedil was shown to be capable of altering the rate of α-methylparatyrosine-induced decline of DA in the ME (Moore *et al.*, 1980).

Further peculiarities in the neurochemical regulation of TIDA neurons are evident. The striatal concentration of dihydroxyphenyl acetic acid (DOPAC), the oxidatively deaminated product of DA metabolism, represents the amount of DA released and then recaptured by the nerve, and it is used as a biochemical index of nigrostriatal DA nerve activity (Roth *et al.*, 1976). In contrast, the concentration of this metabolite of DA does not reflect the activity of TIDA nerves. This is due to the fact that DA, once released from TIDA nerves, is carried by the hypophysial portal system to the AP, and is not recaptured by the nerve terminal and converted to DOPAC. Consistent with this notion is the lack of a high-affinity amine-uptake system in the TIDA nerves (Moore *et al.*. 1980).

Whereas classical neuronal feedback mechanisms do not control the activity of TIDA nerves, hormonal feedback mechanisms are highly active in this sense. The presence of a short-loop feedback mechanism has been suggested for PRL, a hormone that lacks a specific peripheral target gland or tissue.

Implantation of PRL into the rat hypothalamus was shown to be capable of inhibiting all the physiological processess dependent on PRL secretion; high circulating levels of PRL resulting from implantation of mammosomatotropic tumours were associated with a reduced ability of the host pituitary to synthesize PRL (MacLeod, 1976). A mechanism enabling PRL to control its own secretion via a short-loop feedback might have operated by increasing the activity of TIDA neurons which release the PRL-inhibiting factor DA into the portal blood. A vast series of studies has now shown this to be the case. Exogenous treatments with PRL, given systemically or centrally, are found to accelerate, after a long delay, the turnover of DA in the TIDA neurons, either using semi-quantitative or quantitative histofluorescent techniques or sensitive radioenzy-matic procedures and measuring as an end-point the decline in DA induced by α-methylparatyrosine (α-MPT), the accumulation of DOPA following inhibition of DOPA decarboxylase or DA titers in the hypophysial portal blood. Dopamine antagonists, like haloperidol, or estrogens accelerate the turnover of DA in the TIDA neurons only after a definite period of latency, showing that their action is dependent upon their ability to cause sustained elevation of plasma PRL.

In both cases, in fact, hypophysectomy prevents the increase in turnover. Neither exogenous nor endogenous PRL increased by endocrinological or pharmacological means affects DA turnover in extrapyramidal or limbic structures (Moore *et al.,* 1980).

The implementation of these findings for the understanding of either the physiopathology of neuroendocrine disorders or the mechanism of action of DA-agonist and DA-antagonist drugs will be discussed later.

(b) Anterior Pituitary. Consideration of the anatomical substrate for neuro-transmitter-neurohormonal interactions within the CNS necessitates analysis of the pituitary gland and, particularly, of its anterior lobe. Mention has been made previously of the fact that DA, once released from terminals of TIDA neurons, is vehicled to the AP, where it may contribute directly to neurohormonal control of AP hormones. This concept, which permits the gland to be regarded as a neuroendocrine transducer in the sense categorized by Wurtman, is mainly derived from studies on the inhibitory effect of the amine on pituitary lactotrophs.

Proof for the direct action of DA on pituitary lactotrophs may be seen in that: (1) DA-precursor or DA-mimetic drugs are capable of inhibiting PRL release by AP incubated *in vitro,* an effect which can be prevented by specific DA blockers; (2) suppression of elevated baseline PRL levels may be induced by DA-mimetic drugs administered to hypophysectomized rats bearing an ectopic AP or electrolytic lesions of the ME or to stalk-lesioned rhesus monkeys or humans; (3) direct infusion of DA into the hypophysial portal vessels is competent to inhibit PRL release (MacLeod, 1976).

More definitive proof for the hypophysiotropic mechanism of DA was given with the demonstration in the AP of high affinity, saturable, stereoselective and reversible binding sites for DA agonists and antagonists (Table 3.1). These binding sites represent true receptor structures of physiological meaning for PRL control, since the ability of different ligands to compete for labeled sites is highly corre-lated with the ligand's effect on PRL secretion.

Elegant immunocytochemical studies with the use of an antibody to haloperi-dol showed that the majority of DA receptors are associated with mammotrophs. These cells can therefore be viewed as the postsynaptic element for DA release

Table 3.1: Competition for ^{3}H-DHE and ^{3}H-SPIR Binding to Sheep Anterior Pituitary

	K_i (nM)*	
	^{3}H-DHE	^{3}H-SPIR
D-butaclamol	2.7	1.6
Haloperidol	80	12
Bromergocriptine	80	19
Apomorphine	95	220
Dopamine	1,300	780
L-butaclamol	3,200	17,000
Phentolamine	4,700	26,000
Norepinephrine	5,000	18,000
Epinephrine	5,000	22,000
Clonidine	>40,000	>40,000
Serotonin	>40,000	>40,000
TRH	>40,000	>40,000

*The apparent inhibition constants, K_i, were calculated according to the equation $K_i = \dfrac{IC_{50}}{1 + C/K_d}$. The IC_{50} is the competition concentration that yields 50% of ^{3}H-ligand binding in individual experiments and C is the concentration of the ^{3}H-ligand. The K_d utilized for ^{3}H-DHE (dihydroergocriptine) was 5 nM and that for ^{3}H-SPIR (spiroperidol) was 0.85 nM.
Source: Reproduced, with permission from Weiner *et al.* (1979).

from TIDA neurons; similar to nigrostriatal dopaminergic neurons, they exhibit *in vivo* and *in vitro* supersensitivity (to the PRL-inhibiting effect of DA) when TIDA neurons are destroyed or AP-dopaminergic receptors are chronically blocked (Weiner *et al.*, 1979).

In spite of precise characterization of these receptors, the biochemical mechanism(s) subserving the interaction of DA, DA-mimetic and DA-antagonist drugs with the AP are far from being clear; a DA-sensitive adenylate cyclase (AC) has not been observed in the AP, though there has been one report to the contrary (Ahn *et al.*, 1979), while inhibition of the AC by the addition of the agonist has instead been demonstrated in PRL-secreting pituitary tumours (De Camilli *et al.*, 1979).

Thus, from the existing evidence, it would appear that the DA receptor at AP level is independent from AC (type D-2 of Kebabian and Calne) and shows a selectivity different from that of the classical D-1-DA receptor (linked to AC) to DA-agonist and DA-antagonist drugs, i.e., is highly simulated by DA, apomorphine and ergot drugs, and antagonized by benzamides, i.e., sulpiride and metoclopramide (see Tables 3.2 and 3.3) (Kebabian and Calne, 1978).

More recently, suggestion for the presence of two DA receptors at AP level has been formulated; specific binding with the DA antagonist spiroperidol evidenced a biphasic scatchard plot, and guanosine triphosphate decreased the affinity of DA receptors for most agonists, but not antagonists (Weiner *et al.*, 1979).

Although most of the evidence on AP receptors for DA-agonist and DA-antagonist drugs refers to the control of PRL secretion, proof has also been given of the presence of such receptors on tumorous somatotrophs. In acromegalic, but not in normal subjects, direct DA-agonist drugs act to directly stimulate DA receptors on somatotrophs, with ensuing inhibition of elevated human growth hormone (hGH) levels (see the section on DA-agonist drugs and the medical treatment of acromegaly below). In addition, for thyroid-stimulating hormone (TSH), melanocyte-stimulating hormone (MSH) and gonadotropins, the possibility has been envisaged of a direct effect of DA at the AP level. Thus, L-DOPA has been found capable of blocking TRH-induced TSH release, an effect also shared by systemic DA infusion, and a significant fall in plasma LH but not FSH levels has been described in man after infusion of DA or administration of L-DOPA (Müller *et al.*, 1978b). It cannot be excluded that in these instances the true locus of action of dopaminergic stimulation was the ME.

B. Extrahypothalamic Structures for DA-neurohormonal Interaction

It appears from the foregoing that the ME-AP complex is the area where interactions of DA with the neurohormonal system of crucial importance for endocrine regulation occur. However, there are probably extrahypothalamic structures participating in the control of neuroendocrine activity which receive innervation from ascending dopaminergic pathways. Thus, electrical stimulation of limbic structures, such as corticomedial and basolateral amygdala, which are densely innervated by meso-subcortical DA pathways, results in inhibition or stimulation, respectively, of GH release in the rat (Müller, 1979). In a limbic structure, the n accumbens, as in the ME, DA turnover would increase after systemic administration of PRL (Fuxe *et al.*, 1977), thus implying the intervention of mesolimbic DA pathways in this area to mediate feedback inhibition of PRL on its own secretion.

Proof, albeit inferential, for the occurrence of feedback regulation of plasma hormones in extrahypothalamic structures is also provided by the observation that chronic ovariectomy in the rat significantly reduces DA- and apomorphine-sensitive AC in both the n accumbens and the striatum (Kumakura *et al.*, 1979) and that, conversely, estrogens increase DA-receptor sensitivity in the striatum (Hruska and Silbergeld, 1980). In this context, mention has already been made of a nigral-hypothalamic-ME pathway, which may be important for relaying hormonal feedback signals from the striatum to the MBH.

III. CHEMISTRY AND PHARMACOLOGY OF DA-RECEPTOR AGONISTS AND ANTAGONISTS

A. Directly and Indirectly Acting DA Agonists

(a) Direct DA Agonists. The preferred active conformation of DA is the fully extended *trans* form. Thus, rigid analogs of DA, i.e., apomorphine and ADTN (Table 3.2) which are potent receptor agonists, have their side-chain incorporated into a second ring system in an extended form.

Compounds such as piribedil, which is a noncatechol analog of DA, possess a long-acting DA-receptor-agonist activity for transformation to catechol-metabolites (i.e., S-584).

The most exciting group of DA agonists is the group of ergot derivatives, which from a chemical viewpoint can be divided into: (1) peptide-containing ergot alkaloids; (2) ergolines (simple derivatives of lysergic acid); (3) clavines (Table 3.2).

Bromocriptine, the most clinically useful compound, belongs to the peptide-containing ergot drugs. Bromocriptine reduces DA turnover in various regions of the forebrain, and possesses a high affinity for both the agonist- and antagonist-preferring sites of the DA receptors, indicating that it may be a high-affinity partial agonist. Its effects on the DA-sensitive AC system are unclear, since its efficacy in stimulating AC activity *in vivo* is coupled to inability to affect AC in homogenates of the striatum (Spano and Trabucchi, 1978). The ability of bromocriptine to stimulate pituitary DA receptors probably resides in the fact that these receptors are not linked to the AC system.

Most of the dopaminergic ergot drugs, however, are ergoline derivatives, such as lergotrile, lisuride, metergoline, pergolide and CF 25-397. Finally, ergots of the clavine group have been shown to be potent DA-receptor agonists.

One problem with dopaminergic ergot derivatives is their ability to affect central 5-HT and NE receptors. Thus, bromocriptine induces an early agonistic effect at central NE-receptor sites, while later on its effect is more consistent with postsynaptic NE-receptor blockade; lisuride and especially metergoline possess antagonist activity at central and peripheral 5-HT synapses.

(b) Indirectly Acting DA Agonists. These compounds only indirectly influence DA neurotransmission; in fact, they do not act directly on DA receptors, but rather induce release of DA (and NE) from the nongranular (reserpine-resistant) pool, e.g., amphetamine, methamphetamine, L-ephedrine, or from the nongranular and granular pools, i.e., methylphenidate, mazindol, or else block DA reuptake (amphetamine, nomifensine, benztropine). In addition, nomifensine is a strong inhibitor of NE reuptake (Nicholson and Turner, 1977) (Table 3.2).

Table 3.2.

Nonproprietary name	Trade name	Chemical structure	Observations
Direct DA agonists			
DOPAMINE			Does not cross the brain-blood boundary (BBB)
3,4-Dihydroxyphenylethylamine			
APOMORPHINE			
(R)-5,6,6a,7-Tetrahydro-6-methyl-4H-dibenzo [de,g] quinoline-10-11-diol			
ADTN			Easily crosses the BBB in the pro form, and is then metabolized to the active product
2-Amino-6,7-dihydroxy-tetrahydronaphthalene			

PIRIBEDIL (ET 495)   Trivastal

1-(2-Pyrimidil-4-piperonilpiperazine)

Does not stimulate adenylate cyclase

S 584

[1-3,4-(Dihydroxybenzyl)-4-(2-pyrimidinyl)] piperazine

Active metabolite of piribedil

Peptide-containing ergot alkaloids

2-BR-α-ERGOCRYPTINE   Parlodel
(bromocriptine)

Action depending in part on brain CA stores

Ergolines

LERGOTRILE

2-Chloro-6-methyl-ergoline-8/3-acetonitrile

At high doses blockade of NE and 5-HT receptors. Hepatotoxicity, withdrawn from clinical practice

LISURIDE   Lynsenil

[N-(D-6-)methyl-8-iso-ergolenyl-d]-N', N'-diethyl-carbamide

Central and peripheral antagonist of 5-HT receptors

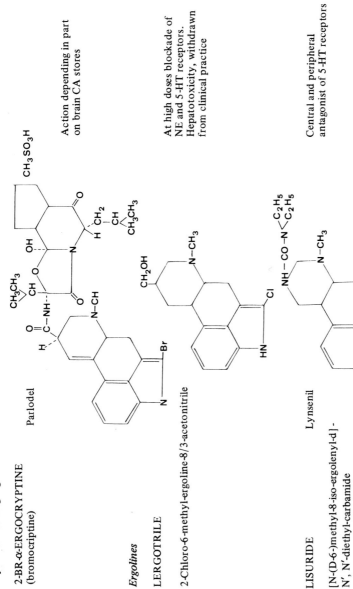

PERGOLIDE

(8β)-8-[Methylthio-methyl]-6-
propylergoline

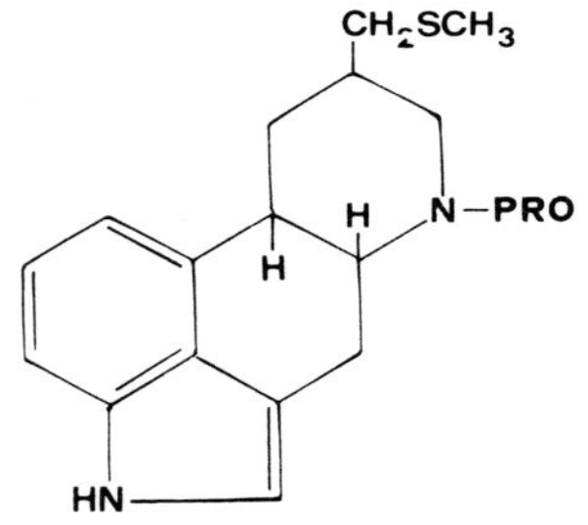

Long-acting pure DA agonist

METERGOLINE   Liserdol

D-8β-[Carboxyamino)methyl]-1,6-
dimethyl-ergoline

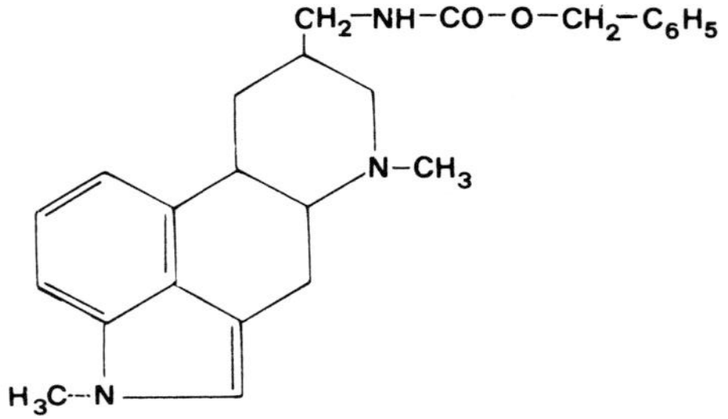

Indirectly acting DA agonists

AMPHETAMINE   Benzedrine

dl-α-Methylphenethylamine

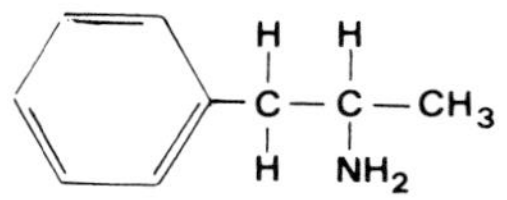

MAZINDOL   Sanorex

5-(4-Chlorophenyl-2, 5-dihydro-3H-
imidazol [2, 1-a] isoindol-5-ol

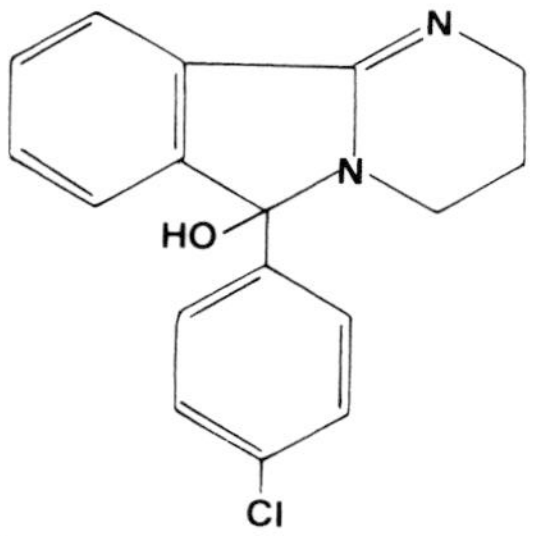

Blockade of NE and 5-HT
receptors

METHYLPHENIDATE   Ritalin

α-Phenyl-2-piperidineacetic acid
methylester

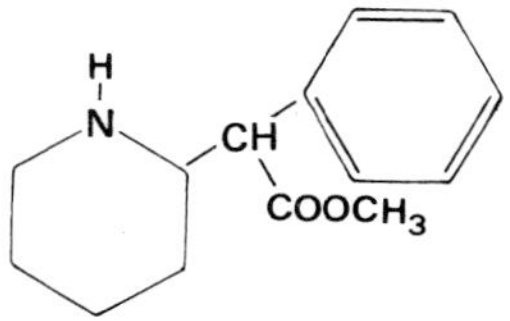

NOMIFENSINE   Alival

(8-Amino-2-methyl-4-phenyl-1, 2, 3, 4-
tetrahydro-isoquinoline

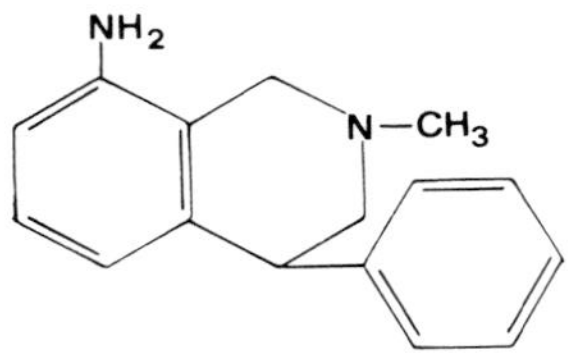

AMINEPTINE   Survector

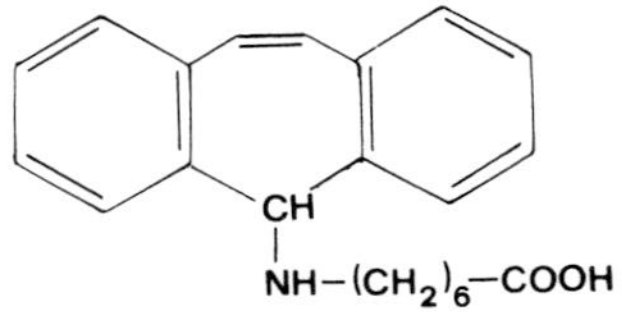

Possible direct stimulation
of DA receptors at high
doses

Table 3.3

Nonproprietary name	Trade name	Chemical structure

CHLORPROMAZINE — Thorazine

2-Chloro-10-(3-dimethylamino-propyl)-phenothiazine

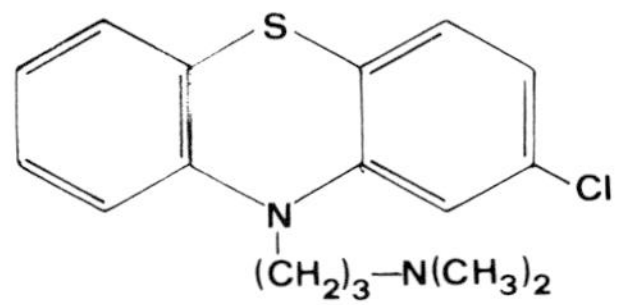

HALOPERIDOL — Haldol

4-[4-(p-Chlorophenyl)-4hydroxy-piperidino]-4'-fluorobutyrophenone

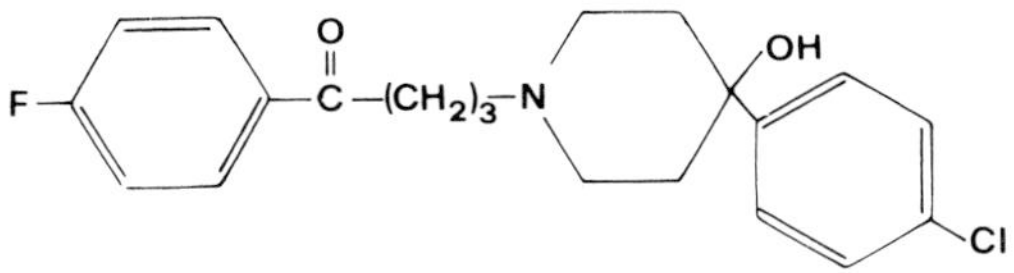

CHLORPROTHIXENE — Taractan

2-Chloro-N, N-dimethylthio-xanthene-98-propylamine

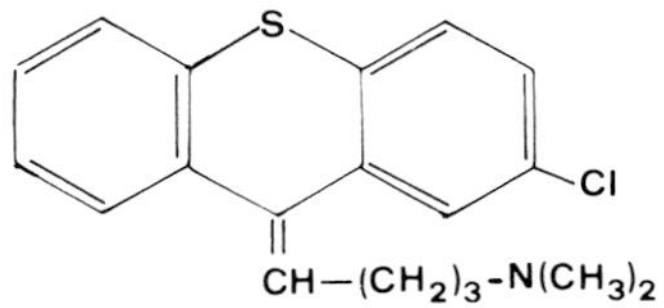

CLOZAPINE — Leponex

8-Chloro-11-(4-methyl-1-piperazinyl) 5H-dibenzo [b, e] [1, 4] diazepine

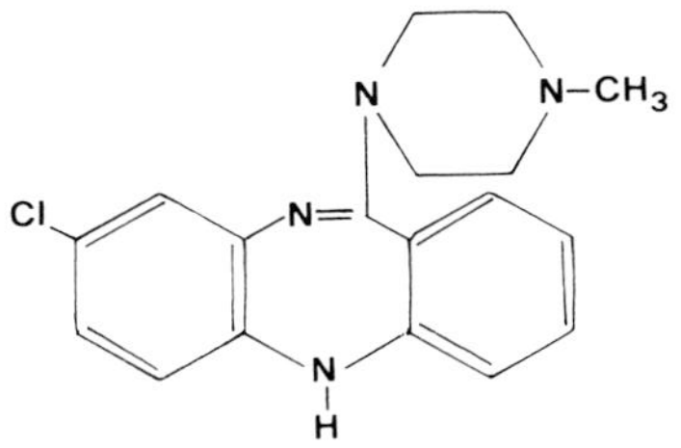

SULPIRIDE   Dobren

5-(Aminosulfonyl)-N-[(1 ethyl-
2-pyrrolidinyl)methyl] -2-methoxybenzamide

METOCLOPRAMIDE   Cerucal

4-Amino-5-chloro-N-[2-(diethylamino)-
ethyl] -2-methoxybenzamide

DOMPERIDONE   Motilium

5-Chloro-1-{1-[3-(1, 3-dihydro-
2-oxo-2H-benzimidazol-1-yl)prolil] -4
piperidynil}-1, 3 dihydro-2H-
benzimidazol-2-one

B. DA-antagonist Drugs

This group of compounds comprises antipsychotic drugs of the phenothiazine, butyrophenone and thioxanthene class (classical neuroleptics) and dibenzodiazepines (clozapine) and substituted benzamides (sulpiride and metoclopramide, MCP, atypical neuroleptics). Drugs representative of the different classes are shown in Table 3.3.

The order of potency of classical neuroleptics with regard to increasing PRL secretion is, in essence, in agreement with their relative potency in inhibiting $[^3H]$-DA and $[^3H]$-haloperidol binding by calf caudate membranes (Creese *et al.*, 1975).

Both clozapine and sulpiride possess a somewhat different neuropharmacological spectrum as compared to classic neuroleptics. They are less, or virtually not, cataleptogenic and produce greater DA turnover changes in the limbic system than in the corpus striatum. In addition, while clozapine is as potent as chlorpromazine in inhibiting the DA activation of AC from both the caudate n. and limbic system, sulpiride is unable to block *in vitro* AC activation induced by both DA and apomorphine and *in vivo* apomorphine-induced adenosine-monophosphate accumulation in rat striatum. However, in common with neuroleptics, sulpiride is competent to stimulate PRL secretion and to displace $[^3H]$-haloperidol from rat striatum membrane preparations (Spano *et al.*, 1979).

Of considerable interest for neuroendocrine research is the recently developed piperidine-benzimidazol derivative, domperidone (Table 3.3). Domperidone has a higher affinity than haloperidol for striatal and pituitary receptors labelled with $[^3H]$-spiroperidol but, in view of its less lipophilic character as compared to neuroleptics, it does not cross the blood-brain barrier (Cocchi *et al.*, 1980).

IV. INDIRECTLY ACTING DA-AGONIST AND DA-ANTAGONIST DRUGS IN THE DIAGNOSIS OF HYPERPROLACTINEMIC STATES

Firm establishment that PRL exists in man as a hormone separate from GH, and the possibility of developing accurate radioimmunoassay procedures for the detection of the hormone in the plasma, have considerably enhanced our understanding of the physiopathology of PRL secretion.

It is now clear that excessive PRL secretion is a common phenomenon and is often implicated in women with menstrual disorders, infertility or galactorrhea and in men with impotence. Hyperprolactinemia (hyperPRL) and associated clinical features may arise in a variety of conditions (Table 3.4). Drug administration is the most common cause of hyperPRL, and it is always important to exclude hypothyroidism, which can so readily be treated. Hyperprolactinemia and associated hypogonadism may also be found in patients with chronic renal

Table 3.4: Hyperprolactinemic States

---

1. Alterations of the Hypothalamopituitary System
   A. Hypothalamic lesions (traumatic, inflammatory and neoplastic disorders)
   B. Stalk section
   C. PRL-secreting tumours (prolactinoma, acromegaly, mixed forms)
   D. Iatrogenic
   E. Primary hypothyroidism
   F. Hyper-hypo-cortisolism
   G. 'Functional' (idiophatic, post-partum, post-pill, etc.)

2. PRL-secreting extrapituitary tumours

3. Chronic infiltrative processes of the mammary gland and the chest

4. Chronic renal failure, liver cirrhosis

---

failure or on hemodialysis and sometimes labels individuals with chronic liver disease.

However, when these conditions have been excluded, the most likely cause of hyperPRL is the presence of a PRL-secreting adenoma and it is important to discern between patients who have discrete PRL-secreting tumours (microadenomas) and those who have not. In the absence of radiologic modifications of the sella turcica or of clinical and endocrine manifestations of hypothalamic disease, the latter cases are lumped under the term 'idiophatic' or 'functional' hyperPRL, which masks our present ignorance of a more proper nosological connotation.

Whatever the etiophatogenesis of the prolactinomas may be, early diagnosis and treatment of PRL-secreting adenomas is of great importance. In fact, according to the 'hypothalamic' hypothesis which postulates that a primary defect in DA neuronal function may result in hyperplasia and then adenomatous transformation of the pituitary lactotrophs, it would seem possible that medical treatment could prevent 'functional' disorders from progressing to tumours. Alternatively, assuming that PRL-secreting tumours may result from a primary pituitary neoplasm (Daughaday, 1977), early diagnosis would allow its selective transsphenoidal removal, with preservation of normal pituitary tissue (Hardy, 1975).

Radiological abnormalities of the pituitary fossa have been judged the most useful diagnostic criterion of a pituitary tumour (Vezina and Sutton, 1974) and advances in diagnostic radiology now make it possible to diagnose pituitary microadenomas (less than 10 mm in diameter) before they expand beyond the confines of the sella turcica. However, small changes revealed by radiology are difficult to interpret and, moreover, minor radiological 'abnormalities' have been revealed in up to 30% of apparently normal subjects (Swanson and Du Boulay, 1975). Therefore, identification of the disease at an early stage is still a challenge for the clinician.

A. Dynamic Tests of PRL Secretion

Several tests have been devised with the aim of discriminating between tumorous and nontumorous hyperPRL. These tests are mainly based on the use of: (1) DA-agonist and DA-antagonist drugs capable of affecting DA neurotransmission or action, and thence of inhibiting or stimulating, respectively, PRL secretion; (2) hypothalamic neurohormones which exert their action directly at the AP level (Faglia *et al.*, 1980b).

On purely theoretical grounds, combined application of neuroactive drugs and hypothalamic hormones should be capable of unravelling the CNS or the pituitary origin of the disorder. Evidence against this assumption, however, is provided by the findings that: (1) most of the DA agonists or antagonists used act at the level of the AP; (2) PRL-secreting adenomas may not be completely autonomous from CNS influences; (3) the altered pituitary tissue may respond normally to hypophysiotropic stimuli.

Based on these premises, it is evident that a neuroactive drug, to be of diagnostic value, should affect central DA neuronal function but without acting directly on DA receptors present on normal or pathological lactotrophs.

(a) Inhibitory Tests. The inhibitory tests first introduced for distinguishing tumorous from nontumorous hyperPRL relied on the use of DA, DA-precursor (L-DOPA) or DA-agonist (bromocriptine) drugs (Kleimberg *et al.*, 1977; Faglia *et al.*, 1980b). None of these tests allowed accurate discrimination between hyperPRL subjects, the main reason being that these drugs act directly at AP level (L-DOPA after its peripheral conversion to DA) and both normal and adenomatous lactotrophs possess DA receptors (Weiner *et al.*, 1979). In view of the foregoing, these agents are unable to detect a possible impairment of hypothalamic (dopaminergic) control of PRL secretion and, therefore, the source of hyperPRL.

In the study of Faglia *et al.* (1980b), L-DOPA (500 mg po) and bromocriptine (2.5-5.0 mg po) failed to inhibit serum PRL in 44% and 30%, respectively, of patients with PRL-secreting tumours and in 25% and 20% of patients without sella abnormalities (Figure 3.2). Similarly, piribedil (100 mg po), while competent to lower plasma PRL in peurperal hyperPRL, did not evidence clear-cut differences in its PRL-lowering effect between patients with or without radiological evidence of pituitary tumours (Müller *et al.*, 1977). Administration of ergots with 5-HT antagonistic properties, such as metergoline and methysergide, also proved to be of no value in the differential diagnosis of hyperPRL (Ferrari *et al.*, 1978).

(b) L-DOPA plus Carbidopa and Nomifensine Tests. More recently, inhibitory tests have been devised based on the use of drugs or drug combinations capable of investigating selective aspects of neurotransmitter function. Since L-DOPA is converted to DA by an L-amino acid decarboxylase, its administration in

Figure 3.2: Incidence (%) of normal responses to inhibitory or stimulatory tests for prolactin secretion in 93 patients with radiological evidence of pituitary tumours and in 25 patients without similar alterations. Here SLP denotes sulpiride, and CB 154 bromocriptine.

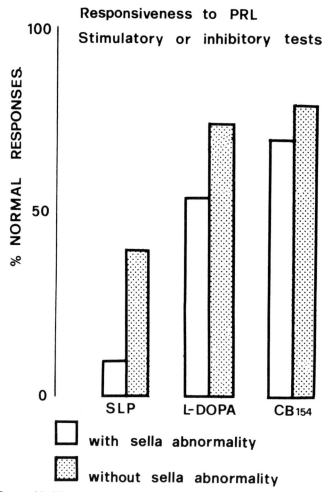

Source: Modified and reproduced, with permission, from Faglia *et al.* (1980b).

combination with an enzymatic inhibitor of the decarboxylase which does not cross the blood-brain barrier (e.g., carbidopa, CD) would result in an increased dopaminergic tone in the CNS, but not in the periphery.

Since direct dopaminergic stimulation of the lactotrophs was prevented by this manipulation, lack of PRL inhibition following administration of L-DOPA in combination with CD would indicate a pituitary adenoma as the cause of the PRL excess. In their report, Fine and Frohman (1978) demonstrated that, while

L-DOPA alone was competent to significantly decrease plasma PRL both in normal subjects and in eight patients bearing PRL-secreting adenomas, administration of L-DOPA plus CD did so only in normal subjects. Unfortunately, these authors did not examine patients with nontumoural hyperPRL.

Another pharmacological approach aimed at isolating those hyperPRL subjects who bear a PRL-secreting tumour relies on the use of nomifensine (Nom), an antidepressant drug which activates DA neurotransmission mainly by inhibiting DA reuptake and by promoting DA release (see Table 3.2). The initial study on the diagnostic use of Nom reported that oral administration of 200 mg po of the drug lowered plasma PRL in ten subjects with puerperal hyperPRL and in seven hyperPRL patients without evidence of pituitary tumour; the drug was ineffective in ten patients in whom a prolactinoma was diagnosed at surgery (Müller *et al.,* 1978a). These results have been confirmed and expanded by the same authors in a further investigation. In 27 hyperPRL subjects in whom the existence of a tumour was established at surgery (25 prolactinomas, one ependymoma and one cholesteatoma) and in 13 subjects with roentgenographic alterations of the sella that were unequivocal for the presence of pituitary tumours, Nom did not significantly lower baseline plasma PRL levels. In 46 hyperPRL subjects without or with only equivocal alterations of the sella, Nom was effective in lowering plasma PRL in 14 subjects, and ineffective in the remaining 32 subjects. Interestingly, while 13 out of 14 Nom responders presented with a radiologically normal sella, 16 out of 32 Nom nonresponders had alterations, though minor, of the sella (Figure 3.3).

Based on the response pattern to Nom of patients with proven or probable tumours, it may also be inferred that the Nom 'nonresponders' bearing an intact sella turcica may harbour a PRL tumour in an early stage of development. However, the Nom test is only suggestive of a discrimination potential and only longitudinal studies of this group of subjects may provide support or otherwise for its infallibility in individual subjects.

The main reason for using Nom in the diagnosis of prolactinoma rested on the proven inability of the drug to directly affect DA receptors located on lactotrophs (Cocchi *et al.,* 1979). However, to account fully for the PRL nonresponsiveness to Nom of tumour-bearing subjects, an additional factor has to be considered, namely, the existence of a central defect of DA transmission or DA delivery in the hypothalamic-pituitary portal capillary system (Bergland, 1979). In fact, by blocking DA reuptake into TIDA neurons or, most likely, by releasing DA from TIDA neurons into the portal vasculature (Apud *et al.,* 1980), Nom should be otherwise competent to inhibit PRL secretion from the tumorous lactotrophs which maintain their responsiveness to DA (Weiner *et al.,* 1979).

Also supporting the existence of a central defect in DA neurotransmission are the results of the test with L-DOPA plus CD (Fine and Frohman, 1978) and the finding that patients who harbour a prolactinoma, unlike normal subjects, do not depress baseline plasma DA titers after bromocriptine administration (Van Loon *et al.,* 1980).

Figure 3.3: Pattern of plasma PRL changes in 86 patients with pathologic hyper-prolactinemia undergoing Nom (200 mg orally) testing. Also shown are data pertaining to a group of puerperal subjects (postpartum day 2), who received acute administration of the same dose of nomifensine. In the ordinate, values are expressed as percentage change from baseline PRL values. Each symbol is the mean of percentage changes from baseline recorded in individual subjects between 120 and 240 min after drug administration.

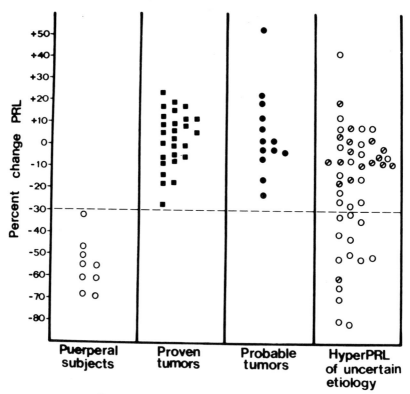

⊘ *Equivocal sella alterations*

However, Faglia *et al.* (1980b) in a study with L-DOPA plus CD in 16 patients with proven PRL-secreting adenomas, and in ten subjects without evidence of tumour, found that five patients with tumour significantly reduced their serum PRL levels and, conversely, only four of the presumptive nontumoural cases consistently reduced their serum PRL. Moreover, in two patients with hyperPRL associated with other endocrine or metabolic disorders (one woman with poly-cystic-ovary syndrome, one diabetic man with sexual impotence), L-DOPA plus CD administration did not lower serum PRL. Similar discrepancies have been reported by Faglia *et al.* (1980b) and Ferrari *et al.* (1980) when evaluating the validity of the Nom test. However, the latter authors were able to discriminate patients with PRL-secreting tumours as a group from hyperPRL patients without demonstrable pituitary adenomas and from normoprolactinemic subjects.

This has never been accomplished by the PRL-lowering drugs used previously, showing that neuropharmacological tests based on the assessment or the stimulation of preexisting CNS-DA function hold promise for widening our knowledge of the physiopathological events underlying inappropriate PRL secretion. Accurate follow-up studies in a large number of subjects may provide support for the diagnostic accuracy of these tests in individual cases and their predictive value.

(c) Stimulatory Tests. The proposition that a defect in DA neuronal function may underlie the development of a PRL-secreting tumour provides an objective rationale to the diagnostic utilization of antidopaminergic drugs in hyperPRL states. These drugs for releasing PRL act by blocking a pituitary receptor which is under modulation by DA delivered mainly through the hypophysial portal circulation. Denervation supersensitivity of the mammotrophs' responsiveness to DA (Weiner *et al.*, 1979) directly implies that DA is continually bathing the AP.

If one admits that in patients prone to developing or harboring a prolactinoma there may be, for an intrinsic defect in DA neurotransmission, an insufficient modulation of the PRL cells by DA transported via the portal blood, unresponsiveness to the PRL-releasing effect of the DA antagonists may be anticipated. However, up to now, blockers of DA receptors, such as chlorpromazine, sulpiride and MCP have not shown discriminatory capacity. Thus, as reported by Faglia *et al.* (1980b), sulpiride (100 mg i.m.) did not elicit any significant serum-PRL increase in 90% of patients with sella alterations and in about 60% of the patients with normal sella (Figure 3.2). Kleimberg *et al.* (1977) observed that chlorpromazine did not produce a positive PRL response, defined as a doubling of the baseline value, in any of nine patients with tumour, but the test was also abnormal in 19 of 36 patients with galactorrhea of nontumoural origin.

Recently, encouraging results have been obtained with the use of domperidone (Dom), a novel antidopaminergic compound (see Table 3.3), which, on the basis of both distribution studies and the inability to enhance DA turnover in the CNS after systemic administration (Cocchi *et al.*, 1980), is thought not to cross the blood-brain barrier.

Previous studies had shown that Dom is a strong PRL releaser in rats (Cocchi *et al.*, 1980), normoprolactinemic subjects and subjects with puerperal hyperPRL (Camanni *et al.*, 1980). Studies in which an evaluation was made of the PRL-releasing effect of Dom in hyperPRL subjects also evaluated with Nom provided interesting results.

As shown in Figure 3.4, a homogeneity in the PRL response to Nom and Dom was present in 36 of the 41 patients investigated. In fact, in only two patients with proven or probable tumour and in three patients with hyperPRL of uncertain etiology, unresponsiveness to Nom was associated with a response of plasma PRL to Dom (at least doubling of baseline PRL levels).

Figure 3.4: Pattern of plasma PRL changes in 41 patients with pathologic hyper-prolactinemia undergoing Nom (upper part) and Dom (lower part) testing. Nom was administered at a dose of 200 mg orally; Dom was administered as a bolus injection of 4 mg i.v. Also shown are data pertaining to two different groups of puerperal subjects (postpartum day 2) receiving acute administration of either drug. In the ordinate, values are expressed as percentage change from baseline PRL values. In the Nom test each symbol is the mean of percentage changes from baseline recorded in individual subjects between 120 and 240 min after administration; in the Dom test, each point refers to the maximum percentage increase recorded. In parentheses, two values exceeding those reported in the ordinate are indicated.

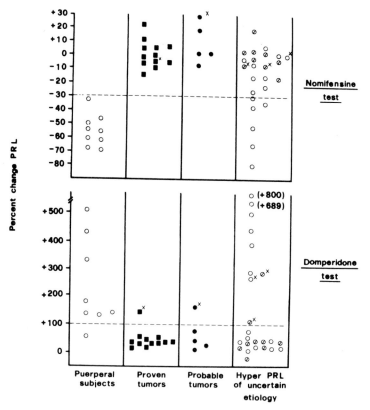

⊘   equivocal sellar alterations

×   lack of homogeneity in response

Domperidone holds promise of being an effective diagnostic agent in hyperPRL states. It has to be noted, however, that responsiveness to Dom occurred in two patients with surgically proven or probable tumour and in three subjects with hyperPRL of uncertain etiology who, on the basis of radiological and clinical findings, were poorly distinguishable from the Dom 'nonresponders'. All these subjects were Nom 'nonresponders.' Close follow-up studies with assessment of the functional and clinical evolution of the latter cases is needed before the real discrimination potential of Dom may be determined. However, it would now appear that Nom is a more reliable indicator than Dom of the presence of adenomatous changes of pituitary lactotrophs. Nevertheless, with these reservations in mind, tests in which Nom and Dom are associated are suitable for differentiating patients with PRL tumours. In this vein, successful application in hyperPRL states of tests based on the combined administration of MCP and TRH has also been reported by Cowden *et al.* (1979).

V. DA-AGONIST AND DA-ANTAGONIST DRUGS: TESTS OF GROWTH-HORMONE RESERVE IN CHILDREN

In 1970 Boyd and coworkers reported high hGH levels in patients with Parkinson's disease who had been treated with L-DOPA, a substance which acts to stimulate hGH release by virtue of its transformation into DA and NE. Since then, L-DOPA has been introduced as a new method for the investigation of hypothalamopituitary function in man. The results of experiments in which the effects of L-DOPA (125-500 mg orally) on hGH release were studied in both normal and short-stature children indicated that it is as reliable a stimulus for hGH release as are the more time-consuming agents such as arginine and insulin (Weldon *et al.,* 1973).

However, the presence of L-DOPA 'nonresponders' among nonhypopituitary subjects (13-40%) (Gomez-Sanchez and Kaplan, 1972) and the unpredictable timing of the response (30-120 min) after drug administration limit its use as a diagnostic test in the evaluation of short stature (Hayek and Crawford, 1972). To inhibit peripheral degradation of L-DOPA and thus facilitate its penetration into brain parenchyma, the drug has been administered along with CD. A combination of CD (50 mg) and L-DOPA (250 mg) resulted in a fivefold potentiation of plasma L-DOPA and a twofold increase in hGH levels (Mars and Genuth, 1973). In a study performed in 75 short-stature patients, this pharmacological cocktail proved to be a reliable screening method for hGH deficiency. Only 16% of the subjects tested were true 'nonresponders' to L-DOPA plus CD (Schönberger *et al.,* 1977). The number of 'nonresponders' among nonhypopituitary children is even reduced when oral administration of L-DOPA (0.5 g/1.73 m^2) is associated with propranolol (0.75 mg/kg), a β-receptor blocker and, hence, an activator of hGH release (Collu *et al.,* 1975). In this vein, the association of L-DOPA with an indirectly acting DA agonist, i.e., Nom, which by blocking both DA and NE reuptake is expected to potentiate the L-DOPA effect, also seems promising. In children with nonendocrine short stature, Nom (12.5 mg i.v.) given alone or in combination with L-DOPA (12.5 mg i.v.) was more effective as an hGH releaser than insulin hypoglycemia or L-DOPA alone (Pintor and Müller, unpublished results).

Since activation of DA neurotransmission causes the release of GH in normal adults (Müller *et al.*, 1978b), it is expected that its blockade would inhibit hGH release or, on considering the low resting concentrations of the hormone, have no effect. In normal adult males, the DA-antagonist drug MCP does not release GH (Sowers *et al.*, 1976) but when administered to adolescent males with non-endocrine short stature, elevates GH in plasma (Cohen *et al.*, 1979b). In this study, the degree of GH response after MCP was similar to that following insulin-induced hypoglycemia and was unrelated to stressful mechanisms, alterations in blood glucose or erratic fluctuations in hGH release. That adolescent but not adult males respond to MCP may indicate that a difference in sex-hormone balance is of importance, a hypothesis consistent with the finding that MCP also evokes GH release in adult males with hypogonadism (Cohen *et al.*, 1979a).

VI. DA-AGONIST DRUGS AND THE TREATMENT OF HYPERPROLACTINEMIC STATES

It is now clear that excessive PRL secretion is a common phenomenon and is often implicated in women with menstrual disorders, galactorrhea and infertility and in men with impotence. Whatever the reason for the exuberant PRL secretion may be (Table 3.4), it is evident on theoretical grounds that DA-mimetic drugs, acting either indirectly by an activation of the endogenous DA function or directly through stimulation of pituitary DA receptors, should be competent to restore to normal the enhanced PRL levels and the associated clinical features. However, there is no sound reason for the use of indirectly acting DA agonists in the treatment of hyperPRL states. In fact, it must be remembered that: (1) the activation of the DA tone induced by these drugs at the level of PRL-secreting mammotrophs cannot be equated to that produced by direct DA agonists (Cocchi *et al.*, 1979); and (2) in many instances, coexistence of a defect in brain dopaminergic function would obliterate the PRL-lowering effect of these drugs. These considerations provide the rationale for the introduction of powerful direct DA agonists into clinical practice. It must be added for the sake of exactness, however, that the discovery that bromocriptine, the most clinically useful compound, possesses the behavioural, biochemical and histochemical properties of a DA-receptor agonist was made subsequent to its use as a PRL-lowering agent.

Historically, in the 1970s the clinical availability of L-DOPA provided the opportunity of studying the effect of a DA precursor on PRL secretion and represented the first pharmacological approach to hyperPRL states. In women with nonpuerperal lactation and high PRL levels due either to PRL-secreting adenomas or 'functional' pituitary disorders, L-DOPA (500 mg orally) resulted in a marked fall in serum PRL in 90 minutes. Sustained suppression of PRL could not be achieved; however, when the drug was given four times daily, more prolonged treatment did not change basal serum PRL or milk production (Malarkey *et al.*, 1971). L-DOPA acts after conversion to DA and its efficacy after administration of an oral dose depends on the concurrence of many factors. In endocrine patients, its therapeutic application is also limited by the high incidence of side effects and the reported impairment of glucose tolerance. Similarly, the short duration of action and the profound side effects of apomorphine and DA limit their therapeutic applicability (Müller *et al.*, 1978b).

The development of ergot compounds which have specific antigalactic action and are free of the uterotonic and vascular properties of the traditional ergot alkaloids resulted in the selection of bromocriptine (Flückiger and Wagner, 1968) and has opened a new era in the medical treatment of hyperprolactinemia.

In addition to peptide-containing ergot alkaloids, a number of simple synthetic ergoline derivatives (e.g., lergotrile, lisuride, metergoline and pergolide; see Table 3.2) have undergone successful clinical trial.

A. Puerperal and Pathological Lactation

Unlike L-DOPA, bromocriptine acts at the level of DA receptors with a much longer duration of action, which appears to be due to a persisting effect at the receptor site. The drug is well adsorbed after oral administration, and has a duration in suppressing PRL of six to twelve hours after a single oral dose (Figure 3.5).

Bromocryptine (2.5 mg orally two or three times a day) can either prevent the initiation of lactation or promptly suppress it once already established. Prolactin levels fall to normal, breast tenderness and engorgement do not occur and there is no rebound milk production if treatment is pursued for two to three weeks (Varga *et al.*, 1972). Metergoline too (4 mg orally t.i.d.) is capable of preventing lactation or rapidly suppressing it when administered after lactation. The drug is well tolerated, and virtually free of side effects (Crosignani *et al.*, 1978a). However, while there is unanimous consent on the beneficial antigalactic effect of metergoline, this is not true for its mechanism of action. Metergoline, though developed as a potent antiserotoninergic agent (see Table 3.2), has also shown dopaminergic properties. Nevertheless, some differences which may reflect a different mechanism of action can be envisaged between metergoline and bromocriptine with respect to the clinical effect both on lactation and on circulating PRL levels. Unlike bromocriptine, which necessitates prolonged administration, metergoline has a persistent antigalactic effect even after as short a time as five days, despite the fact that, in contrast to the abrupt decrease induced by bromocriptine, plasma PRL falls only slowly and progressively after metergoline (Crosignani *et al.*, 1978a). Both drugs appear superior and safer in the inhibition of puerperal lactation than estrogens and represent an advisable alternative method for obtaining this goal.

Dopamine-agonist drugs suppress not only physiological but also pathological lactation, an event which occurs in about 80% of all subjects with hyperPRL. Whatever the cause of galactorrhea (i.e., idiophatic without amenorrhea, resulting from intense DA-receptor blockade due to neuroleptics, or associated with a PRL-secreting adenoma), it is found that 10 mg daily of bromocriptine, 12 mg

Figure 3.5: Plasma PRL values in patients with pathologic hyperprolactinemia (baseline values fall in the range 27-600 ng/ml) after administration of bromocriptine (CB 154, 2.5 mg po), L-DOPA (500 mg po) or infusion of DA (280 μg/min/120 min). Values (mean ± SEM) are expressed as a percentage of the baseline. The number of patients is given in parentheses.

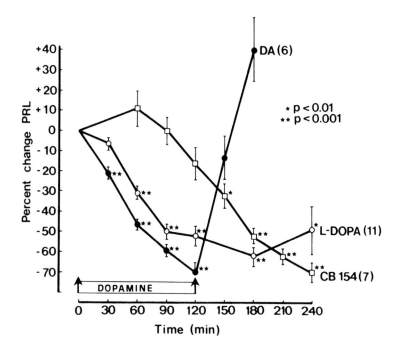

metergoline and 0.6-1.2 mg of lisuride, given in three divided doses, usually restores PRL levels to normal and stops milk production within a week. Bromocriptine has to be given at first in slowly increasing doses and with food in order to avoid side effects such as postural hypotension, nausea and vomiting. Starting from a dose of 1.25 mg on retiring, the dose is increased every three to four days until a dose of 2.5 mg three or four times a day is reached. Higher dosages are necessary (i.e., for bromocriptine, up to 50 mg daily) if initial PRL levels are very high in the presence of PRL-secreting adenomas.

When galactorrhea is associated with amenorrhea in women or decreased libido and potency in men, cessation of milk secretion and suppression of PRL levels are accompanied, respectively, by regular menstruation and restored libido and potency. If treatment with ergot alkaloids is stopped, hyperPRL and galactorrhea usually recur. A list of ergot alkaloids which have been used successfully in hyperPRL states is presented in Table 3.5.

Table 3.5: Ergot Drugs Used in Clinical Practice for Inhibiting PRL in Hyper-prolactinemic Subjects

Drugs	Authors
Ergonovine	
Methylergonovine	Müller *et al.* (1978b)
2-Br-α-ergocryptine	
Lergotrile	Thorner *et al.* (1978)
Lisuride	Liuzzi *et al.* (1978)
Metergoline	Crosignani *et al.* (1978b)
Methysergide	Crosignani *et al.* (1978b)
Pergolide*	Lemberger and Crabtree (1979)

* Effect observed in normoprolactinemic subjects

B. Infertility

Galactorrhea with high PRL levels is usually associated with hypogonadism, whether the hyperPRL is physiological during the post-partum period or pathological due to pituitary or hypothalamic disease. The majority of patients have normal basal gonadotropin levels, and indeed normal responses to the gonadotropin-releasing hormone (LH-RH), although a disparity in the releasable pituitary LH and FSH with a profound reduction of the LH/FSH ratio has been evidenced in some patients (Lachelin *et al.*, 1977a). A defect in pulsatile release of gonadotropins appears to be present, since hyperPRL patients fail to show the normal increase in gonadotropin after the administration of exogenous estrogens; that is, the positive feedback response is blocked. These events may be explained by the feedback action of high PRL levels on TIDA neurons, with increased DA turnover and ensuing dopaminergically mediated inhibition of LH-RH discharge from the hypothalamus. Supporting this proposition is the finding that, in rats bearing a transplantable PRL-secreting tumour, hypothalamic LH-RH titers are increased (Gil-Ad *et al.*, 1978). Important peripheral actions of PRL cannot be neglected; increased PRL levels may impair the action of the gonadotropins on the gonads and disrupt the ovarian production of progesterone. Chronic failure of cyclical estrogen secretion from the ovary, being the gonadotropin blocked at this level, may contribute to the impairment of the hypothalamic cycling center.

Whatever may be the precise mode of action of the antifertility effect of hyperPRL, ergot drugs are astonishingly successful in lowering PRL levels and restoring fertility within four to twelve weeks of starting treatment in most cases. Since, as previously noted, DA seems to be inhibitory to LH release in man (Lachelin *et al.*, 1977b), the antifertility action of dopaminergic ergot drugs must rely principally on their PRL-lowering capacity.

Restoration of ovarian function by metergoline, in view of the reported serotoninergic inhibition of gonadotropin release (Müller *et al.*, 1978b), has also been ascribed to direct stimulation of LH-RH centers by this antiserotoninergic drug (Crosignani *et al.*, 1978b), although no effect was shown by metergoline on plasma LH levels in normoprolactinemic women (Pontiroli *et al.*, 1980). The different pharmacological profile of bromocriptine and metergoline would explain why the former is more uniformly effective than the latter in suppressing PRL secretion and restoring ovarian function.

Ergot drugs appear to be effective not only in patients with hyperPRL, but also in amenorrheic normoprolactinemic subjects (Seppälä *et al.*, 1976). However, in studies comparing the efficacy of bromocriptine and metergoline to that of a placebo, no significant difference was evident in the number of subjects who menstruated following either treatment (Crosignani *et al.*, 1980).

Ergot alkaloids may also be effective against infertility present in women with the clinical picture of polycystic-ovary syndrome, when this is accompanied by high PRL levels (about 20% of patients). Chronic administration of bromocriptine reduces elevated PRL levels, stops galactorrhea and lowers the increased levels of testosterone and dehydroepiandrosterone. In these subjects, restoration of fertility with return of a normal menstrual cycle is associated with amelioration of peripheral signs such as hirsutism and acne (Besser, 1978).

Short luteal phase and low progeterone may be present with elevated plasma PRL in some infertile women; in this instance, bromocriptine normalizes the hyperhermic phase of the cycle by lowering plasma PRL, and pregnancy ensues (del Pozo *et al.*, 1976). In view of the permissive effect of PRL on luteal-cell function (McNatty *et al.*, 1977), reduction of plasma progesterone during the luteal phase has also been shown in normally menstruating women whose normal PRL levels have been reduced following the institution of bromocriptine (del Pozo and Wolf, 1978). Thus, PRL 'oversuppression' may also endanger normal progesterone production.

C. Prolactin-secreting Adenomas

Aside from drug-induced hyperPRL, the most frequent cause of excessive PRL secretion is due to PRL-secreting adenomas, which represent the majority of pituitary tumours. In fact, it has been found that PRL hypersecretion is present in about 60-70% of pituitary tumours previously classified as functionless adenomas and in about 30-40% of GH-secreting adenomas (Faglia *et al.*, 1980a). Prolactin adenomas account for between 20 and 30% of otherwise unexplained secondary amenorrhea. Detection of these tumours in an early stage requires sophisticated technologies such as polytomography and computerized axial tomography, as well as the demonstration of a particular response pattern to functional tests of PRL secretion.

What accounts for the relatively large number of newly discovered cases, whether the availability of new technologies or a real change in the history of the disorder, is an unanswered question. It has been suggested that the apparent increased incidence may be due to widespread use of oral contraceptives, which, because of their estrogen content, may stimulate the growth of adenomas (Sherman *et al.,* 1978). Other studies ascribe the increasing incidence to advances in diagnostic and surgical technology instead (Coulman *et al.,* 1979).

Even though little is known about the pathogenesis of PRL-secreting tumours, there is evidence to suggest that the abnormalities governing the hypersecretion of PRL may be due to a primary defect in central dopaminergic neurotransmission, ultimately leading to hyperplastic and then neoplastic changes of pituitary lactotrophs. Estrogens might also be related to the physiopathology of PRL-secreting adenomas in this way. A single injection of estrogen in the rat induces specific cytopathological changes in the ARC n (Brawer *et al.,* 1978), an area crucial for TIDA functioning, and decreases DA concentration in the hypophysial portal blood (Cramer *et al.,* 1979).

It cannot be excluded, however, that the pituitary may be the locus of the primary etiologic abnormality and that PRL hypersecretion *via* the PRL-DA short-loop feedback may result in secondary alteration of CNS neurotransmitter function. Restoration of the pathological condition to normal after selective removal of the hypersecreting cells (Hardy, 1975) would tend to support the 'pituitary' hypothesis. Whatever the primary pathophysiological event in the development of a PRL-secreting adenoma, administration of ergot derivatives causes immediate and sustained PRL suppression, with rapid restoration of fertility in women and normalization of hyperPRL hypogonadism in men (von Werder *et al.,* 1980).

These events are mediated by ergot-induced inhibition of PRL secretion from tumorous lactotrophs and ensuing suppression of PRL antifertility effects. In addition, evidence is now emerging that ergots for an antiproliferative effect on pituitary adenoma growth (Quadri *et al.,* 1972; Gil-Ad *et al.,* 1978) may lead to shrinkage of the prolactinoma, resolution of visual-field defects and changes in bone structure of the sella turcica (Thorner *et al.,* 1980; von Werder *et al.,* 1980) (Figure 3.6). As a result of the increased mitotic rate and signs of enhanced proliferation of macro- as against microprolactinoma (Peillon *et al.,* 1978), the antiproliferative effect of ergot seems to be specific for the PRL-secreting cells in macroprolactinomas with extremely elevated PRL levels and high PRL turnover (von Werder *et al.,* 1980). Ergot therapy thus offers promise for the transformation of an invasive macroprolactinoma into a surgically resectable tumour.

D. Pregnancy

There is major concern about the use of ergot drugs to induce ovulation in patients who desire to become pregnant and who harbour a prolactinoma. The concern relates to the possible acceleration of prolactinoma growth due to increased estrogen production during pregnancy (Bergh *et al.,* 1978). Surgery is indicated for large tumours, but there is no consensus about management of small adenomas. Prophylactic surgery, external radiation or yttrium implants

Figure 3.6: Visual-acuity and visual-field plots in a patient bearing a macropro-lactinoma before and during six months of bromocriptine therapy. The visual fields were plotted by one observer using the Goldman apparatus under identical conditions with a 0.25-mm² object at two different light intensities, 1000 apostilb (I₄) and 100 apostilb (I₂). The black periphery indicates a normal visual field for comparison. Before therapy, a complete temporal hemianopsia in the left eye and an incomplete hemianopsia in the right eye were present, with reduced visual acuity in both. After three days the left visual field had improved, and visual acuity was restored to 20/20. Thereafter, progressive improvement in the visual fields were noted; the only abnormality at six months was an equivocal superior bitemporal quadrantic defect to the low-intensity object.

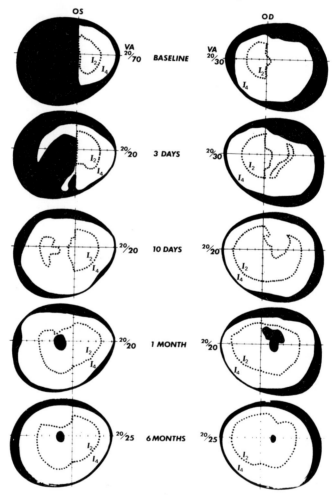

Source: Reproduced, with permission from Thorner *et al.* (1980).

have been recommended prior to administration of bromocriptine when pregnancy is anticipated (Besser, 1978). However, some endocrinologists feel that infertile patients with PRL-secreting microadenomas can be allowed to become pregnant on bromocriptine alone, provided that they are carefully evaluated and supervised during pregnancy (Zarate *et al.*, 1979).

Bromocriptine has been given alone or after radiotherapy or surgery to many thousands of women with successful deliveries; its use should be stopped once pregnancy is confirmed, although the drug is not teratogenic and uneventful pregnancies have been registered when its use has been continued (Griffith *et al.*, 1978).

VII. DA-AGONIST DRUGS AND THE MEDICAL TREATMENT OF ACROMEGALY

Irrefutable evidence has been produced that, in contrast to the stimulatory effect exerted in normal subjects (Müller *et al.*, 1977), application of dopaminergic compounds in acromegaly induces a consistent suppression of the elevated hGH levels. In 1972, Liuzzi *et al.* made the crucial observation that acute administration of L-DOPA lowered hGH levels in the plasma of some acromegalics, a finding which has subsequently been extended to include other DA-agonist drugs (i.e., apomorphone, piribedil) and the ergot drugs, bromocriptine, lisuride, lergotrile and metergoline (Müller *et al.*, 1978b) (Figure 3.7).

Also, direct infusion of DA was shown to be capable of inducing a clear-cut decrease of baseline hGH levels, which was even more marked than that elicited by L-DOPA administration (Camanni *et al.*, 1977). Moreover, it became apparent that, unlike direct DA agonists, indirectly acting DA agonists such as amantadine, amphetamine and nomifensine, despite their ability to release hGH in healthy subjects, were ineffective in lowering high resting hGH levels in acromegalics (Müller, 1979).

In addition, the competence of L-DOPA to inhibit hGH secretion in 'responder' acromegalics is related to peripheral conversion of the precursor to DA and not to activation of central DA neurotransmission. In fact, peripheral inhibition of DOPA decarboxylase by CD prevented the hGH-lowering effect of administered L-DOPA (Camanni *et al.*, 1978). In summary, these data and others not reported here indicate that the effectiveness of a given dopaminergic compound to lower basal hGH levels in acromegaly encompasses direct stimulation of DA receptors present on the hyperplastic or tumorous somatotroph. Indirectly acting DA agonists which require a pool of DA available for release are instead ineffective (Müller, 1979).

Why DA receptors become unmasked and operative on tumoral somatotrophs is poorly understood, although this event is probably related to the elusive physiopathology of the disease. It has been noted that, in this condition, GH-secreting cells behave as PRL-secreting cells; they also release GH after TRH

Figure 3.7: GH-lowering effect of acute administration of direct DA-agonist drugs in 'responder' acromegalic patients, showing the pattern of GH decrease in the plasma after administration of bromocryptine (CB 154, 2.5 mg po), L-DOPA (500 mg po), apomorphine (0.75 mg sc), metergoline (4 mg po) and lisuride (0.2 mg po). In the L-DOPA and bromocryptine experiments, patients also received acute administration of placebo. Values are expressed as the ratio of suppressed (S) to baseline (B) ± SEM.

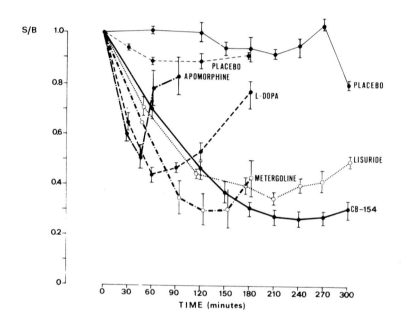

Source: Redrawn and reproduced, with permission, from Liuzzi *et al.* (1976, 1978) and Chiodini *et al.* (1974, 1976).

(Irie and Thushima, 1972), and the hypothesis has been formulated that there exists a mammosomatotropic cell which synthesizes and releases both GH and PRL (Liuzzi *et al.,* 1976). Clinical histological findings may support this view (Guyda *et al.,* 1973). Accepting that, in some acromegalic 'responder' patients, there may be GH-secreting cells with receptorial properties similar to those of the PRL-secreting cells, it is still not clear whether this is due to a primary pituitary or, alternatively, hypothalamic disturbance. Certainly, brain dopaminergic function is altered in acromegalic 'responder' patients as indicated by: (1) failure of directly and indirectly acting DA agonists to stimulate hGH release; (2) the ability of DA infusion to restore PRL responsiveness to sulpiride when hyperPRL coexists (Moriondo *et al.,* 1980).

However, altered CNS dopaminergic function may be merely secondary to the negative-feedback effect of GH hypersecretion on neural tissue. In hypophysectomized rats, huge doses of GH increased DA turnover in the TIDA neurons (Fuxe *et al.*, 1974). If this mechanism is also operative in the human, progressive desensitization of dopaminergic synapses by increased DA input may be envisioned (Beck *et al.*, 1978). Supporting this 'pituitary' hypothesis of acromegaly is the finding of normalization of the GH response to TRH and dopaminergic drugs in some patients after selective adenomectomy (Faglia *et al.*, 1980a).

Despite our present ignorance of the natural history of acromegaly, the inhibitory effect of dopaminergic agonists on GH secretion has provided the rationale for the medical treatment of this disease. Of the compounds tested so far, the long-acting DA agonist bromocriptine (up to 20 mg daily) and lisuride (up to 1.6 mg daily) are the most effective in achieving a stable prolonged suppression of hGH levels. Responsiveness to acute administration of a single dose of either compound is a reliable criterion to predict responsiveness to long-term treatment (Belforte *et al.*, 1977).

Among acromegalics, 20-30% of patients are 'nonresponders,' even to high doses of the compound. Responder patients exhibit a different degree of suppression, i.e., according to different authors for bromocriptine from 80 ng/ml to < 1 ng/ml, or from > 200 ng/ml to 6 ng/ml, or from 197 ng/ml to 2 ng/ml, the average suppression being proportional to the initial values (Köbberling *et al.*, 1980).

Like dopaminergic ergot compounds, metergoline causes a clear-cut inhibition of plasma hGH levels in some of the acromegalic patients studied, although the inhibitory effect of this drug is less sustained and prolonged than that of bromocriptine. The close concordance existing in individual subjects between the hGH-inhibitory effect of bromocriptine and that of metergoline and the abolition induced by pimozide suggests that, in both instances, a dopaminergic mechanism underlies the neuroendocrine effect (Müller, 1979).

Apart from lowering plasma hGH levels, clinical and metabolic improvements have been reported following the institution of the medical therapy. Headache, high blood pressure, excessive sweating, soft-tissue swelling, libido and sexual performance are all improved. Resumption of menses and pregnancy and cessation of galactorrhea have been described, but they may be due to concomitant reduction of hyperPRL. In addition, improved glucose tolerance, as well as reduction of hydroxyprolinuria, can be observed. Plasma hGH levels and disability return if treatment is stopped. Clinical improvement has also been reported in 'nonresponder' or poor 'responder' patients. This dissociation between the clinical and hormonal response may result from a proportionally greater fall in the biologically active form of hGH, with the radioimmunologically competent but less biologically active form remaining in the circulation. Alternatively, since metabolic and/or clinical improvement can also be seen in nonacromegalic patients on bromocriptine treatment, it may be that they do not reflect constantly true amelioration of the acromegalic disease.

Growth-hormone responses to TRH or insulin hypoglycemia are usually preserved during bromocriptine therapy; moreover, drug withdrawal is quickly followed by reinstatement of hGH levels to pretherapy values (Köbberling *et al.*, 1980). These facts tend to indicate an inhibitory action of dopaminergic ergots of hGH release but not synthesis. However, reduction of tumour size has been recently reported in eight of 43 acromegalic patitents under chronic bromocriptine treatment, but only two of them had not received radiotherapy in addition to bromocriptine (Wass *et al.*, 1979). Thus, to assess the antiproliferative effect of this drug in acromegaly, further studies are needed.

In conclusion, while it is indisputable that there are acromegalic patients who benefit from, although they are not cured by, chronic treatment with ergot drugs, it is apparent that these drugs have to be considered as a useful adjunct to, but not a replacement for, more radical approaches. They should be especially directed to: (1) patients after unsuccessful surgery and/or radiation; (2) patients waiting for surgery or for radiation to take effect; (3) elderly, surgically high-risk patients; (4) patients who refuse other types of therapy (Köbberling *et al.*, 1980).

VIII. BROMOCRIPTINE AND THE MEDICAL TREATMENT OF CUSHING'S DISEASE

Information on the action of ergot derivatives on the secretion of ACTH is still scanty (Table 3.6). Earlier studies in the rat had demonstrated that chronic treatment with both ergotamine and bromocriptine are capable of stimulating ACTH secretion. A reverse picture was present when ergotamine was administered to rats which bore a tumour secreting ACTH (and PRL) since in this instance it decreased the elevated serum ACTH levels without affecting the pituitary ACTH concentration (MacLeod and Krieger, 1976). Similarly, in patients with Cushing's disease and Nelson's syndrome, a marked diminution in plasma ACTH was reported after a single oral dose of 2.5 mg bromocryptine (Lamberts *et al.*, 1980a). In other studies, however, no effect of bromocryptine on plasma ACTH was observed in three out of three patients with Cushing's disease and in 13 patients with Cushing's disease after bilateral adrenalectomy. In four of five patients with Cushing's disease who responded to acute drug administration by lowering plasma ACTH, chronic treatment with bromocriptine (up to 20 mg/daily for one year) consistently reduced the cortisol secretory rate and/or urinary 17-OHCS secretion; however, long-term therapy was not always effective in these patients because of the 'escape' from therapy which was observed in two of them (Lamberts *et al.*, 1980a).

It remains debatable whether the primary source of Cushing's disease is the hypothalamus (Krieger and Glick, 1972) or the AP gland, even though clinical remission following selective adenomectomy would support the 'pituitary' hypothesis (Tyrrel *et al.*, 1978). The only question yet to be answered is whether selective removal of the adenoma results in permanent cure of Cushing's disease. Only careful, long-term follow-up of patients treated by transsphenoidal

Table 3.6: Ergot Drugs and ACTH Secretion

Drug	Experimental model	Effect on plasma ACTH levels***	Authors
Ergotamine*	Normal rat	↑	
Bromocriptine	Normal rat	↑	MacLeod and Krieger (1976)
Bromocriptine	Dispersed cells from ACTH-secreting tumour	↓	Lamberts *et al.* (1980b)
Ergotamine*	Rats with ACTH-PRL secreting tumour	↓	MacLeod and Krieger (1976)
	Healthy man	→	
Bromocriptine**	Pituitary-dependent Cushing's syndrome	↓	Lamberts *et al.* (1980a)
	Nelson's syndrome	↓	

  * Seven-day treatment
  ** Single treatment
  *** A vertically upward arrow denotes stimulation and a downward arrow inhibition, while a horizontal arrow indicates no effect.

microsurgery will resolve this question.

In patients with Cushing's disease, the ACTH and cortisol responses to insulin-induced hypoglycemia remained absent after acute bromocriptine administration, while the absent GH response to hypoglycemia returned partially (Lamberts *et al.*, 1980a). This discrepant finding makes it unlikely that a dopaminergic depletion is present in the hypothalamus of patients with Cushing's disease (Lamberts *et al.*, 1977), since in this instance both the ACTH and the GH response to hypoglycemia should have been restored after bromocriptine. Rather, it may be that suppression of plasma cortisol by the drug more rapidly restores the responsiveness to hypoglycemia of hypothalamic centers for GH than for ACTH secretion (Lamberts *et al.*, 1980a).

Suppression of ACTH and cortisol secretion by bromocriptine would occur though direct inhibition of neoplastic corticotroph cells, a view supported by the notion that bromocriptine added *in vitro* to cells from a rat ACTH-secreting tumour induced a significant suppression of ACTH production (Lamberts *et al.*, 1980b).

It is of note that patients who respond to bromocriptine are usually those with higher resting ACTH levels and larger size of the pituitary tumours, an extra argument for an action of bromocriptine directed at the pituitary These patients also respond paradoxically to hypothalamic hormones, since somatostatin, TRH and LH-RH have been reported to suppress or increase, respectively, ACTH secretion in these patients (Lamberts *et al.*, 1980a). An action of ergot drugs and hypothalamic hormones exerted at the level of dedifferentiated pituitary cells may be envisioned in these circumstances.

IX. SUMMARY AND CONCLUSIONS

New knowledge on neurotransmitter control of AP hormone secretion and, particularly, on the major role played by TIDA neurons, permits the fruitful exploitation of DA-agonist and DA-antagonist drugs for diagnostic and therapeutic purposes in hypothalamo-pituitary dysfunction.

Application of indirectly acting DA agonists such as nomifensine, which probes selective aspects of CNS-DA function, holds promise for the development of simple, inexpensive and rapid tests for discriminating between individuals with and without PRL-secreting pituitary adenoma, a task never accomplished by PRL-lowering drugs used previously. In addition, these compounds may widen our understanding of the physiopathological events underlying excess PRL secretion. Since individuals harbouring a prolactinoma may have a defect in CNS dopaminergic function, DA antagonists, which, if they are to act, require sufficient modulation of lactotrophs by DA released into the portal blood, may also be of diagnostic value in hyperPRL states. In view of the stimulatory effect exerted on hGH secretion, DA-mimetic drugs can also be profitably used for testing GH reserve in short-stature children, a goal achieved, rather paradoxically, by the DA-antagonist drug metoclopramide.

Direct DA-agonist drugs, namely, potent and long-acting peptide-ergot derivatives and ergolines, are valid pharmacological weapons for the suppression of puerperal lactation and for the treatment of pathological hyperPRL, of hGH and, probably, of ACTH hypersecretion.

Hyperprolactinemia is a common marker of gonadal dysfunction, infertility, amenorrhea or menstrual disorders in women, or impotence in man. Galactorrhea also occurs in some of these patients. The most widely used ergot compound (i.e., bromocriptine) acts as a DA-receptor agonist directly on the pituitary cells to lower PRL, restoring circulating PRL levels to normal in both female and male patients. Normalization of PRL levels is associated with return of gonadal function to normal: women have normal ovulatory menstruation and potency is restored in men. Plasma PRL and clinical hypogonadism usually return if the drug is stopped but, in some patients bearing macroprolactinomas, PRL secretion is constantly suppressed after long-term treatment with bromocriptine. This effect is probably consequent to reduction in tumour size for the antiproliferative action of the drug.

Direct DA-agonist drugs do not directly affect somatotrophs of an intact AP, but inhibit GH release in many acromegalics, acting directly on DA receptors unmasked on tumorous somatotrophs. Recognition of this mechanism has provided the possibility of an effective medical approach to therapy, especially with bromocriptine. Medical treatment of acromegaly with ergot drugs must be viewed as a useful adjunct to, even though not a replacement for, more radical approaches.

Finally, at least in some patients with ACTH hypersecretion due to Cushing's disease or Nelson's syndrome, chronic application of bromocriptine consistently reduces adrenal cortisol production, although 'escape' from therapy seems to be more frequent than in states of PRL excess. Even in this instance, suppression of ACTH and cortisol secretion by ergots would occur through direct inhibition of neoplastic-corticotroph cells.

ACKNOWLEDGEMENTS

This chapter includes previously unpublished results of experiments supported in part by a grant from the CNR research project on the 'Biology of Reproduction', provided to E.E. Müller and A.R. Genazzani. The secretarial help of Miss Isabella Zago is gratefully acknowledged.

REFERENCES

Ahn, H.S., Gardner, E. and Makman, M.H. (1979). *Erop. J. Pharmacol., 53,* 313
Apud, J., Cocchi, D., Iuliano, E., Müller, E.E. and Racagni, G. (1980). *Brain Res., 168,* 226
Beck, W., Hancke, J.L. and Wuttke, W. (1978). *Endocrinology 102,* 837
Belforte, L., Camanni, F., Chiodini, P.G., Liuzzi, A., Massara, F., Molinatti, G.M., Müller, E.E. and Silvestrini, F. (1977). *Acta Endocrinol.* (Kbh.) *85,* 235
Ben-Jonathan, N., Oliver, C., Weiner, H.J., Mical, R.S. and Porter, J.C. (1977). *Endocrinology 100,* 452
Bergh, T., Nillius, S.J. and Wide, L. (1978). *Br. Med. J., 1,* 875
Bergland, R.M. (1979). *Lancet ii,* 1270
Besser, G.M. (1978). *Med. J. aust., 2* (suppl.), 14
Boyd, A.E., Lebovitz, H.G. and Pfeiffer, J.G. (1970). *N. Eng. J. Med., 283,* 1425
Brawer, J.R., Naftolin, P., Martin, J. and Sonneschein, C. (1978). *Endocrinology 103,* 501
Camanni, F., Massara, F., Belforte, L., Rosatello, A. and Molinatti, G.M. (1977). *J. Clin. Endocrinol. Metab., 44,* 465
Camanni, F., Picotti, G.B., Massara, F., Molinatti, G.M., Mantegazza, P. and Müller, E.E. (1978). *J. Clin. Endocrinol. Metab., 47,* 647
Camanni, F., Genazzani, A.R., Massara, F., La Rosa, R., Cocchi, D. and Müller, E.E. (1980). *Neuroendocrinology 30,* 2
Carlsson, A. (1975). In *Pre- and Postsynaptic Receptors,* ed. E. Usdin and W.E. Bunney (Dekker, New York) p. 49
Chiodini, P.G., Liuzzi, A., Botalla, L., Cremascoli, G. and Silvestrini, F. (1974). *J. Clin. Endocrinol. Metab., 38,* 200
Chiodini, P.G., Liuzzi, A., Müller, E.E., Botalla, L., Cremascoli, G., Oppizzi, G., Verde, G. and Silvestrini, F. (1976). *J. Clin. Endocrinol. Metab., 43,* 356
Cocchi, D., Locatelli, V., Picciolini, E., Genazzani, A.R. and Müller, E.E. (1979). *Proc. Soc. Exp. Biol. Med.* (NY) *162,* 38
Cocchi, D., Gil-Ad, I., Parenti, M., Stefanini, E., Locatelli, V. and Müller, E.E. (1980). *Neuroendocrinology 30,* 65
Cohen, H.N., Hay, I.D., Beastall, G.H. and Thompson, J.A. (1979a). *Clin. Endocrinol., 11,* 95
Cohen, H.N., Hay, I.D., Thompson, J.A., Logue, F., Racliffe, W.A. and Beastall, G.H. (1979b). *Clin. Endocrinol., 11,* 89
Collu, R., Leboeuf, G., Letarte, J. and Ducharme, J.R. (1975). *Pediatrics 56,* 262
Coulman, C.B., Annegers, J.F., Abbond, C.F., Laws, E.R. and Kurland, L.T. (1979). *Fert. Ster., 32,* 28

92   *Dopamine Agonist and Antagonist Drugs*

Cowden, E.A., Thomson, J.A., Doyle, D., Ratcliffe, J.G., Macpherson, P. and Teasdale, G.M. (1979). *Lancet i,* 1155
Cramer, O.M., Parker, C.R. and Porter, J.C. (1979). *Endocrinology 105,* 419
Creese, I., Burt, D.R. and Snyder, S.R. (1975). *Life Sci., 17,* 993
Crosignani, P.G., Lombroso, G.C., Caccamo, A., Reschini, E. and Peracchi, M. (1978a). *Obst. Gynec., 51,* 113
Crosignani, P.G., Peracchi, M., Lombroso, G.C., Reschini, E., Mattei, A., Caccamo, A. and D'Alberton, A. (1978b). *Am. J. Obst. Gynec., 132,* 307
Crosignani, P.G., Ferrari, C., Fadini, R., Meschia, M., Picciotti, M.C. and Reschini, E. (1980). In *Neuroactive Drugs in Endocrinology,* ed. E.E. Müller (Elsevier/North-Holland, Amsterdam) p. 341
Daughaday, W.H. (1977). *New Eng. Med. J., 297,* 12
De Camilli, P., Macconi, D. and Spada, A. (1979). *Nature* (Lond.) *287,* 252
Faglia, G., Beck-Peccoz, P., Ambrosi, B., Travaglini, P., Moriondo, P., Elli, R. and Rondena, M. (1980a). In *Pituitary Microadenomas,* ed. G. Faglia, M. Giovannelli and R.M. MacLeod (Academic Press, New York) p. 277
Faglia, G., Moriondo, P., Beck-Poccoz, P., Travaglini, P., Ambrosi, B., Spada, A. and Nissim, M. (1980b). In *Neuroactive Drugs in Endocrinology,* ed. E.E. Müller (Elsevier/North-Holland, Amsterdam) p. 263
Ferrari, C., Caldara, R., Rampini, P., Telloli, P., Romussi, M., Bertazzoni, A., Polloni, G., Mattei, A. and Crosignani, P.G. (1978). *Metabolism 27,* 1499
Ferrari, C., Crosignani, P.G., Caldara, R., Picciotti, M.C., Malinverni, A., Barattini, B., Rampini, P. and Telloli, P. (1980). *J. Clin, Endocrinol. Metab., 50,* 23
Fine, S.A. and Frohman, L.A. (1978). *J. Clin. Invest., 61,* 973
Flückiger, E. and Wagner, H.R. (1968). *Experientia* (Basel) *24,* 1130
Fuxe, K., Hökfelt, T., Jonsson, G. and Löfström, A. (1974). In *Neurosecretion − The Final Neuroendocrine Pathways,* ed. F. Knowles and L. Volrath (Springer Verlag, Berlin) p. 269
Fuxe, K., Eneroth, P., Gustafson, J.A., Lofström, A. and Skett, P. (1977). *Brain Res., 122,* 177
Gil-Ad, I., Locatelli, V., Cocchi, D., Carminati, R., Arezzini, C. and Müller, E.E. (1978). *Life Sci., 23,* 2245
Gomez-Sanchez, C. and Kaplan, N.M. (1972). *J. Clin. Endocrinol. Metab., 34,* 1105
Green, J.D. and Harris, G.W. (1947). *J. Endocrinol., 5,* 136
Griffith, R.W., Turkaly, I. and Brown, P. (1978). *Brit. J. Clin. Pharm., 5,* 227
Guyda, H., Robert, F., Colle, E. and Hardy, J. (1973). *J. Clin. Endocrinol. Metab., 36,* 531
Hardy, J. (1975). In *Progress in Neurological Surgery,* ed. H. Krayenbühl, P.E. Maspes and W.H. Sweet (S. Karger, Basel) p. 200
Hayek, A. and Crawford, J.D. (1972). *J. Clin. Endocrinol. Metab., 34,* 764
Hökfelt, T. (1974). *Fed. Proc., Fed. Am. Soc. Exp. Biol., 33,* 2177
Hruska, R.E. and Silbergeld, E.K. (1980). *Europ. J. Pharmacol., 61,* 397
Irie, M. and Thushima, T.J. (1972). *J. Clin. Endocrinol. Metab., 35,* 97
Kebabian, J.W. and Calne, D.B. (1978). *Nature* (Lond.) *277,* 93
Kizer, J.S., Palkovits, M. and Brownstein, M. (1976). *Brain Res., 108,* 363
Kleimberg, D.L. Noel, G.L. and Frantz, A.G. (1977). *N. Eng. J. Med., 296,* 589
Köbberling, J., Schwinn, G., Mayer, G. and Dirks, H. (1980). In *Neuroactive Drugs in Endocrinology,* ed. E.E. Müller (Elsevier/North-Holland, Amsterdam) p. 315
Krieger, D.T. and Glick, S.M. (1972). *Am. J. Med., 53,* 25
Kumakura, K., Hoffman, M., Cocchi, D., Trabucchi, M., Spano, P.F. and Müller, E.E. (1979). *Psychopharmacology 71,* 13
Lachelin, G.C.L., Abu-Fadil, S. and Yen, S.S.C. (1977a). *J. Clin. Endocrinol. Metab., 44,* 1163
Lachelin, G.C.L., Leblanc, H. and Yen, S.S.C. (1977b). *J. Clin. Endocrinol. Metab., 44,* 728
Lamberts, S.W.J., Timmermans, H.A.T., de Jong, F.H. and Birkenhäger, J.C. (1977). *Clin. Endocrinol.* (Oxf.) *7,* 183
Lamberts, S.W., Klijn, J.C.M., de Quijada, M., Timmermans, H.A.T., Uitterlinden, P. and Birkenhäger, J.C. (1980a). In *Neuroactive Drugs in Endocrinology,* ed. E.E. Müller (Elsevier/North-Holland, Amsterdam) p. 371
Lamberts, S.W.J., MacLeod, R.M. and Krieger, D. (1980b). In *Pituitary Microadenomas,* ed. G. Faglia, M. Giovanelli and R.M. MacLeod (Academic Press, New York) p. 151

Lemberger, L. and Crabtree, R.E. (1979). *Science 205,* 1151

Liuzzi, A., Panerai, A.E., Chiodini, P.G., Secchi, C., Cocchi, D., Botalla, L., Silvestrini, F. and Müller, E.E. (1976). In *Growth Hormone and Related Peptides,* ed. A. Pecile and E.E. Müller (Excerpta Medica, Amsterdam) p. 236

Liuzzi, A., Chiodini, P.G., Oppizzi, G., Botalla, L., Verde, G., De Stefano, L., Colussi, G., Graf, K.J. and Horowski, R. (1978). *J. Clin. Endocrinol. Metab., 46,* 196

MacLeod, R.M. (1976). In *Frontiers in Neuroendocrinology,* ed. L. Martini and W.F. Ganong (Raven Press, New York) p. 169

MacLeod, R.M. and Krieger, D.T. (1976). Abstract 317, *Fifty-eighth Annual Meeting of the Endocrinological Society,* (San Francisco)

McNatty, K.P., McNeilly, A.S. and Sawers, R.S. (1977). In *Prolactin and Human Reproduction,* ed. P.G. Crosignani and C. Robyn (Academic Press, New York) p. 109

Malarkey, W.B., Jacobs, L.S. and Daughaday, W.H. (1971). *New Engl. J. Med., 285,* 1160

Mars, H. and Genuth, S.M. (1973). *Clin. Pharmacol. Ther., 14,* 390

Moore, K.E., Demarest, K.T., Johnston, C.A. and Alper, R.H. (1980). In *Neuroactive Drugs in Endocrinology,* ed. E.E. Müller (Elsevier/North-Holland, Amsterdam) p. 109

Moriondo, P. Travaglini, P., Rondena, M., Beck-Peccoz, P., Conti-Puglisi, F., Ambrosi, B. and Faglia, G. (1980). In *Pituitary Microadenomas,* ed. G. Faglia, M. Giovanelli and R.M. MacLeod (Academic Press, New York) p. 247

Müller, E.E. (1979). In *Hormonal Proteins and Peptides,* Vol. 7, ed. C.H. Li (Academic Press, New York) p. 123

Müller, E.E., Genazzani, A.R., Murru, S. and Fioretti, P. (1977). *Acta Endocrinol.* (Kbh.) *86,* 33

Müller, E.E., Genazzani, A.R. and Murru, S. (1978a). *J. Clin. Endocrinol. Metab., 47,* 1352

Müller, E.E., Nisticò, G. and Scapagnini, U. (1978b). *Brain Neurotransmitters and Anterior Pituitary Function,* (Academic Press, New York)

Nicholson, P.A. and P. Turner (eds) (1977). *Nomifensine. Br. J. Clin. Pharmacol.,* suppl. 2, p. 53S

Peillon, F., Racadot, J., Moussy, D., Vila-Porcile, E., Olivier, L. and Racadot, O. (1978). In *Treatment of Pituitary Adenomas,* ed. R. Falbusch and K. von Werder (Thieme, Stuttgart) p. 114

Pontiroli, A.E., Alberetto, M., Pellicciotta, G., De Castro e Silva, E., De Pasqua, A., Girardi, A.M. and Pozza, G. (1980). *Acta Endocrinol.* (kbh.) *93,* 271

del Pozo, E. and Wolf, A. (1978). *Triangolo 16* (suppl.), 43

del Pozo, E., Wyss, H., Lancranjan, L., Obobensky, W. and Varga, L. (1976). In *Ovulation in the Human,* ed. P.G. Crosignani and D.R. Mishell (Academic Press, London)

Quadri, S.K., Lu, K.H. and Meites, J. (1972). *Science 176,* 417

Roth, R.H., Marrin, L.C. and Walters, J.R. (1976). *Europ. J. Pharmacol., 36,* 163

Schönberger, W., Grimm, W. and Ziegler, R. (1977). *Europ. J. Pediatr., 127,* 15

Seppälä, M., Ranta, T. and Hirvonen, E. (1976). *Lancet ii,* 1154

Sherman, B.M., Schlechte, J., Halmi, N.S., Chapler, F.K., Harris, C.E., Duello, T.M. Van Gilder, J. and Granner, D.K. (1978). *Lancet ii,* 1019

Sowers, J.B., McCallum, R.W., Hershman, J.M., Carlson, H.E., Sturdevant, R.L. and Meyer, N. (1976). *J. Clin. Endocrinol. Metab., 43,* 679

Spano, P.F. and Trabucchi, M. (eds) (1978). *Ergot Alkaloids. Pharmacology 16,* suppl. 1 (Karger, Basel)

Spano, P.F., Trabucchi, M., Corsini, G.U. and Gessa, G.L. (eds) (1979). *Sulpiride and Other Benzamides* (Italian Brain Research Foundation Press, Milan).

Swanson, H.A. and du Boulay, G. (1975). *Brit. J. Radiol., 48,* 366

Thorner, M.O., Ryan, S.M., Wass, J.A.H., Jones, A., Bouloux, P., Williams, S. and Besser, G.M. (1978). *J. Clin. Endocrinol. Metab., 47,* 372

Thorner, M.O., Martin, W.H., Rogol, A.D., Morris, J.L. Perryman, R.L., Conway, B.R., Howards, S.S., Wolman, M.G. and MacLeod, R.M. (1980). *J. Clin. Endocrinol. Metab., 51,* 438

Tyrrel, J.B., Brooks, R.M., Ritzgerald, P.A., Cofold, P.B., Forshman, P.H. and Wilson, C.B. (1978). *New Eng. J. Med., 298,* 753

Van Loon, G.R., Appel, N.M., Kim, C. and Ho, D. (1980). In *Neuroactive Drugs in Endocrinology,* ed. E.E. Müller (Elsevier/North-Holland, Amsterdam) p. 279

Varga, L., Lutterbeck, P.M., Pryor, J.S., Wenner, R. and Erg, H. (1972). *Br. Med. J., 2*, 743

Vezina, J.L. and Sutton, T.J. (1974). *Am. J. Roentgenol. Radium Ther. Nucl. Med., 120*, 46

Wass, J.A.H., Thorner, M.O., Charlesworth, M., Moult, P.J.A., Dacie, J.E., Jones, A.E. and Besser, G.M. (1979). *Lancet ii*, 66

Weiner, R.J., Cronin, M.J., Cheung, C.T., Annunziato, L., Faure, N. and Goldsmith, P.C. (1979). In *Neuroendocrine Correlates in Neurology and Psychiatry*, ed. E.E. Müller and A. Agnoli (Elsevier/North-Holland, Amsterdam) p. 41

Weldon, V.Y., Gupta, S.K., Haymond, M.W., Pagliara, A.S., Jacobs, L.S. and Daughaday, W.H. (1973). *J. Clin, Endocrinol. Metab., 36*, 42

von Werder, K., Eversmann, T., Falbusch, R. and Rjosk, H.K. (1980). In *Neuroactive Drugs in Endocrinology*, ed. E.E. Müller (Elsevier/North-Holland, Amsterdam) p. 347

Wurtman, R. (1973). In *Frontiers in Catecholamine Research*, ed. E. Usdin and S.H. Snyder (Pergamon Press, New York) p. 781

Zarate, A., Canales, E.S., Alger, M. and Forsbach, G. (1979). *Acta Endocrinol.* (Kbh.) *92*, 407

4   SOME ASPECTS OF THE NEUROENDOCRINE REGULATION OF MAMMALIAN SEXUAL BEHAVIOUR

Knut Larsson and Carlos Beyer

TABLE OF CONTENTS

I	INTRODUCTION	97
II	FEMININE SEXUAL BEHAVIOUR	97
III	MASCULINE SEXUAL BEHAVIOUR	98
IV	HORMONES INVOLVED IN THE ORGANIZATION OF THE BRAIN SUBSTRATE FOR SEXUAL BEHAVIOUR	99
V	HORMONES RELATED TO THE DISPLAY OF FEMALE SEXUAL BEHAVIOUR	101
VI	EFFECTS OF REMOVING ENDOCRINE GLANDS AND HORMONE REPLACEMENT ON FEMININE SEXUAL BEHAVIOUR	104
VII	NEURAL REGULATION OF FEMALE SEXUAL BEHAVIOUR	106
VIII	HORMONAL CONTROL OF MASCULINE SEXUAL BEHAVIOUR	108
IX	NEURAL CONTROL OF MASCULINE SEXUAL BEHAVIOUR	109
X	CELLULAR MECHANISMS IN THE ORGANIZATIONAL AND ACTIVATIONAL ACTIONS OF SEX STEROIDS	111
XI	MECHANISM OF ORGANIZATIONAL ACTION OF STEROIDS	111
XII	MECHANISMS OF ACTIVATIONAL ACTION OF SEX STEROIDS	112
XIII	SUMMARY & CONCLUSIONS	114
	REFERENCES	116

I. INTRODUCTION

Hormones influence mammalian sexual behaviour by two distinct processes: (a) by controlling the development of the neural nets necessary for sexual behaviour during the fetal or infantile stage; and (b) by stimulating these neural nets at adulthood to facilitate the display of sexual behaviour. Young (1961) proposed the term *organizational* for the early hormonal action and *activational* for the effect exerted by hormones at adulthood.

This chapter deals with the hormones involved in the organization and stimulation of sexual behaviour, the neural structures on which they act and the cellular and molecular mechanisms through which they exert their behavioural actions. Knowledge in these areas has mainly been obtained by experiments performed on rodents, and this explains the apparent bias towards laboratory animals in this chapter.

Heterosexual behaviour is the result of a continual interaction of male and female individuals, where the behaviour of one partner is dependent upon the behaviour of the other. However, it is more convenient to focus attention on the behaviour of one sex at a time and to express measures of behaviour in terms of that sex. Patterns of sexual behaviour are usually stereotyped within species, but are variable across species (Dewsbury, 1979).

II. FEMININE SEXUAL BEHAVIOUR

Beach (1976) proposed three concepts representing characteristics of oestrous females: *attractivity, proceptivity* and *receptivity*. *Attractivity* is related to the female's stimulus value for eliciting sexual behaviour in the male and includes morphological and physiological changes such as the appearance of the genital area (genital swelling, color changes) and the production of odoriferous substances. *Proceptivity* includes the repertory of female reactions directed towards the male and constitutes 'her assumption of initiative in establishing or maintaining sexual interaction' (Beach, 1976). For example, the estrous rat exhibits various proceptive behaviour patterns; she displays a presentation posture which may be accompanied by ear wiggling caused by rapid head shaking. When presenting, she assumes a crouching position differing from sitting by its tenseness and extension of the legs. Presentation is preceded by a hopping reaction and a zigzagging running behaviour called *darting*. Changes in locomotor behaviour directed to searching for males or remaining in their vicinity are proceptive behaviour patterns occurring in many different species. Presentation of the genital region

97

towards the male, grooming of the male and being in proximity to him are the three most striking components of proceptivity in some female primates. *Receptivity* involves adoption of a posture facilitating penile insertion and ejaculation. Many subprimate mammals, including rodents and carnivores, show lordosis, a posture consisting of arching of the back and elevation of the pelvis, frequently accompanied by tail deviation. Lordosis is usually triggered by male mounting. Recent studies in this laboratory have shown that the male rabbit, when mounting the female, exhibits a rhythmic pattern of pelvic thrusting which stimulates lordosis behaviour in the female (Contreras and Beyer, 1979). Changes in the rhythmicity of this pattern, much too small to be recognized by the naked eye of the observer, are perceived by the doe, who may not respond with lordosis.

Besides displaying feminine sexual behaviour, females of many species show total or partial performance of the masculine copulatory pattern, sometimes called *pseudomale* behaviour (Morris, 1955). This behaviour is oriented towards both females and males.

III. MASCULINE SEXUAL BEHAVIOUR

The genital contact that characterizes copulation is usually preceded by sequences of precopulatory or 'courtship' behaviour, which serve to bring together the male and the female at a time when they are likely to mate and conceive. A detailed analysis of the male copulatory pattern in a large number of mammalian species has recently been made by Dewsbury (1979). Therefore, we will limit ourselves to describing copulation in some representative mammals in which quantitative or semiquantitative analysis of mating is available.

By using an accelerometric technique, Contreras and Beyer (1979) measured several aspects of the motor copulatory pattern in rabbits and rats. As shown in Figure 4.1, mounting in the rabbit consists in performance of pelvic oscillations at a relatively constant frequency (13 per second) until the animal achieves intromission, when pelvic thrusting ceases. Ejaculation occurs approximately half a second after intromission. In the rat, pelvic-thrusting frequency during copulation is higher than in the rabbit (19-21 per second), and mating consists of a series of four to twelve mounts with intromissions, each separated by a brief intercopulatory interval, culminating in ejaculation. Ejaculation in the rat, as in most species, is followed by a postejaculatory interval, during which the male is refractory to sexual stimulation. In contrast with the highly stereotyped pattern of copulatory behaviour in lower mammals, the copulatory pattern of higher primates has a certain degree of flexibility. Thus, the gorilla may use more than one position for copulating: i.e., dorsal–ventral, ventral–ventral (Nadler, 1975).

Figure 4.1: (A) Typical accelerometric record of a copulation in an experienced New Zealand white rabbit. Upper trace carries a time mark (1 s), and a signal that was operated by an observer. Narrow signal indicates time during which mounting occurred; wide signal indicates the moment in which the observer recorded intromission. Note that during intromission pelvic thrusting stopped. (b) Mounting train by an experienced rabbit. Note the constant frequency of pelvic thrusting, in spite of variation in amplitude. (c) Mounting train performed by an inexperienced rabbit. Note that the mounting pattern is similar to that of experienced rabbits. (d) Mounting trains performed by a female rabbit. Note that the pattern of mounting varies considerably from that of a male.

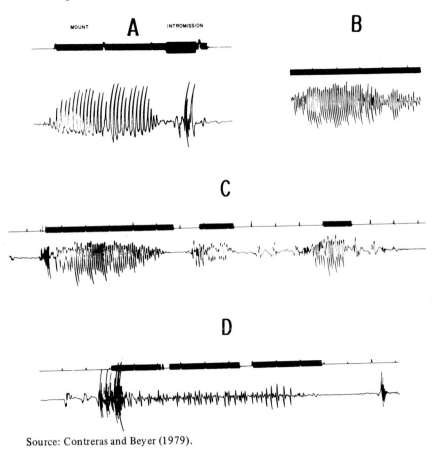

Source: Contreras and Beyer (1979).

IV. HORMONES INVOLVED IN THE ORGANIZATION OF THE BRAIN SUBSTRATE FOR SEXUAL BEHAVIOUR

There is evidence that the differentiation of the brain substrates for sexual behaviour occurs at an early stage of development (the fetal stage in the guinea pig, the first week of life in the rat), and that this process is achieved by the action of testicular secretions on the developing brain. In the absence of testicular hormones, as in the normal female or in the neonatally castrated male rat, the subject develops a female type of sexual behaviour. Conversely, administration of testosterone (T) during this critical stage results in virilization of the sexual behaviour, irrespective of the genetic sex of the animal. The major hormone produced by the fetal or infantile testis is T, and it was long assumed that this was the hormone involved in producing sexual differentiation. Recently, however, it was suggested that the active hormone may be an estrogen rather than an androgen (McEwen, 1976, 1979; Naftolin *et al.*, 1976). This hypothesis is related to the problem of the metabolism of T in the brain. In the developing brain, as in the adult animal, T is steadily metabolized. One of the principal pathways for T metabolism involves ring A reduction through a widely occurring enzyme, called 5α-reductase (Figure 4.2). A second principal pathway for T metabolism involves aromatization of ring A, resulting in the conversion of T to estradiol (E_2). The sexual differentiation of the genital tract appears to be accomplished by T and 5α-reduced steroids, while brain sexual differentiation may be dependent upon estrogen. Several experimental data support this idea. Thus, treatment of neonatally castrated male rats with estradiol benzoate ($E_2 B$), or with the synthetic estrogen RU 2858, augmented the display of masculine sexual behaviour in adulthood. By contrast, dihydrotestosterone (DHT), a 5α-reduced steroid, was ineffective in promoting behavioural masculinization. Administration of aromatization inhibitors preventing the conversion of androgen to estrogen interfered with the virilizing action of T in neonatally castrated rats, and increased the propensity of males to exhibit female sexual behaviour (McEwen, 1979). Summarizing the above findings, sexual differentiation of the rat brain may occur, at least in part, as a result of T-derived estrogenic action upon the central nervous system during the perinatal period.

The role of sex steroids in the organization of the neural substrate for sexual behaviour in primates is not as clear as in rodents. Female rhesus monkeys which were virilized as the result of T treatment to their pregnant mothers showed a slight, but not statistically significant, tendency to display more mounting behaviour than control animals. On the other hand, these monkeys were significantly more aggressive, a male characteristic, than normal females (Eaton *et al.*, 1973).

These data suggest that the rhesus brain is also modified by prenatal androgen, though to a much lesser extent than the rodent brain. This result recalls that reported in girls virilized *in utero* as a result of excessive androgen production by their mothers. During their development these girls showed a clear shift towards maleness, as evidenced by their preference for masculine games and clothing. However, no drastic alteration in their sexual behaviour was noted (Money and Ehrhard, 1972). The failure of sex steroids to determine the type of sexual behaviour to be displayed in adult primates is congruent with the idea that

Figure 4.2: Structural formulas and metabolic pathways for testosterone and related steroids. The enzymatic sequences established in brain tissue are indicated with arrows.

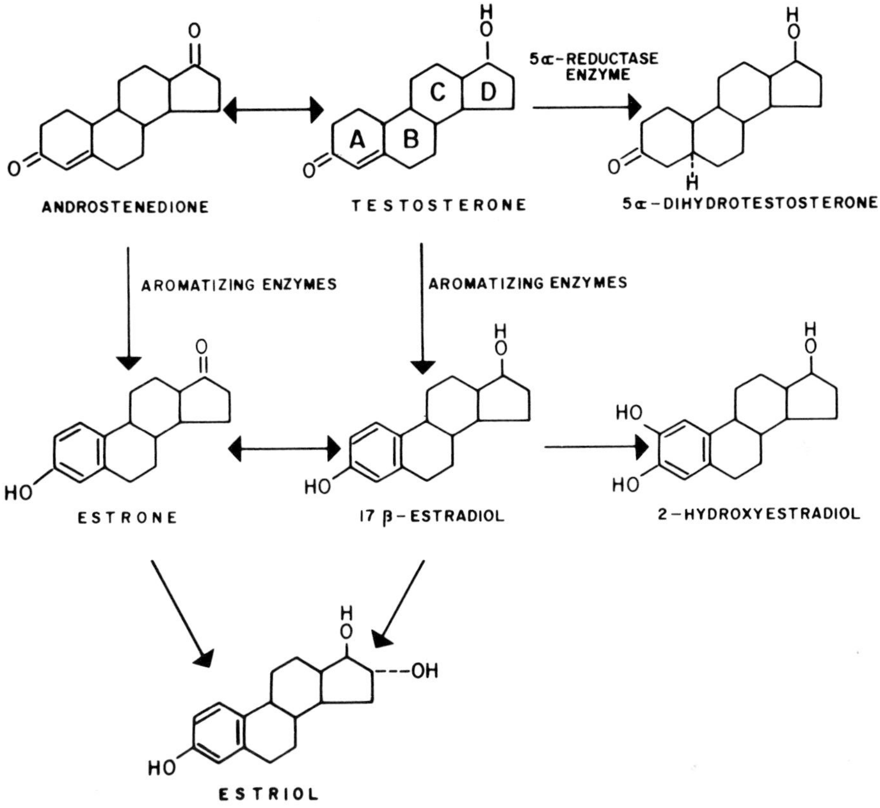

ANDROSTENEDIONE   TESTOSTERONE   5α-DIHYDROTESTOSTERONE

AROMATIZING ENZYMES   AROMATIZING ENZYMES

ESTRONE   17 β-ESTRADIOL   2-HYDROXYESTRADIOL

ESTRIOL

psychological factors are the main determinants of sexual behaviour in primates.

V. HORMONES RELATED TO THE DISPLAY OF FEMALE SEXUAL BEHAVIOUR

Information on the nature of the hormones involved in female sexual behaviour basically comes from two experimental approaches: (a) experiments correlating the display of estrous behaviour with either the condition of the ovaries and genital tract or plasma levels of some hormones; and (b) administration of hormones to gonadectomized or hypophysectomized animals. Females of most mammalian species show distinct periods of sexual activity (estrus) alternating with periods of sexual quiescence (diestrus). Variations in ovarian functioning underlie these behavioural fluctuations. Besides variations in sexual receptivity during the estrous cycle, there are seasonal variations, particularly in species living in the wild. In these cases, periods of ovarian activity (estrus) alternate with periods of ovarian quiescence during which no sexual activity is displayed (anestrus).

Figure 4.3 presents data on the relative duration of the periods of sexual receptivity and the relative plasma levels of gonadal hormones in various mammalian species. As seen in this figure, the apex of receptivity in most species is preceded by high levels of estrogen, occurring around the time of ovulation. Androgen also reaches its highest values around ovulation in some species and may also facilitate sexual behaviour. A periovulatory progesterone (P) rise coincides with onset of estrus in rats and guinea pigs. This rise is relevant for the normal display of estrus in rodents, since P administration facilitates sexual behaviour in estrogen-primed rodents. Receptivity usually ends shortly after either spontaneous or induced ovulation and, in some species, its cessation can be temporarily related to a rise of P produced by the luteinized follicles.

A peak of pseudomale behaviour occurs around ovulation in many species, suggesting the involvement of gonadal steroids in this behaviour (Beach, 1968). However, there is not always a close relationship between the display of lordosis and mounting, suggesting that other nonhormonal factors determine the occurrence of pseudomale behaviour.

Old-world monkeys have a menstrual cycle, and under free-ranging conditions their sexual activity appears to be more or less restricted to a few days in the middle of this cycle. In the laboratory, where the investigator sets the conditions for mating, sexual activity may take place at any time during the cycle but is most frequent near midcycle.

In a study of five female rhesus monkeys during 33 menstrual cycles, Michael and Bonsall (1979) showed that one day after the peak of E_2 levels (i.e., at the time of ovulation), the male reaches a peak of ejaculatory activity, and this is synchronized with shorter times taken by females to gain access to males and longer periods spent in the proximity of males. In general, they found that high E_2 levels were associated with short access times and short ejaculation times (time from first mount in the series of the ejaculatory mounts), while high P levels were associated with long access times and long ejaculation times.

Interestingly, the changes in the hormonal status of the female accompanying the menstrual cycle are reflected not only in the behaviour of the female, but in that of the male as well (Michael and Bonsall, 1979). This indicates the importance of communication of the hormonal status between the partners.

In women, a higher frequency of intercourse and orgasm is reported to occur around the time of ovulation (Udry and Morris, 1968), a situation similar to that of other primates.

Sexual behaviour diminishes during pregnancy, pseudopregnancy and lactation in most mammals. The low level of sexual activity observed during pregnancy may either be due to the high levels of P secreted by corpora lutea or to low circulatory levels of estrogen resulting from absence of follicular development. In women, no consistent changes in sexual behaviour are observed during pregnancy,

Figure 4.3: Patterns of estradiol (dashed curve) and progesterone (solid curve) secretion during the reproductive cycle of some reflex and spontaneous ovulators. The profile of 20-α-hydroxypregnenone (dotted curve) secretion around coitus is also shown for the rabbit. Time zero corresponds to coitus in reflex ovulators (rabbit, cat). In primates, day zero corresponds to the occurrence of the LH peak, whereas in the guinea pig, dog, mare and ewe day zero is defined as the onset of behavioural estrus.

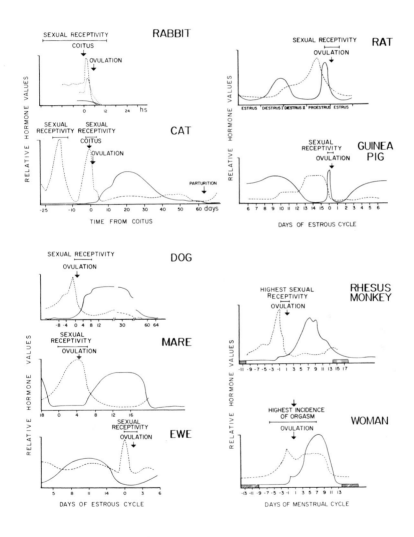

Source: Moralí and Beyer (1979).

though a diminution in the frequency of coitus usually occurs in the last trimester, probably as a result of cultural rather than biological factors.

VI. EFFECTS OF REMOVING ENDOCRINE GLANDS AND HORMONE REPLACEMENT ON FEMININE SEXUAL BEHAVIOUR

Prepubertal ovariectomy prevented the appearance of sexual behaviour in most subprimate mammals studied (Young, 1961; Komisaruk, 1978). However, prepubertally ovariectomized rabbits may display lordosis in adulthood, indicating that sexual behaviour can be initiated in the absence of ovarian hormones in this species (Beyer and McDonald, 1973). In general, postpubertal ovariectomy gradually abolishes sexual motivation and most components of estrous behaviour in all subprimate female mammals. In most species, proceptive behaviour rapidly disappears after ovariectomy. Darting and hopping are not displayed by ovariectomized rats and, in female rhesus, ovariectomy reduces the number of grooming and copulation invitations. Some species retain receptive behaviour after ovariectomy, though at a low level. Thus lordosis persists long after castration in some spayed rabbits (Beyer and McDonald, 1973). Adrenal steroids may play a role in this effect, since adrenalectomy further reduces sexual behaviour in ovariectomized rabbits. In women, ovariectomy usually does not lead to loss of libido and sexual gratification but, when it is combined with adrenalectomy, libido disappears (Bancroft, 1978). A similar result is observed after hypophysectomy.

Treatment with estrogen restores sexual behaviour in ovariectomized females of most species. In rats and guinea pigs, E_2 is the most potent estrogen in inducing lordosis, followed in decreasing order by estrone (E_1) and estriol (E_3) (Beyer *et al.,* 1971; Feder and Silver, 1974). Attractiveness and the various components of estrous behaviour have different thresholds for estrogen stimulation. In spayed rats and hamsters, proceptivity requires much larger doses of E_2 than lordosis. The intensity of some proceptive patterns (e.g., ear wiggling and hopping in the rat) are directly correlated with the dosages of estrogen administered. Some characteristics of the lordosis response, such as the degree of arching during lordosis, the latency of lordosis in response to male mounting and the duration of the lordotic posture, also vary with the amount of estrogen given (Komisaruk, 1978; Moralí and Beyer, 1979). Treatment of ovariectomized rhesus monkeys with estrogen restores not only the female's sexual behaviour but the male partner's sexual activity as well (Michael and Bonsall, 1979). Small doses of estrogen given intravaginally were more active than subcutaneously administered ones in restoring mounting attempts by the male, suggesting that this effect is mainly achieved through the production of a pheromone by the vagina. In contrast to the stimulating effect of estrogen on female sexual behaviour in all species studied, women usually do not increase their libido in response to estrogen administration.

Androgen stimulates lordosis in ovariectomized females of several species

(Moralí and Beyer, 1979). Estrous behaviour induced by androgen is similar to that produced by estrogen, though much larger amounts are required to obtain a similar level of receptivity. However, since androgens are secreted in large amounts by the adrenals and the gonads in females, its physiological role cannot be excluded. The role of androgen in the control of primate female sexual behaviour is controversial.

In rhesus monkeys, extirpation of the adrenal glands, which are the main source of female androgen, decreased the number of female sexual invitations, and increased the proportion of refusals to male mounting attempts (Everitt *et al.*, 1972). Similarly, dexamethasone administration, which inhibits the adrenal production of androgens by suppressing corticotropin secretion, antagonized the stimulatory action of estrogen on the female's sexual behaviour (Everitt and Herbert, 1971). In ovariectomized female rhesus monkeys, treatment with T induces high levels of invitational behaviour and stimulates operant behaviour for access to males. Administration of androstenedione or T restores the behaviour of adrenalectomized or dexamethasone-treated females. All these data point to an important role of androgen in the normal display of sexual behaviour in primates. However, Michael and Bonsall (1979) found that normal female sexual behaviour can be induced in ovariectomized rhesus monkeys by chronic administration of E_2 and P, without any androgen, indicating that androgen is not essential for the induction of female sexual behaviour in this species. In women, though estrogen fails to stimulate sexual behaviour, increased libido is reported as a side effect of androgen therapy (Salmon and Geist, 1943).

It has been proposed that androgen elicits female sexual behaviour through its conversion to estrogen. This idea is supported by the following findings: (a) aromatizable androgens are much more potent in inducing sexual receptivity in rabbits and rats than nonaromatizable androgens; and (b) antiestrogens such as MER-25 inhibit the estrous response produced by daily administration of T to spayed rabbits and rats (Beyer *et al.*, 1976a).

In several species (mouse, rat, hamster), estrous behaviour normally occurs as the joint effect of estrogen and P. Facilitation of lordosis by P only occurs when the female is primed with estrogen for a minimal period of time. P then promotes attractiveness in estrogen-primed rats, as well as the appearance of proceptive behavioural components such as ear wiggling and darting. By contrast, P does not appear to influence the nature and intensity of the lordosis response. Provided the period of estrogen priming has been long enough, P exerts its effects very rapidly. When injected intravenously or directly into the brain, P facilitates sexual behaviour within ten to 90 minutes (Lisk, 1973). This is in striking contrast to estrogen, which does not exert any behavioural action until 18 hours after its injection.

Besides its facilitatory effect, P has an inhibitory effect on estrous behaviour (Feder and Marrone, 1977; Morin, 1977). Experiments performed on rats, hamsters and guinea pigs have shown that, when P is given to ovariectomized animals close to a priming estrogen injection, the subsequent P injection fails to elicit lordosis behaviour ('concurrent inhibition'). This type of inhibitory effect of P, in which it counteracts the possible action of circulating estrogen, may explain the decrease or absence of sexual behaviour during pregnancy in some species. P also exerts a 'sequential' inhibitory effect on sexual behaviour following its facilitatory action on estrogen-primed rodents. After a first sequential treatment of estrogen and P, the female requires a recovery interval before she again responds to hormonal stimulation.

VII. NEURAL REGULATION OF FEMALE SEXUAL BEHAVIOUR

Electrolytic lesions in various parts of the hypothalamus suppress feminine sexual behaviour (Figure 4.4). In most species investigated (rat, guinea pig, cat), the integrity of the anterior-ventromedial hypothalamus is essential for the display of estrous behaviour (Pfaff *et al.*, 1972; Lisk, 1973; Morali and Beyer, 1979). A possible exception is the rabbit, in which lesions in the premammillary area interrupt estrous behaviour even when exogenous estrogen is administered (Sawyer, 1959). Lesions outside the hypothalamus, such as the habenula nuclei or the stria medullaris, reduce soliciting behaviour and receptivity in ovariectomized estrogen-treated rats, and electrical stimulation in or near the habenulo-interpeduncular tract produces movements of the tail, rump and hindlimbs resembling components of the female mating pattern (Pfaff *et al.*, 1972). Other workers, however, have found that rats with lesions in the medial habenula become more receptive, suggesting an inhibitory role of these nuclei.

In close agreement with the lesioning studies, implants of crystalline estrogen into the anterior hypothalamus, the ventromedial or the premammillary region have been reported to stimulate sexual behaviour in ovariectomized females of various species (rat, rabbit, guinea pig, cat). T or TP implants in the same regions may also induce sexual behaviour (Lisk, 1973). Additional indirect evidence for the participation of the ventromedial part of the hypothalamus in estrous behaviour came from autoradiographic studies showing a selective concentration of $^3HE_2$ in nuclei from this region after administration of radioactive estrogen (Sar and Stumpf, 1973).

Paradoxically, lesions in some brain areas may facilitate the expression of lordosis behaviour. Thus, lesions in the medial preoptic area (mPOA) or in septum, or dorsal-anterior deafferentation of the mPOA, reduced the amount of estrogen required to induce lordosis in ovariectomized rats (Nance *et al.*, 1974; Komisaruk, 1978). The location of these lesions suggests the existence of an inhibitory pathway originating in the telencephalon and passing to the anterior hypothalamus through the POA. The inhibitory system may include various other telencephalic structures, such as the neocortex and the basomedial amygdaloid nucleus, since lesions in the cerebral cortex or induction of cortical spreading depression by potassium chloride increased the lordosis quotient of ovariectomized rats receiving subthreshold doses of estrogen (Clemens and Christensen, 1975).

Figure 4.4: Sagittal (A) and transverse (B) section of a representative mammalian brain, showing areas in which lesions or electrical stimulation affect receptivity and location of areas in which steroid implants facilitate receptivity (C). Abbreviations: AC, anterior commisure; AH, anterior hypothalamus; AR, arcuate nucleus; CC, corpus callosum; Cx, cerebral cortex; H, habenula; V, lateral ventricle; MB, mammillary bodies; MFB, medial forebrain bundle; MRF, mesencephalic reticular formation; OB, olfactory bulb; OC, optic chiasma; PH, posterior hypothalamus; POA, preoptic area; S, septum; SO, supraoptic nucleus; T, temporal lobe including amygdala; VMN, ventromedial nucleus; III, third ventricle.

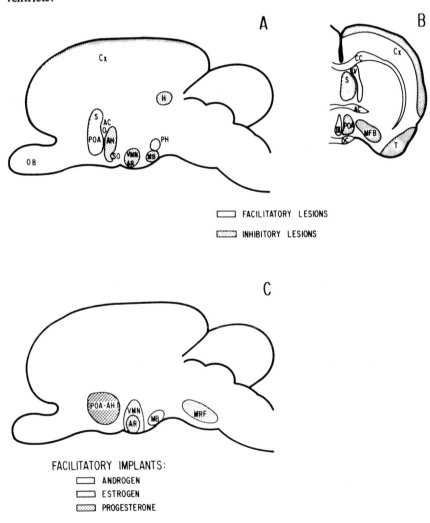

Source: Moralí and Beyer (1979).

Female rats with lesions in the basomedial amygdaloid nucleus and female deermice with lesions in the basolateral amygdala have been found to mate in diestrus (Clemens and Christensen, 1975). Female rabbits with temporal-lobe lesions did not show a change in their lordosis quotient, but exhibited a striking increase in mounting activity, which could be directed even toward animals of other species (Beyer and McDonald, 1973).

To summarize, lesion experiments indicate a stimulatory role of the anterior-ventromedial part of the hypothalamus and the occurrence of a widespread, ill-defined telencephalic inhibitory system for lordosis behaviour. A neural system involved in the recognition of the adequate sexual partner may exist in both male and female animals within the entorhinal-amygdala system.

In contrast to the good correspondence between the results of lesion studies and estrogen or androgen implants, the results with P implants are somewhat more variable. Some workers have reported facilitation of lordosis by P implants in the medial basal hypothalamus of estrogen-primed ovariectomized females (Morín, 1977), while others found P effective only when implanted into the mesencephalic reticular formation (Gorski, 1976).

VIII. HORMONAL CONTROL OF MASCULINE SEXUAL BEHAVIOUR

The reduction of sexual activity after castration in animals and man has been known since antiquity. Yet in some species and individuals, the sexual activity of the male may be maintained long after castration. Dogs and goats may copulate for a year or more after castration without showing any reduced activity. In one strain of mice, the complete mating pattern persisted indefinitely after removal of the gonads (McGill and Manning, 1976). Primates, including man, show an even greater individual variation in the effect of castration (Luttge, 1971). Thus, some men stop copulating almost immediately after castration, while others continue to display sexual activity for a decade or more. In intact man, sexual activity may be temporarily decreased by treatment with the antiandrogen, cyproterone acetate (Neumann *et al.*, 1970).

Although T readily restores sexual activity in castrated animals, it might not be the active hormone in stimulating sexual behaviour since it is subjected to active metabolism. Moreover, the active androgen in stimulating male sexual accessories is dihydrotestosterone (DHT), which results from the 5α-reduction of T. In order to investigate the role of T metabolites (estrogens and 5α-reduced androgens) in the activation of sexual behaviour, Beyer *et al.* (1973) treated castrated male rats with ten different naturally occurring androgens. Only those androgens which could be aromatized stimulated masculine sexual behaviour, namely T, androstenedione and androstenediol, whereas 5α-reduced androgens such as DHT and androstenediol, although highly potent in stimulating peripheral genital tissues, were unable to activate sexual behaviour. Moreover, 19-OH-T and 19-OH-androstenedione, very weak androgens which are intermediate compounds in the aromatization of T and androstenedione to E_2 and E_1, respectively, restored copulatory behaviour in castrated rats (Luttge, 1979; Parrot, 1974). If the action of T is mediated by E_2, one might think that estrogen alone may promote sexual behaviour. However, treatment of castrated rats with E_2 does not usually induce sexual activity unless fairly large doses of the estrogen are administered. Since E_2 alone does not readily induce masculine sexual behaviour, we undertook an experiment in which castrated rats were treated with low doses of E_2 together with a 5α-reduced androgen, speculating that this condition better corresponds to the natural situation where T is metabolized both to estrogen and 5α-reduced androgens (Larsson *et al.*, 1973). In the doses used, none of these hormones provoked sexual behaviour when given alone but, when combined, they were as active as T. A similar synergism between estrogen and DHT has been reported in other species such as the rabbit, hamster and mouse, but not in the guinea pig (Beyer *et al.*, 1976a). Aromatization of T can be blocked by a number of compounds. Administration of any one of the three aromatization inhibitors, 1, 4, 6-androstatriene-3, 17-dione, 4-OH-androstenedione or aminoglutethimide, or infusion of aminoglutethimide into the POA blocked T-induced sexual activity in castrated male rats (Christensen and Clemens, 1975; Beyer *et al.*, 1976b; Moralí *et al.*, 1977).

To summarize, the observations that only aromatizable androgens are effective in activating sexual behaviour in castrated male rats, that a combination of DHT and an estrogen is as potent in activating sexual behaviour as T, and that inhibition of aromatization prevents T from activating sexual behaviour support the idea that, in some species, estrogen interacts with androgen in stimulating libido. What mechanisms accomplish this interaction between an estrogen and an androgen remains an unsolved question. Studies on the role of aromatization in the regulation of male sexual behaviour must be extended to other species, particularly to primates.

IX. NEURAL CONTROL OF MASCULINE SEXUAL BEHAVIOUR

Extensive lesions encompassing the entire medial preoptic-anterior hypothalamic continuum (mPOA) permanently abolished masculine sexual behaviour of male rats, while smaller lesions, damaging only parts of the mPOA, temporarily suppressed sexual activity (Heimer and Larsson, 1966/7). These effects were obtained without causing any clear gonadal dysfunction. Moreover, T failed to restore sexual behaviour in MPO-lesioned rats. Similar findings have been reported in cats, dogs and rhesus monkeys (Larsson, 1979). Rhesus monkeys continued to display masturbatory behaviour following mPOA lesions, but were otherwise sexually inactive (Slimp *et al.*, 1978).

As could be expected from the above data, TP implants in the mPOA restored copulatory behaviour in castrated male rats (Davidson, 1966; Lisk, 1973). Electrical stimulation of the mPOA may also result in increased sexual activity (van Dis and Larsson, 1971).

The mPOA may be viewed as the origin of a final common pathway for sexual behaviour that is activated by a combination of arousing stimuli from the olfactory and vomeronasal systems, from the gustative, auditive, visual and tactile sensory systems, and by circulatory hormones. The relative contribution of these various sources of sexual stimulation varies in different species. Olfaction in macrosmatic species and vision in primates are assumed to be particularly important for sexual arousal.

There is evidence that the mPOA is regulated by the activity of other brain structures. Lesions at various levels along the olfactory pathways produce deficits in the mating behaviour of male rats. After severance of the olfactory peduncle, effects similar to those following lesions of the olfactory bulb or destruction of the olfactory mucosa were obtained (Larsson, 1971). Since under all these conditions, both the vomeronasal and olfactory mucosa systems were destroyed, the relative contribution of the two olfactory systems to the effect obtained remains unresolved. The mPOA receives projections from the corticomedial division of the amygdaloid nuclei via stria terminalis. Lesions in the corticomedial amygdaloid nuclei as well as in the bed nucleus of stria terminalis resulted in a slower rate of copulation, an effect resembling that shown by rats having impaired olfaction (Emery and Sachs, 1976). The medial forebrain bundle is a massive, polysynaptic fiber system conveying impulses from various telencephalic and diencephalic structures. Lesions in this bundle resulted in a reduction or total suppression of sexual activity. By contrast, electrical stimulation of the lateral preoptic area in the rostral end of the medial forebrain bundle facilitated expression of sexual activity (Madlafousek *et al.*, 1970).

Extensive lesions in the mesodiencephalic junction in the male rat resulted in a drastic increase in sexual activity (Heimer and Larsson, 1964). The animals ejaculated after fewer intromissions than usual, and with shorter latencies. Most remarkable was that postejaculatory intervals were shortened to one or two minutes. Similar changes in the coital pattern have been reported following restricted lesions in the rostral ventral midbrain (Barfield *et al.*, 1975; Clark *et al.*, 1975). These findings suggest the existence of a neuronal system encompassing structures in the mesencephalon and diencephalon, exerting a modulatory influence upon sexual behaviour. Recently, we observed that lisuride, a drug that is believed to interfere with dopamine and serotonin neurotransmission, partly mimicked the effect of the mesencephalic-diencephalic lesions, suggesting that the behavioural effects observed after these lesions may be due to interference with monoamine neurotransmission (Ahlenius *et al.*, 1980).

Changes in masculine sexual behaviour have also been reported to follow bilateral lesions in the temporal lobe of monkeys and cats (Kluver and Bucy, 1939). Animals did not show any disturbances in the normal mating pattern following these lesions, but displayed 'bizarre' sexual behaviour including attempts to copulate with animals of other species and even with inanimate objects. Cats, which normally show territoriality, copulated outside their familiar environment when lesioned in the entorhinal cortex (Green *et al.*, 1959).

Summarizing the data obtained by lesion and stimulation techniques, male sexual behaviour seems to be dependent on the activity of two opposite mechanisms in the brain, one exerting an excitatory and the other an inhibitory influence on the behaviour. The excitatory mechanism is centered around an area encompassing the mPOA, and extends to structures including the bed nucleus of stria terminalis. This area receives input from various sensory systems, of which olfaction appears to be particularly important in rodents, the most studied order in this field of research. The inhibitory mechanism relies on still badly identified neural pathways that pass through medial portions of caudal diencephalon and rostral mesencephalon.

X. CELLULAR MECHANISMS IN THE ORGANIZATIONAL AND ACTIVATIONAL ACTIONS OF SEX STEROIDS

Sex steroids both organize and activate the neural substrate for sexual behaviour. The nature of these two responses differs, the organizational effect most likely requiring growth changes while the activational effect only involves molecular rearrangements. The time scale of these effects is also dissimilar. However, in spite of these differences, the initial cellular events triggered by sex steroids are probably similar in both types of actions.

The general model of steroid-hormone action involves the following processes (Jensen and Jacobson, 1962): (a) sex steroids penetrate the cell and bind to a receptor protein (cytosol receptor) inducing its transformation into an activated form (nuclear receptor); (b) the steroid-receptor complex is translocated to the nucleus, where it modifies genome expression and stimulates RNA synthesis; and (c) mRNAs are transported to the cytoplasm and code for the synthesis of specific proteins.

XI. MECHANISM OF ORGANIZATIONAL ACTION OF STEROIDS

Much work has been devoted to localizing, in the fetal or infantile brain, androgen or estrogen receptors similar to those observed in nonneuronal tissues (Kato *et al.*, 1971; McEwen, 1976; Kato, 1977). Most workers have failed to observe a preferential uptake of T by the hypothalamus (McEwen, 1976). Therefore, high-affinity, low-capacity T receptors may not be involved in the induction of brain masculinization by androgen. Existence of estrogen receptors in the neonatal brain has also been debated since no specific accumulation of radioactivity is found in the brain following (^3H)-E_2 injections before two to three weeks of age (McEwen, 1976). However, a fetoneonatal estrogen-binding protein has been found to sequester small-to-moderate doses of (^3H)-E_2, thus preventing the estrogen from reaching its intraneuronal receptor sites (McEwen,

1976; Plapinger and McEwen, 1978). Moreover, small quantities of (^3H)-E_2 associated with nuclei from limbic regions are found when (^3H)-T is administered to neonatal rats (McEwen, 1976). These results support the existence of E_2 receptors in the neonatal brain, and point to the importance of androgen aromatization for brain virilization. The role of estrogenic receptors in brain virilization is further supported by the finding that antiestrogens protect from the virilizing effect of neonatal T (McDonald and Doughty, 1973).

The role of *de novo* protein synthesis in brain masculinization has been studied using various antibiotics. Both actinomycin-D, a transcription blocker, and puromycin, a translational inhibitor, block TP-induced sterility in neonatal rats (Gorski, 1971), indicating that protein synthesis is essential for masculinization. However, no studies on the effect of antibiotics on the virilization of sexual behaviour have yet been made.

A group of chemicals including tranquillizers, barbiturates, and progestins have been reported to protect against neonatal T virilization (Gorski, 1971). All these drugs can induce membrane stabilization, a process interfering with secretion. Therefore, it has been proposed that neonatal androgen exerts its organizational effect by stimulating the synthesis and secretion of 'inducer' molecules regulating the growth and direction of afferent axons to the mPOA (Beyer *et al.*, 1979). Membrane stabilizers may protect from T brain virilization by interfering with the secretion of inducers. In support of this idea there is evidence that brain sexual differentiation involves growth changes in certain brain areas. Raisman and Field (1973) found that neonatal T treatment, as well as the testicular secretion of androgen by the neonatal male rat, caused a decrease in the number of presynaptic endings of nonamygdaloid origin on dendritic spines of preoptic neurons. Other workers have noted sexual differences in the arrangement of dendritic arborizations in hamsters (Greenough *et al.*, 1977) or in the staining of some areas within the medial preoptic area (Gorski, 1979).

XII. MECHANISMS OF ACTIVATIONAL ACTION OF SEX STEROIDS

As previously mentioned, the nature of the hormonal stimuli for inducing female sexual behaviour in mammals shows species variation. Thus the rabbit requires only estrogen, the rat estrogen plus P, and the ewe P plus estrogen (Moralí and Beyer, 1979). Similarly in the male, there are species in which androgen is sufficient to activate full copulatory behaviour (guinea pig), while others may require estrogen in addition (rat). It is logical to believe that differences in cellular mechanisms underlie these variations in the hormonal requirements for the expression of sexual behaviour. However, it also appears that a common mechanism involving the stimulation of the synthesis of specific proteins is the essential process for the activation of sexual behaviour. This basic mechanism involves, as in other steroid effectors, the binding of the steroid with specific receptors and its transport to the nucleus.

Receptors with similar physicochemical characteristics have been identified in all estrogen effector tissues including the brain, and it has been assumed that the same molecule mediates the action of estrogen in the brain as in other estrogen-dependent organs (Kato, 1977).

However, this molecule may not be involved in the behavioural effects of estrogen since it binds E_1 and E_3 poorly. Both E_1 and E_3 stimulate female sexual behaviour in ovariectomized rats and guinea pigs (Beyer *et al.*, 1971; Feder and Silver, 1974) and synergize with DHT in stimulating male sexual behaviour in castrated male rats (Larsson *et al.*, 1976). A single receptor has been proposed to mediate the various effects of T. However, androstenedione, which binds weakly with this molecule, is as efficient in restoring sexual behaviour as T. All these data suggest that the brain receptors implicated in sexual behaviour differ from those reported in other effectors. Obviously the search for steroid receptors involved in sexual behaviour must be oriented toward molecules with adequate binding affinities for all steroids of a series with the property of stimulating sexual behaviour.

E_2, which attaches to cytosol estrogen-binding sites, is transferred into the cell nuclei, and substantial amounts of (^3H)-E_2 can be recovered in isolated nuclei of brain neurons. There is strong evidence that brain protein synthesis is stimulated by sex hormones, and that this process is related to the activation of sexual behaviour. Actinomycin-D, a transcription blocker, interferes with female sexual behaviour when administered immediately before, simultaneously with or within twelve hours after the injection of E_2 (Ho *et al.*, 1973). Similarly, cycloheximide, a translation blocker, blocks lordosis when given around the time of E_2 administration (Quadagno and Ho, 1975). The observation that actinomycin-D was still active twelve hours after E_2 injection is surprising, since little nuclear E_2 remains at this time. It may be, however, that the 'nuclear' receptors which may be directly related to genome activation have a longer residence in the nucleus than E_2. Moreover, continuous RNA synthesis is not essential for the continuous manifestation of estrous behaviour since intrabrain infusion of antibiotics into estrous rats neither interrupts sexual behaviour nor shortens its duration. This suggests that mRNAs or proteins required for sexual behaviour are synthesized within a few hours after E_2 injection, and that they remain in the neurons for a period during which sexual behaviour can be displayed.

Specific proteins produced in response to sex steroids and stimulating the expression of sexual behaviour have not been identified. Neither have we any information on the nature of the cellular or molecular events that follow the stimulation of protein synthesis by estrogen or androgen in their target neurons. From the complexity of sexual behaviour, it can be inferred that its neural substrate comprises a large series of interconnected neurons, and that the functioning of this neural net involves a series of distinct cellular events. Pharmacological studies suggest that synapses of various chemical types are related to the facilitation and inhibition of sexual behaviour. Changes in neurotransmitter levels or in enzymes regulating their synthesis, degradation or transport have been observed following E_2 or T administration, and might be relevant to sexual behaviour. We have recently shown that castration of postpubertal but not prepubertal male rats is followed by an increased synthesis of catecholamines and serotonin in the limbic system and in the diencephalon, and that these effects can be counteracted by treatment with T (Engel *et al.*, 1979). Other evidence suggesting a monoaminergic link in the neural control of sexual behaviour is that parachlorophenylalanine (PCPA), a drug that inhibits the synthesis of 5-HT, may induce full masculine sexual behaviour in otherwise sexually inactive, castrated male rats (Sodersten *et al.*, 1976). Monoamines may activate adenylate cyclase postsynaptically and cAMP changes have been reported to be related to the expression of masculine sexual behaviour (Christensen and Clemens, 1974). Sex steroids, including estrogen, may activate sexual behaviour through other means than modifying genome expression. Recently we observed that 4-mercuric-E_2, an estrogen that does not penetrate the nucleus or does not remain there, elicits lordosis behaviour in ovariectomized rats (Beyer, Canchola and Cruz, unpublished observations).

It seems likely that P acts at other sites than the nucleus to facilitate sexual behaviour. The fact that P only acts in animals primed with E_2, together with the short latency of its effect on sexual behaviour, suggests that P probably acts at an extranuclear site, interacting with the chemical signals generated by the previous E_2 treatment. These signals could be stable mRNAs or proteins which need to be translated or activated to exert their effects on lordosis behaviour. Sexual behaviour may also be activated in the estrogen-primed ovariectomized rat by administration of nonsteroidal agents such as LH-RH (Moss and McCann, 1973), and prostaglandin E_2 (Rodríguez-Sierra and Komisaruk, 1977). Interestingly, these hormones act at the membrane level to raise cyclic AMP in their target cells, suggesting that this nucleotide, through the activation of protein kinases, activates the protein already synthesized by previous estrogen treatment. This idea is supported by our observation that either systemic or intrahypothalamic administration of db-cAMP induces lordosis behaviour in estrogen-primed ovariectomized rats (Beyer *et al.*, 1980). Inhibition of sexual behaviour by P is probably achieved through the synthesis of a protein, since cycloheximide, a translation blocker, interferes with this effect in the guinea pig (Wallen *et al.*, 1972).

XIII. SUMMARY AND CONCLUSIONS

The preceding review reveals the great diversity of neuroendocrine mechanisms related to sexual behaviour and, consequently, the difficulty in proposing concepts that are valid for all mammalian species. An example of this is the question of the role of fetal or neonatal hormones in the development of the brain substrate for sexual behaviour. Brain virilization by early androgen administration has been demonstrated in some rodents, but not in rabbits and primates. Moreover, psychological rather than hormonal factors appear to determine behavioural sexual differentiation in the human. The hormonal requirements for the display of sexual behaviour in females also vary greatly among species. Thus, estrous behaviour is induced by estrogen alone in cats and rabbits, by estrogen and P in rodents and probably by androgen in women and other primates. It appears that P is required in those species in which the heat period is short but intense, this hormone acting by regulating both the initiation and duration of estrus.

The anatomical localization of the neural substrate for sexual behaviour has been mapped by lesions and by implantation of sex steroids in the brain. All these studies point to the basal hypothalamus as the region essential for the integration of lordosis behaviour. However, the exact localization of these 'sex centers' varies from species to species. Moreover, there is some discussion about the site of action of P, which, according to some workers, acts at the same site as estrogen, while others consider that it acts on the midbrain.

The nature of the specific hormonal stimulus for male sexual behaviour is still the subject of some controversy, and probably important species variations may exist even within the rodents. Thus, the rat apparently requires estrogen for the display of sexual behaviour but the guinea pig does not. The role of aromatization for normal male sexual behaviour has not been established in primates and many other groups, though it would be no surprise to find that it is as variable in these species as in rodents. The neural substrate for male copulatory behaviour appears to involve a complex net of telencephalic and diencephalic structures. Lesion studies as well as implantation of androgen in the brain suggest that preoptic neurons are involved in the initiation of copulatory behaviour in several species including primates. The excitability of these neurons depends on a variety of factors such as hormones and the afferent input, both excitatory and inhibitory, from various brain regions.

The cellular mechanisms involved in the activation of sexual behaviour remain largely unknown, but several solid facts have recently emerged. There is strong evidence that protein synthesis is required for the induction of estrous behaviour, at least in rodents. Blockers of transcription or transduction inhibit the stimulatory effect of estrogen on lordosis behaviour. The nature of the proteins stimulated by estrogen is unknown, but they appear to be related to the synthesis or availability of neurotransmitters. Some recent data suggest that most of these proteins are in an inactive form or state, and that passage to an active form requires their phosphorylation through cAMP-dependent kinases. Activation of adenyl cyclases resulting in higher intraneuronal levels of cAMP is produced directly or indirectly by P, or by other hormones such as LH-RH or prostaglandin E_2. Some evidence exists that facilitation of estrous behaviour involves the activation of noradrenergic synapses. Much less information on the molecular basis of male sexual behaviour exists in the literature. However, it appears that protein synthesis is also involved in this process. Recent data also suggest a role of cAMP on the hormonal induction of male sexual behaviour in rodents. Practically no studies on the biochemical basis of sexual behaviour have been made in species other than rodents. Since large species variations exist regarding the neuroendocrine regulation of sexual behaviour, biochemical studies must be done in species representative of other groups before the present cellular models, based on studies performed in the rat, can be generalized to other species.

REFERENCES

Ahlenius, S., Larsson, K. and Svensson, L. (1980). *Eur. J. Pharmacol.* (in press)

Bancroft, J. (1978). In *Biological Determinants of Sexual Behaviour,* ed J.B. Hutchison (Wiley, Chichester-New York) p. 493

Barfield, R.J., Wilson, C. and McDonald, P.G. (1975). *Science 189,* 147

Beach, F.A. (1968). In *Reproduction and Sexual Behaviour,* ed M. Diamond (Indiana University Press, Bloomington) p. 83

Beach, F.A. (1976). *Horm. Behav., 7,* 105

Beyer, C. and McDonald, P.G. (1973). *Adv. Reprod. Physiol., 6,* 185

Beyer, C., Moralí, G. and Vargas, R. (1971). *Horm. Behav., 2,* 273

Beyer, C., Larsson, K., Pérez-Palacios, G. and Moralí, G. (1973). *Horm. Behav., 4,* 99

Beyer, C., Moralí, G., Larsson, K. and Sodersten, P. (1976a). *J. Steroid Biochem, 7,* 1171

Beyer, C., Moralí, G., Naftolín, F., Larsson, K. and Pérez-Palacios, G. (1976b). *Horm. Behav., 7,* 353

Beyer, C., Larsson, K. and Cruz, M.L. (1979). In *Endocrine Control of Sexual Behavior,* ed. C. Beyer (Raven Press, New York) p. 365

Beyer, C., Canchola, E., Cruz, M.L. and Larsson, K. (1980). In *Endocrinology 1980* (Australian Academy of Science, Canberra) p. 615

Christensen, L.W. and Clemens, L.G. (1974). *J. Endocrinol., 61,* 159

Christensen, L.W. and Clemens, L.G. (1975). *Endocrinology 97,* 1545

Clark, T.K., Caggiula, A.R., McConell, R.A. and Antelman, S.M. (1975). *Science 190,* 169

Clemens, L.G. and Christensen, L.W. (1975). In *The Behaviour of Domestic Animals,* 3rd edn, ed. E.S.E. Hafez (Bailliere Tindall, London) p. 108

Contreras, J.L. and Beyer, C. (1979). *Physiol. Behav., 23,* 939

Davidson, J. (1966). *Endocrinology 79,* 783

Dewsbury, D.A. (1979). In *Endocrine Control of Sexual Behavior,* ed. C. Beyer (Raven Press, New York) p. 3

van Dis, H. and Larsson, K. (1971). *Physiol. Behav., 6,* 85

Eaton, G.G., Goy, R.W. and Phoenix, C.H. (1973). *Nature New Biology 242,* 119

Emery, D.E. and Sachs, B.D. (1976). *Physiol. Behav., 17,* 803

Engel, J., Ahlenius, S., Almgren, O., Carlsson, A., Larsson, K. and Sodersten, P. (1979). *Pharmacol. Biochem. Behav., 10,* 149

Everitt, B.J. and Herbert, J. (1971). *J. Endocrinol., 51,* 575

Everitt, B.J., Herbert, J. and Hamer, J.D. (1972). *Physiol. Behav., 8,* 409

Feder, H.H. and Marrone, B.L. (1977). *Ann. NY Acad. Sci., 286,* 331

Feder, H.H. and Silver, R. (1974). *Physiol. Behav., 13,* 251

Gorski, R. (1971). In *Frontiers in Neuroendocrinology,* ed. L. Martini and W.F. Ganong (Oxford University Press, New York) p. 237

Gorski, R.A. (1976). *Psychoneuroendocrinology 1,* 371

Gorski, R.A. (1979). In *Ontogeny of Receptors and Reproductive Hormone Action,* ed. T.H. Hamilton, J.H. Clark and W.A. Sadler (Raven Press, New York) p. 371

Green, J.D., Clemente, C.D. and De Groot, J. (1959). *J. Comp. Neurol., 108,* 505

Greenough, W.T., Carter, S.C., Steermon, C. and De Voogd, T.J. (1977). *Brain Res., 126,* 63

Heimer, L. and Larsson, K. (1964). *Experientia 20,* 460

Heimer, L. and Larsson, K. (1966/7). *Brain Res., 3,* 248

Ho, G.K.W., Quadagno, D.M., Cooke, P.H. and Gorski, R.A. (1973). *Neuroendocrinology 13,* 47

Jensen, E.V. and Jacobson, H.I. (1962). *Recent Prog. Horm. Res., 18,* 387

Kato, J. (1977). In *Receptors and Mechanisms of Action of Steroid Hormone,* Part II, ed. J.R. Pasqualini (Marcel Dekker, New York) p. 603

Kato, J., Sugimura, N. and Kobayashi, T. (1971). In *Hormones in Development*, ed. M. Hamburgh and E.J.W. Barrington (Appleton-Century-Crofts, New York) p. 689

Kluver, N. and Bucy, P.C. (1939). *Arch. Neurol. Psychiatry 42*, 979

Komisaruk, B.R. (1978). In *Biological Determinants of Sexual Behavior*, ed. J.B. Hutchison (Wiley, New York) p. 349

Larsson, K. (1979). In *Endocrine Control of Sexual Behaviour*, ed. C. Beyer (Raven Press, New York) p. 77

Larsson, K., Sodersten, P. and Beyer, C. (1973). *Horm. Behav., 4,* 289

Larsson, K., Sodersten, P., Beyer, C., Moralí, G. and Pérez-Palacios, G. (1976). *Horm. Behav., 7,* 379

Lisk, R.D. (1973). In *Handbook of Physiology*, Sec. 7, Vol. II, ed. R.O. Greep, E.B. Astwood and S.R. Geiger (American Physiolocal Society, Washington, D.C.) p. 223

Luttge, W.G. (1971). *Arch. Sex Behav., 1,* 61

Luttge, W.G. (1979). In *Endocrine Control of Sexual Behavior*, ed. C. Beyer (Raven Press, New York) p. 341

McDonald, P.G. and Doughty, C. (1973). *Neuroendocrinology 13,* 182

McEwen, B.S. (1976). In *Subcellular Mechanisms in Reproductive Neuroendocrinology*, ed. F. Naftolin, K.J. Ryan and I.J. Davies (Elsevier, Amsterdam) p. 277

McEwen, B.S. (1979). In *Endocrinology*, Vol. I, ed. L.J. DeGroot (Grune and Stratton, New York) p. 35

McGill, T.E. and Manning, A. (1976). *Anim. Behav., 24,* 507

Madlafousek, J., Freund, K. and Grotova, I. (1970). *J. Comp. Physiol. Psychol., 72,* 28

Michael, R.P. and Bonsall, R.W. (1979). In *Endocrine Control of Sexual Behavior*, ed. C. Beyer (Raven Press, New York) p. 279

Money, J.M. and Ehrhard, A.A. (1972). *Man and Woman, Boy and Girl. The Differentiation and Dimorphism of Gender Identity from Conception to Maturity* (Johns Hopkins University Press, Baltimore)

Moralí, G. and Beyer, C. (1979). In *Endocrine Control of Sexual Behavior*, ed. C. Beyer (Raven Press, New York) p. 33

Moralí, G., Larsson, K. and Beyer, C. (1977). *Horm. Behav., 9,* 203

Morín, L.P. (1977). *Physiol. Behav., 18,* 701

Morris, D. (1955). *Behaviour 8,* 46

Moss, R.L. and McCann, S.M. (1973). *Science 181,* 177

Nadler, R.D. (1975). *Science 189,* 813

Naftolin, F., Ryan, K.J. and Davies, I.J. (1976). In *Subcellular Mechanisms in Reproductive Neuroendocrinology*, ed. F. Naftolin, K.J. Ryan, and I.J. Davies (Elsevier, Amsterdam) p. 347

Nance, D.M., Shryne, J. and Gorski, R. (1974). *Horm. Behav., 5,* 73

Neumann, F., von Berswordt-Wallrabe, R., Elger, W., Steinbeck, H., Hann, J.D. and Kramer, M. (1970). *Recent Prog. Horm. Res., 26,* 337

Parrot, R.F. (1974). *J. Endocrinol., 61,* 105

Pfaff, D.W., Lewis, C., Diakow, C. and Keimer, M. (1972). In *Progress in Physiological Psychology*, ed. E. Stellar and J. Sprague (Academic Press, New York) p. 253

Plapinger, L. and McEwen, B.S. (1978). In *Biological Determinants of Sexual Behaviour*, ed. J.B. Hutchison (Wiley, Chichester-New York) p. 153

Quadagno, D.M. and Ho, G.K.W. (1975). *Horm. Behav., 6,* 19

Raisman, G. and Field, P.M. (1973). *Brain Res., 54,* 1

Rodríguez-Sierra, J.F. and Komisaruk, B.R. (1977). *Horm. Behav., 9,* 281

Sar, M. and Stump, W.E. (1973). *Endocrinology 92,* 251

Salmon, V.J. and Geist, S.H. (1943). *Endocrinology 3,* 235

Sawyer, C.H. (1959). *Anat. Rec., 124,* 358

Slimp, J.C., Hart, B.L. and Goy, R.W. (1978). *Brain Res., 142,* 105

Sodersten, P. (1974). *Horm. Behav., 5,* 111

Sodersten, P., Larsson, K., Ahlenius, S. and Engel, J. (1976). *Pharmacol. Biochem. Behav., 5,* 319

Udry, J.R. and Morris, N.M. (1968). *Nature 220,* 593

Wallen, K., Goldfoot, D.A., Joslin, W.D. and Paris, C.A. (1972). *Physiol. Behav., 8,* 221

Young, W.C. (1961). In *Sex and Internal Secretions*, Vol. II, ed. W.C. Young (Williams and Wilkins, Baltimore) p. 1173

PART TWO

BRAIN ENDORPHINS, PSYCHOTROPIC DRUGS
AND NEUROENDOCRINE REGULATION

5 OPIATES AND NEUROENDOCRINE REGULATION

André Dupont, Nicholas Barden, Fernand Labrie,
Louis Ferland and Lionel Cusan

TABLE OF CONTENTS

I	INTRODUCTION	123
II	PROLACTINE AND GROWTH-HORMONE SECRETION	123
III	PHYSIOLOGICAL ROLE IN STRESS AND SUCKLING	126
	A Ether stress	127
	B Suckling	128
IV	ROLE OF DOPAMINE AND SEROTONIN	129
	A Changes of hypothalamic DA turnover	129
	B Studies with antidopaminergic drugs	129
	C Role of serotonin	129
V	DISTRIBUTION OF β-ENDORPHIN AND ENKEPHALINS	131
VI	CHANGES OF β-ENDORPHIN AND MET-ENKEPHALIN LEVELS	135
	A Estrogen and haloperidol treatment	135
	B Estrous cycle	137
VII	SUMMARY AND CONCLUSIONS	137
	REFERENCES	139

I. INTRODUCTION

Following reports of the presence of endogenous opiate activity in the brain (Pasternak *et al.*, 1977; Terenius and Wahlstrom, 1975), two pentapeptides having the following structures: H-tyr-gly-gly-phe-met-OH (met-enkephalin) and H-tyr-gly-gly-phe-leu-OH (leu-enkephalin) have been isolated from porcine (Bradbury *et al.*, 1976) and calf (Simantov and Snyder, 1976) brain and shown to possess high opiate-agonist activity. The sequence of met-enkephalin is the same as the N-terminus of β-endorphin (Smith *et al.*, 1976), the C-fragment (β-LPH$_{61-91}$) of β-lipotropin first isolated from sheep pituitary gland (Smith *et al.*, 1976). It has been possible to isolate and characterize β-endorphin from camel pituitary glands (Li and Chung, 1976), as well as from many other species including the human, and both *in vitro* (Bradbury *et al.*, 1976; Cox *et al.*, 1976; Li and Chung, 1976) and *in vivo* (Loh *et al.*, 1976) methods have shown it to be an analgesic agent.

On the other hand, the opiates morphine and methadone are well known to be potent stimuli of growth hormone and prolactin secretion in the rat (Kokka *et al.*, 1973; Ferland *et al.*, 1977b). Moreover, morphine stimulates prolactin release in the human (Tolis *et al.*, 1975). Since met-enkephalin and β-endorphin bind to the opiate receptor and have potent morphine-like activity in various biological assays, the possibility was raised that the endogenous opioid peptides, besides their well-known analgesic potency and activity as behaviour modulators, could also be involved in the neuroendocrine control of GH and prolactin secretion.

This chapter will attempt to summarize current knowledge on the role of endogenous opiates in the control of neuroendocrine functions.

II. PROLACTIN AND GROWTH-HORMONE SECRETION

As illustrated in Figure 5.1A the intraventricular injection of 0.5-25 μg of β-endorphin led to a rapid and important stimulation of PRL release in unanesthetized freely moving rats. With the 0.5-μg dose, a significant rise was already measured five minutes after injection of the peptide and a maximal stimulation (approximately sevenfold) was measured after ten minutes, with a slow decrease of plasma hormone levels at later time intervals. The higher doses of β-endorphin (2, 5 and 25 μg) led to a progressive increase of PRL release, a 30-60-fold increase being measured between 20 and 60 minutes after injection of 25 μg of the peptide.

Although inactive at 0.5 μg, doses of 2 μg and higher of β-endorphin led to a

Figure 5.1: **Effect of increasing doses of β-endorphin on plasma PRL (A) and GH (B) levels in the rat.** Male rats bearing intraventricular and intrajugular cannulas were injected intravenously with 0.2 ml of sheep somatostatin antiserum five minutes before the intraventricular injection of the indicated amounts of synthetic β-endorphin. PRL and GH concentrations were measured at the indicated time intervals after administration of β-endorphin to between eight and ten animals per group. Data are expressed as means ± SEM.

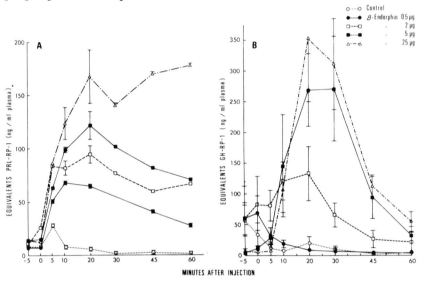

Source: Dupont *et al.* (1977a, b).

significant stimulation of plasma GH release. With the 2-μg dose, a six to ten-fold stimulation of the plasma GH concentration was measured ten and 20 minutes after β-endorphin injection with a progressive decrease to basal levels reached at 45 minutes. The two higher doses (5 and 25 μg) of β-endorphin led to a 20-30-fold stimulation of plasma GH levels measured 20 and 30 minutes after injection of the peptide.

As shown in Figure 5.2, met-enkephalin, the NH_2-terminal pentapeptide of β-endorphin, was much less potent than β-endorphin in stimulating PRL and GH release. In fact, at doses of 500 and 1,000 μg, met-enkephalin led to approximately four-and sixfold increases of plasma PRL levels, respectively. Maximal stimulation was measured ten to 20 minutes after injection of the pentapeptide, with a rapid return to basal levels between 30 and 45 minutes. It can be seen in Figure 2B that stimulation of GH release was observed only at the 1,000-μg dose, thus indicating a greater sensitivity of PRL than GH responses to the opioid peptide.

Specificity of the stimulatory effect of β-endorphin on PRL and GH release is

Figure 5.2: Effect of 0.5 or 1.0 mg of met-enkephalin on plasma PRL (A) and GH (B) levels in the rat. The experiment was performed as described in Figure 5.1.

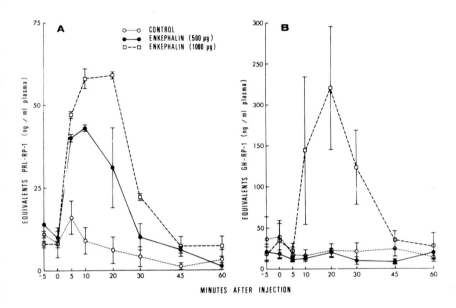

indicated by the finding that naloxone, a specific opiate antagonist, completely blocked the stimulation of GH release (Figure 3B) at a dose of 0.5 mg/kg, whereas the highest dose of naloxone (12.5 mg) was required to completely abolish the rise of PRL secretion (Figure 3A). This higher resistance of the PRL than GH response to naloxone might be due to a greater sensitivity of PRL than GH release to the stimulatory action of opiate peptides. This possibility is supported by the data of Figures 5.1 and 5.2, where changes of PRL release were observed earlier and at lower doses compared to GH after β-endorphin and met-enkephalin injection. It is also possible that these effects could be secondary to differential access of β-endorphin and/or naloxone to their sites of action on PRL and GH release.

The stimulatory effect of met-enkephalin on PRL and GH release is much lower than that of β-endorphin. In fact, our data demonstrate that β-endorphin is 2,000 times more potent than met-enkephalin in stimulating prolactin and GH secretion. The analgesic action of met-enkephalin is also much weaker than that of β-endorphin when the agents are injected intracerebrally. These data indicate that the higher potency of β-endorphin is probably due to its higher resistance to degradation. It is of interest, however, to note that met-enkephalin is rapidly inactivated by plasma and brain tissue.

The stimulatory effect of morphine (Ferland *et al.*, 1977b), β-endorphin (Dupont *et al.*, 1977a, b), met-enkephalin (Dupont *et al.*, 1977a, b) and the endorphin analogs (Cusan *et al.*, 1977) on GH release is observed in animals

Figure 5.3: Effect of increasing doses of naloxone on plasma PRL and GH release induced by intraventricular injection of 5 μg of β-endorphin in the rat. The experiment was performed as described in Figure 5.1, except that naloxone was injected in the indicated groups ten minutes before β-endorphin administration.

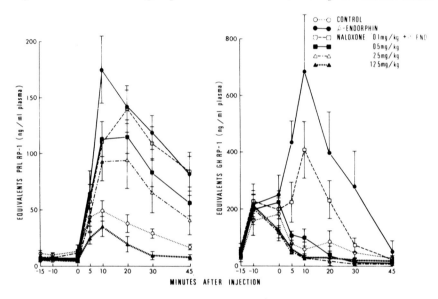

where circulating somatostatin was neutralized by excess somatostatin antiserum, thus suggesting that the endogenous tetradecapeptide does not play an important role in opiate-induced GH release (Ferland *et al.*, 1977a). Since opiate peptides cannot stimulate GH release from pituitary cells *in vitro*, it is more likely that the observed GH release is due to stimulated release of hypothalamic GH-releasing activity (GH-RH).

As mentioned earlier and in agreement with the data obtained in the rat, morphine has been found to stimulate prolactin secretion in man (Tolis *et al.*, 1975) and the potent enkephalin analog FK 33-824 stimulates both prolactin and growth-hormone secretion in normal men (Von Graffenried *et al.*, 1978), this effect being reversible by the opiate antagonist, naloxone.

III. PHYSIOLOGICAL ROLE IN STRESS AND SUCKLING

Although, as described above, endorphins and their analogs can stimulate growth-hormone and prolactin secretion in experimental animals and man, such data do not prove the physiological role of these peptides in the control of neuroendocrine functions. To investigate such a role, advantage can be taken of the availability of a pure opiate antagonist, naloxone. Finding naloxone blockage of a physiological change of hormone secretion would strongly support the role of endogenous opiate peptides in this process.

A. Ether Stress

In agreement with previous data (Neill, 1970), it can be seen in Figure 5.4 that exposure to ether vapor led to a rapid stimulation of plasma PRL release. A maximal effect was already seen after five minutes with a progressive decrease toward control levels after 30 minutes. When the opiate antagonist naloxone (10 mg/kg) was injected ten minutes before exposure to ether, the stimulation of prolactin release was completely abolished ($p < 0.01$).

Figure 5.4: Inhibitory effect of naloxone on prolactin release induced by ether stress.

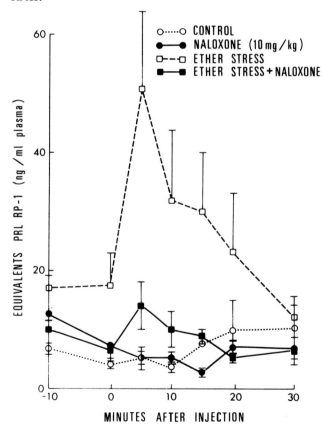

Source: Ferland *et al.* (1978).

B. Suckling

As illustrated in a representative experiment shown in Figure 5.5, a single injection of naloxone (5 mg/kg) five minutes before return of the pups led to a 50-95% inhibition of the marked rise of plasma prolactin induced by suckling up to the last time interval studied (90 minutes). In complementary experiments where naloxone was injected after ten, 45 and 90 minutes, plasma prolactin levels were still reduced from 600 ± 95 to 115 ± 40 ng/ml two hours after the start of suckling (data not shown).

The present data clearly indicate that endogenous opiate peptides could be involved as mediators of the stimulatory effect of stress and suckling on prolactin release in the rat.

Figure 5.5: Inhibitory effect of naloxone on prolactin release induced by suckling.

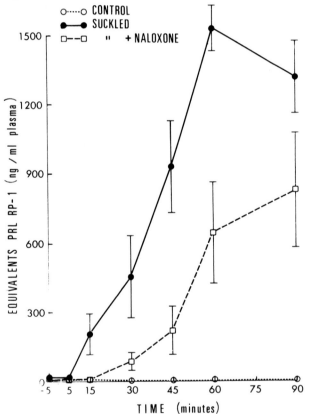

Source: Ferland *et al.* (1978).

IV. ROLE OF DOPAMINE AND SEROTONIN

A. Changes of Hypothalamic DA Turnover

Not only met-enkephalin but also β-endorphin and morphine markedly reduce DA turnover within the lateral and medial palisade zones of the median eminence (Ferland *et al.*, 1977b, 1980). This marked effect on the tuberoinfunibular DA neurons suggests that several of the neuroendocrine actions of opioid peptides and morphine may be mediated via inactivation of the median-eminence DA system.

B. Studies with Antidopaminergic Drugs

Following the findings of an opiate-induced inhibition of DA turnover in the tuberoinfundibular neurons, we have used antidopaminergic drugs as a second approach for measuring the role of DA in the effect of endorphins and opiates on prolactin secretion. Finding that morphine could not further stimulate prolactin secretion after administration of a maximal dose of an antidopaminergic drug would thus indicate that DA is involved in the action of the opiate on prolactin secretion. That the potent stimulatory effect of morphine on prolactin release is secondary to inhibition of dopaminergic activity is clearly indicated by the observation that the acute response of plasma prolactin induced by a maximal dose of morphine (40 mg/kg) is not further increased by simultaneous administration of a high dose of haloperidol (1 mg/kg) (Figure 5.6).

It can also be seen in Figure 5.6 that the acute release of prolactin induced by the first injection of morphine is followed by a refractory period, since no change of plasma prolactin levels is observed when the same dose of morphine is administered 30 and 60 minutes later. On the other hand, when haloperidol is injected during this refractory period to morphine (45 minutes after the injection of the opiate), a marked stimulation of prolactin release is observed, thus suggesting that the rapid decrease of plasma prolactin levels following acute stimulation by morphine is due to a return of inhibitory dopaminergic activity and a loss of responsiveness to morphine.

C. Role of Serotonin

Since a role of serotonin has been proposed in the stimulation of prolactin induced by stress (Goodman *et al.*, 1976) and evidence has recently been obtained for a role of endorphins in the stress- and suckling-induced rise of prolactin release, we next used an inhibitor (parachlorophenylalanine) and a precursor (5-hydroxytryptophan) of serotonin biosynthesis to study a possible role of serotonin in the stimulatory effect of endorphins on prolactin release.

Inhibition of serotonin biosynthesis with parachlorophenylalanine (PCPA) reduced by 40-45% the rise of plasma prolactin levels induced by morphine in the absence (Figure 7A) or presence (Figure 7B) of simultaneous treatment with 5-hydroxytryptophan (5-HTP). Treatment with PCPA alone had no significant effect on the already low plasma levels of prolactin in control animals. It is of

Figure 5.6: **Effect of haloperidol or repeated injection of morphine on plasma prolactin levels in male rats already treated with morphine. Morphine was injected as a single dose or given at times of 0, 30, and 60 minutes. In another group, haloperidol was injected 45 minutes after morphine.**

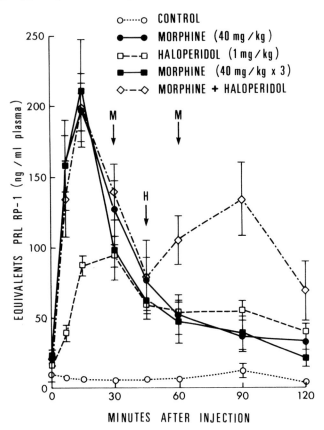

interest that the marked rise of prolactin secretion induced by 5-HTP was superimposable in control and PCPA-treated rats (Figure 7B), thus indicating that lowering of endogenous serotonin by PCPA does not affect the sensitivity of the serotonergic response mechanisms.

The present data clearly indicate that the acute release of prolactin induced by morphine can be accounted for by an inhibition of the inhibitory hypothalamic dopaminergic influence on prolactin secretion. This is well illustrated by the absence of further increase of prolactin release by a high dose of haloperidol in animals treated with morphine (Figure 5.6). These data are in agreement with the inhibition of dopamine turnover observed in the medial palisade zone of the median eminence in rats injected intraventricularly with met-enkephalin (Ferland *et al.*, 1977b). These effects are probably due to the activation of presynaptic

Figure 5.7: Effect of morphine, 5-hydroxytryptophan (5-HTP) or parachloro-phenylalanine (PCPA) alone or in combination on plasma prolactin levels in the rat. Morphine was injected s.c. at time 0, whereas 5-HTP was injected s.c. 45 minutes before the experiment and PCPA was injected i.p. 48 and 24 hours before time 0.

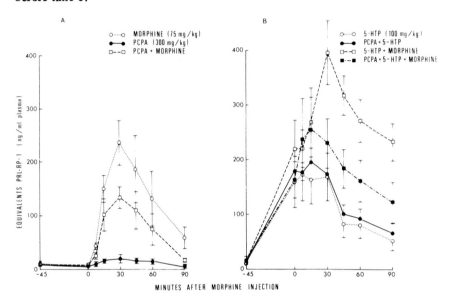

inhibitory opiate receptors located on tuberoinfunibular dopaminergic nerve endings.

The present observations suggest that presynaptic inhibitory opiate receptors are present on dopaminergic neurons of the tuberoinfundibular system. Since the stimulatory effect of morphine on prolactin secretion is partially suppressed after inhibition of serotonin biosynthesis with PCPA, it is possible that presynaptic stimulatory opiate receptors are present on serotonergic nerve endings in contact with the tuberoinfundibular dopaminergic system. It is also possible that effects of opiates on dopaminergic and serotonergic activity are not direct but are instead mediated by one or more other neurotransmitters.

V. DISTRIBUTION OF β-ENDORPHIN AND ENKEPHALINS

Although β-LPH could be the precursor of β-endorphin and met-enkephalin, recent data suggest the existence of two separate systems: an enkephalin system and a β-LPH/β-endorphin/ACTH system (Bloom *et al.*, 1978; Watson *et al.*, 1978).

The localization of β-LPH, endorphins and enkephalins has been studied in the pituitary gland and the central nervous system using radioimmunoassay and immunohistochemical techniques. In the pituitary gland, enkephalin immunoreactivity was found mainly in the pars intermedia-neurohypophysis, whilst β-endorphin was concentrated in the intermediate lobe and was the predominant peptide in the adenohypophysis (Rossier *et al.*, 1977; Lissitzky *et al.*, 1978). In the brain, immunohistochemical techniques indicate that β-endorphin-reactive neurons and fibers are distributed in the basal hypothalamic area and project into the preoptic area, the medial dorsal thalamus and the periaqueductal gray.

The concentration of met-enkephalin-like and β-endorphin-like immunoreactivities was determined in 42 discrete areas of rat brain (Table 5.1). The highest concentrations of β-endorphin are found in the median eminence and the hypothalamic nuclei, followed by the medial preoptic nucleus, the nucleus interstitialis striae terminalis, periaqueductal gray, nucleus medialis thalami and nucleus amygdaloideus centralis. The highest concentration of met-enkephalin immunoreactivity is found in the globus pallidus, followed by the nucleus interstitialis striae terminalis, medial preoptic nucleus, nucleus lateralis hypothalami, nucleus amygdaloideus centralis and paraventricular nucleus.

The regional distribution of β-endorphin, met- and leu-enkephalin-like material in bovine brain is shown in Table 5.2. As observed previously in the rat (Rossier *et al.*, 1977; Bloom *et al.*, 1978; Watson *et al.*, 1978), enkephalins and β-endorphin display quite different distributions. It seems that β-endorphin-like immunoreactivity is mainly concentrated in the hypothalamic area, preoptic area, habenula, lateral pontine area, nucleus interpeduncularis and septum, while the highest concentration of enkephalin-like material is found in the basal ganglia and certain structures of the olfactory brain, namely the nucleus accumbens septi. The tuberculum olfactorium also contains high amounts of leu- and met-enkephalin-like immunoreactivity. Lesser amounts of enkephalin-like material can be measured in the median eminence, the septum, the preoptic area, the hypothalamic area and the habenula.

In this study, the ratio of met- to leu-enkephalin varies from 1.14 to 9.54 (Table 5.2). It is interesting to note that the ratio ranges from 6.15 to 7.8 for the structures of the basal ganglia. Except for the median eminence, the structures having a low concentration of peptides are those associated with a low ratio of met- to leu-enkephalin.

The distribution of β-endorphin and met-enkephalin in 42 discrete areas of rat brain and in 30 structures of bovine brain supports the view that enkephalins and endorphins form distinct systems in the brain. This mutual exclusivity of β-endorphinergic and enkephalinergic systems has been previously suggested by immunohistochemical data (Rossier *et al.*, 1977; Simantov *et al.*, 1977; Watson *et al.*, 1978). In the rat, the endorphin system is mainly concentrated within the hypothalamic area and five extrahypothalamic sites, namely the medial preoptic area, nucleus interstitialis striae terminalis, medial dorsal thalamus, nucleus paraventricularis thalami and periaqueductal gray. Such an extrahypothalamic distribution of β-endorphin might have some relevance to the electrophysiological studies of Renaud (1976), which indicate that axons of basomedial hypothalamic

Table 5.1: Regional Distribution of Met-enkephalin and β-Endorphin in the brain*

	Met-enkephalin	β-Endorphin
	(ng/mg of protein)	
Frontal cortex	0.36 ± 0.07	ND
Septum	1.43 ± 0.36	0.77 ± 0.08
Tractus diagonalis	2.07 ± 0.33	1.06 ± 0.33
Medial preoptic nucleus	10.99 ± 0.54	21.93 ± 0.96
Lateral preoptic nucleus	6.38 ± 1.66	2.90 ± 0.13
Nucleus interstitialis striae terminalis	16.22 ± 1.11	15.46 ± 1.57
Nucleus accumbens	1.88 ± 0.29	ND
Tubercle olfactorium	1.41 ± 0.33	ND
Striatum A 9000	2.33 ± 0.68	ND
Striatum A 7000	0.85 ± 0.44	— —
Globus pallidus	18.25 ± 3.07	— —
N. amygdaloideus medialis A 4250	1.28 ± 0.35	4.28 ± 1.14
N. amygdaloideus medialis A 3500	1.35 ± 0.30	3.47 ± 0.29
N. amygdaloideus centralis A 4250	7.13 ± 1.81	5.76 ± 1.25
N. amygdaloideus lateralis, pars post	0.50 ± 0.13	3.09 ± 0.91
N. amygdaloideus basalis, pars lat	0.50 ± 0.06	4.43 ± 1.73
N. amygdaloideus corticalis	0.49 ± 0.05	ND
Nucleus anterior hypothalami	7.21 ± 0.62	21.17 ± 1.80
Nucleus lateralis hypothalami	9.16 ± 2.95	5.81 ± 1.32
Nucleus suprachiasmaticus	4.80 ± 0.95	25.76 ± 2.83
Nucleus periventricularis	1.88 ± 0.11	30.79 ± 3.16
Paraventricular nucleus	6.42 ± 0.83	13.18 ± 0.22
Median eminence	2.69 ± 0.51	41.78 ± 0.77
Arcuate nucleus I-II-III	2.45 ± 0.61	29.67 ± 0.37
Nucleus ventromedialis hypothalami	5.44 ± 1.79	21.84 ± 1.80
Nucleus dorsomedialis hypothalami	3.26 ± 0.65	25.68 ± 4.92
Nucleus arcuate IV-V	3.6 ± 0.84	26.39 ± 8.01
Nucleus premamillaris ventralis	4.00 ± 0.77	9.60 ± 0.40
Nucleus premamillaris dorsalis	0.67 ± 0.03	4.59 ± 0.95
Nucleus mamillaris medialis, pars medialis	0.80 ± 0.12	6.32 ± 0.94
Nucleus lateralis thalami	0.16 ± 0.04	1.84 ± 0.21
Nucleus medialis thalami	0.33 ± 0.01	6.16 ± 1.41
Nucleus ventralis thalami	0.13 ± 0.03	1.32 ± 0.41
Nucleus paraventricularis thalami	—	30.47 ± 5.29
Hippocampus	0.17 ± 0.02	1.47 ± 0.20
Nucleus habenulae lateralis	1.51 ± 0.07	6.32 ± 0.94
Substantia nigra	0.37 ± 0.07	ND
Area ventralis tegmenti	0.95 ± 0.15	ND
Formatio reticular + lemniscus medialis	0.38 ± 0.07	ND
Interpeduncular nucleus	2.55 ± 0.27	1.33 ± 0.15
Periaqueductal gray	1.93 ± 0.39	8.19 ± 1.77
N. Medianus raphes	0.50 ± 0.10	ND

*The brains were frozen and specific areas were punched. Means ±SEM of three groups of four rats (pool) are presented. ND denotes not detectable.

Table 5.2: Regional Distribution of β-Endorphin, Met-enkephalin and Leu-enkephalin in Various Parts of Bovine Brain*

	β-Endorphin	Met-enkephalin	Leu-enkephalin	met-enk/leu-enk
		(ng/mg of protein)		
Median eminence	1.916 ± 2.500	5.340 ± 0.087	0.560 ± 0.013	9.54
Polus frontalis	−	0.570 ± 0.037	−	−
Cortex temporalis	−	0.430 ± 0.047	0.198 ± 0.060	2.17
N. caudatus	−	15.640 ± 0.400	2.004 ± 0.061	7.80
Putamen	−	11.100 ± 1.330	1.557 ± 0.013	7.13
Tuberculum olfactorium	−	14.990 ± 2.340	2.231 ± 0.267	6.72
N. accumbens septi	−	18.500 ± 1.640	2.036 ± 0.384	9.09
Area preoptica	3.752 ± 0.480	3.280 ± 0.127	0.677 ± 0.110	4.84
Septum	1.036 ± 0.250	4.433 ± 0.395	0.655 ± 0.110	6.77
Globus pallidus	−	37.980 ± 4.070	6.173 ± 1.402	6.15
Cortex parietalis	−	0.274 ± 0.033	−	−
Hypothalamus anterior	2.120 ± 0.060	0.908 ± 0.130	0.362 ± 0.031	2.51
Thalamus anterior	−	0.422 ± 0.036	0.369 ± 0.038	1.14
Corpus amygdaloideum	−	1.387 ± 0.147	0.409 ± 0.010	3.39
Hippocampus	−	0.184 ± 0.020	−	−
Hypothalamus median	3.924 ± 0.032	2.392 ± 0.329	0.516 ± 0.034	4.64
Thalamus median	−	1.067 ± 0.146	0.364 ± 0.069	2.93
Habenula	3.503 ± 0.338	2.764 ± 0.223	0.319 ± 0.055	8.66
Hypothalamus posterior	1.374 ± 0.400	2.401 ± 0.083	1.161 ± 0.250	2.07
N. interpeduncularis	1.653 ± 0.264	1.034 ± 0.192	0.223 ± 0.115	4.64
Tectum mesencephali	−	0.784 ± 0.054	0.276 ± 0.080	1.75
S. grisea centralis	−	1.045 ± 0.085	0.220 ± 0.038	4.75
S. nigra	−	1.112 ± 0.062	0.564 ± 0.300	1.97
Tegmentum	−	2.009 ± 0.132	−	−
Medial pontine area	−	1.429 ± 0.208	0.570 ± 0.250	2.51
Lateral pontine area	1.850 ± 0.040	0.978 ± 0.034	0.216 ± 0.080	4.53
Area postrema	−	0.868 ± 0.080	−	−
Anterior medulla	−	0.547 ± 0.170	−	−
Cerebellar vermis	−	0.057 ± 0.008	−	−
Cerebellar hemisphere	−	0.060 ± 0.010	−	−

*Results are the mean ±SEM of triplicate determinations. A dash means not detectable.

neurones innervate the anterior hypothalamic area, the medial preoptic area, the medial dorsal thalamus and the periaqueductal gray area. Knowing that fibers originating from the amygdala reach the hypothalamus through the stria terminalis, the scattered distribution of β-endorphin in nuclei amygdala and its concentration in the nucleus interstitialis striae terminalis suggest the existence of a β-endorphinergic pathway.

Recently, Larsson *et al.* (1979) have indicated the existence of separate neuronal systems for met-enkephalin and leu-enkephalin and report that leu-enkephalin-containing neurons are only 20-25% as numerous as met-enkephalin-containing neurons in the striatum. In the light of this report, the different ratios of met-enkephalin and leu-enkephalin found in various structures may reflect differences in innervation by the two distinct neuronal systems.

VI. CHANGES OF β-ENDORPHIN AND MET-ENKEPHALIN LEVELS

A. Estrogen and Haloperidol Treatment

There is good evidence for a close association between brain enkephalinergic (and endorphinergic) and dopaminergic systems. This is clearly suggested by the met-enkephalin-induced reduction of hypothalamic dopamine turnover (Ferland *et al.*, 1977b) and by the localization of opiate receptors on dopaminergic neurones in the striatum and nucleus accumbens (Pollard *et al.*, 1977, 1978). It was thus of interest to study a possible effect on met-enkephalin levels of treatments known to interact with dopaminergic systems. For example, estrogen treatment is known to increase the turnover of dopamine in the median eminence (Fuxe *et al.*, 1977). Moreover, the same treatment exerts potent antidopaminergic actions at the level of the pituitary gland and striatum. These data pertain to the reversal of the dopamine-induced inhibition of prolactin secretion (Raymond *et al.*, 1978) and the inhibition by estrogen treatment of the accumulation of striatal acetylcholine (Euvrard *et al.*, 1980) and circling behaviour (Bédard *et al.*, 1978) induced by dopamine agonists in the rat.

The effect of chronic estrogen or haloperidol treatment on the met-enkephalin content of several rat brain areas is shown in Figure 5.8. It can be seen that both treatments lead to differential effects on met-enkephalin content in various brain structures. Thus, while estrogen treatment leads to stimulation ($p < 0.01$) in the nucleus interstitialis striae terminalis, medial preoptic nucleus, lateral preoptic nucleus and periaqueductal gray, an inhibitory effect is found in the nucleus amygdaloideus centralis. Following chronic haloperidol treatment, an increased met-enkephalin concentration is found in the striatum, medial preoptic nucleus, lateral preoptic nucleus and interpeduncular nucleus, while no significant effect is seen in the other areas.

Enkephalinergic pathways have been described between the striatum and globus pallidus (Cuello and Paxinos, 1978) and enkephalin-containing neurones of the amygdala centralis project through the stria terminalis to possibly terminate in the nucleus interstitialis striae terminalis (Gros *et al.*, 1978). It is thus of interest to mention the opposite effects of chronic estrogen treatment on met-enkephalin levels in the nucleus interstitialis striae terminalis and nucleus amygdaloideus centralis. Whilst in the former structure met-enkephalin levels are increased, a significant decrease is evident in the latter structure. In this context, it is interesting to note that changes in the release of DA from nerve terminals and dendrites were simultaneous but in opposite directions following perfusion of the substantia nigra with substance P or antiserum to substance P (Michelot *et al.*, 1979). The involvement of the preoptic area in the regulation of gonadotropin release has been known for a long time and the stimulation of met-enkephalin levels by estrogen or haloperidol treatment in the medial and

Figure 5.8: **Effect of a chronic treatment with 17β-estradiol (1 μg, b.i.d., s.c. for 14 days) or with haloperidol (0.5 mg/kg, b.i.d., s.c. for five days or 1 mg/kg, b.i.d., s.c. for nine days) on the met-enkephalin content of discrete brain nuclei of ovariectomized rats. Data are given as the means ± SEM of three groups of four rats (pool). Abbreviations: cp, striatum; sl, nucleus septi lateralis; st, nucleus interstitialis striae terminalis; pom, nucleus preopticus medialis; pol, nucleus preopticus lateralis; am, nucleus amygdaloideus medialis; ac, nucleus amygdaloideus centralis; ip, nucleus interpeduncularis; SNR, substantia nigra, zona reticulata.**

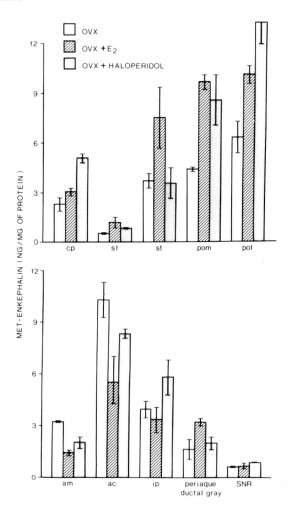

lateral preoptic nuclei agrees well with the effects of estrogens and opiate peptides on LH release (Bruni *et al.*, 1977). It should also be mentioned that changes of met-enkephalin levels can only be used as suggestive evidence for changes of met-enkephalin release. In fact, the effects observed are likely to be time-dependent, and direct measurements of met-enkephalin turnover and release will be required before definitive conclusions can be drawn.

Although both estrogens and haloperidol can exert antidopaminergic actions, the mechanisms involved are likely to be quite different. Different sites of action are suggested by the differential action of estrogens and haloperidol on met-enkephalin levels in structures such as the nucleus interstitialis striae terminalis and interpeduncular nucleus and by the apparent additivity of their inhibition of met-enkephalin levels in the globus pallidus. In support of previous findings, the present data are suggestive of an intimate interrelationship between endogenous opioid peptides and dopaminergic systems of the central nervous system.

B. Estrous Cycle

A role for endogenous opiate peptides in the control of prolactin secretion is suggested by the prevention of stress- or suckling-induced increases in serum prolactin by naloxone administration (Figures 5.4 and 5.5). In order to obtain further evidence for the involvement of endogenous opiate peptides in the control of prolactin secretion, we have measured the β-endorphin content of discrete brain nuclei during the estrous cycle in the rat.

Measurements of the β-endorphin content of 16 specific hypothalamic and extra-hypothalamic nuclei during all stages of the estrus cycle show that significant variations occur only in suprachiasmatic and arcuate nuclei and the median eminence (Figure 5.9). In the arcuate nucleus, which contains the highest β-endorphin content of all brain nuclei, the afternoon of proestrus coincides with a 50% decrease in β-endorphin content compared to the mean content on all other days of the cycle. In contrast, the β-endorphin content of the suprachiasmatic nucleus and median eminence was increased by 100% and 65%, respectively, on the afternoon of proestrus when compared to the mean content on other days. Whilst in the suprachiasmatic nucleus the β-endorphin content remained elevated throughout estrus, in both the arcuate nucleus and median eminence, the values had returned to those seen on other days of the cycle by the day of estrus.

VII. SUMMARY AND CONCLUSION

Since opiate peptides have been demonstrated to reduce the turnover of dopamine in the median eminence (Ferland *et al.*, 1977b), and dopamine levels in the median eminence are decreased on the afternoon of proestrus (Crowley *et al.*, 1978), it is possible that the increased β-endorphin content of median eminence seen on the afternoon of proestrus is related to this effect, and the subsequent increase in prolactin secretion results from removal of the dopamine inhibition at the level of the anterior pituitary. The cell bodies of the brain β-endorphinergic

Figure 5.9: Changes in the β-endorphin content of certain hypothalamic nuclei during the estrus cycle of the rat. Groups of 18 four-day cycling rats were sacrificed at 07.00 h each day of the estrus cycle and also at 15.00 h on the day of proestrus. The β-endorphin content of specific brain nuclei removed by 0.5 mm or 1.0 mm punches of frozen brain slices was measured by radioimmunoassay (Lissitzky *et al.*, 1978) using synthetic β-endorphin as standard and a specific antiserum. Results shown are the means ± SEM of three separate determinations of the β-endorphin content of tissue pooled from six animals. The significance of differences in the means was tested by the Duncan-Kramer multiple-range test following analysis of variance. (An asterisk denotes p < 0.01 as against means on diestrus I, II, proestrus AM and estrus; a double asterisk denotes p < 0.01; and a triple asterisk denotes p < 0.05 as against means on diestrus I, II and proestrus AM).

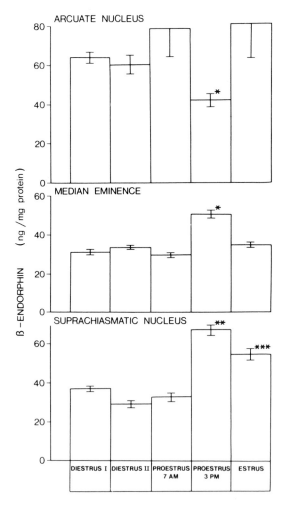

system are believed to be located exclusively in the arcuate nucleus (Bloom *et al.*, 1978) and the decreased β-endorphin content of this nucleus on the afternoon of proestrus may reflect increased axonal transport to the median eminence and the suprachiasmatic nucleus. We are unable to relate the increased β-endorphin content of the suprachiasmatic nucleus on the afternoon of proestrus and during estrus to any physiological effects. However, it is interesting to recall the importance of this nucleus in the regulation of diurnal rhythms and, presumably, the estrus cycle, in view of its acute sensitivity to light/dark cycles. Moreover, a role for endogenous opioid peptides in the regulation of LH secretion has been indicated (Cicero *et al.*, 1979) and recently Arendash and Gallo (1979) reported that electrical stimulation of the suprachiasmatic nucleus can modify episodic LH release in untreated and estrogen-primed ovariectomized rats. It is becoming increasingly clear that peripheral hormones can influence central-nervous-system neurotransmitter systems and the changes in β-endorphin seen during the estrus cycle are probably due to fluctuations in peripheral gonadal steroids, particularly estrogens.

REFERENCES

Arendash, G.W. and Gallo, R.V. (1979). *Endocrinology 104*, 333

Bédard, J., Dankova, J., Boucher, R. and Langelier, P. (1978). *Can. J. Physiol. Pharmacol.*, *56*, 538

Bloom, F., Battenberg, E., Rossier, J., Ling, N. and Guillemin, R. (1978). *Proc. Nat. Acad. Sci. (USA) 75*, 1591

Bradbury, A.F., Smyth, P.G. and Snell, C.R. (1976). *Nature 260*, 793

Bruni, J.F., Van Vugt, D., Marshall, S. and Meites, J. (1977). *Life Sci.*, *21*, 461

Cicero, T.J., Schainker, B.A. and Meyer, E.R. (1979). *Endocrinology 104*, 1286

Cox, B.M., Goldstein, A. and Li, C.H. (1976). *Proc. Nat. Acad. Sci. (USA) 73*, 1821

Crowley, W.R., O'Donohue, T.L. and Jacobowitz, D.M. (1978). *Brain Res.*, *147*, 315

Cuello, A.C. and Paxinos, G. (1978). *Nature 271*, 178

Cusan, L., Dupont, A., Kledzik, G.S., Labrie, F., Coy, D.H. and Schally, A.V. (1977). *Nature 268*, 544

Dupont, A., Cusan, L., Labrie, F., Coy, D.H. and Li, C.H. (1977a). *Biochem. Biophys. Res. Commun.*, *75*, 76

Dupont, A., Cusan, L., Garon, M., Labrie, F. and Li, C.H. (1977b). *Proc. Nat. Sci. (USA) 74*, 358

Euvrard, C., Raynaud, J.P., Boissier, J.R. and Labrie, F. (1980). *Brain Res.*, *169*, 215

Ferland, L., Labrie, F., Arimura, A. and Schally, A.V. (1977a). *J. Mol. Cell. Endocrinol.*, *2*, 247

Ferland, L., Fuxe, K., Eneroth, P., Gustafsson, J.A. and Skett, P. (1977b). *Eur. J. Pharmacol.*, *43*, 89

Ferland, L., Kledzik, G.S., Cusan, L. and Labrie, F. (1978). *J. Mol. Cell. Endocrinol.*, *16*, 267

Ferland, L., Labrie, F., Cusan, L., Dupont, A., Lépine, J., Beaulieu, M., Denizeau, F. and Lemay, A. (1980). In *The Endocrine Function of the Brain*, ed. M. Motta (Plenum Press, New York, in press)

Fuxe, K., Lofstrom, A., Eneroth, P., Gustafsson, J.A., Skett, P., Hökfelt, F., Wiesel, F.A. and Agnati, L. (1977). *Psychoneuroendocrinology 2*, 203

Gallo, R.V., Rabii, J. and Moberg, G.P. (1975). *Endocrinology 97*, 1096

Goodman, G., Lawson, D.M. and Gala, R.R. (1976). *Proc. Soc. Exp. Biol. Med.*, *153*, 225

Gros, C., Pradelles, P., Humbert, J., Dray, F., Le Gal, La Salle, G. and Ben Ari, Y. (1978). *Neurosc. Lett.*, *10*, 193

Kokka, N., Garcia, J.F. and Elliot, H.W. (1973). *Brain Res., 39,* 347

Larsson, L.I., Childers, S. and Snyder, S. (1979). *Nature 282,* 407

Li, C.H. and Chung, D. (1976). *Proc. Nat. Acad. Sci. (USA) 73,* 1145

Lissitzky, J.C., Morin, O., Dupont, A., Labrie, F., Seidah, H.G., Chretien, M., Lis, M. and Coy, D.H. (1978). *Life Sci., 22,* 1715

Loh, H.H., Tseng, L.F., Wei, E.I. and Li, C.H. (1976). *Proc. Nat. Acad. Sci. (USA) 73,* 2895

Michelot, R., Leviel, V., Giorgirieff-Chesselet, M.F., Cheramy, A. and Glowinski, J. (1979). *Life Sci., 24,* 715

Neill, J.D. (1970). *Endocrinology 87,* 1192

Pasternak, G.W., Goodman, R. and Snyder, S.H. (1977). *Life Sci., 16,* 1765

Pollard, H., Llorens-Cortes, J.C., Schwartz, J.C. (1977). *Nature 268,* 745

Pollard, H., Llorens-Cortes, C., Schwartz, J.C., Gros, C. and Dray, F. (1978). *Brain Res., 151,* 392

Raymond, V., Beaulieu, M., Labrie, F. and Boissier, J.R. (1978). *Science 200,* 1173

Renaud, L.P. (1976). *Brain Res., 105,* 59

Rossier, J., Vargo, T.M., Minick, S., Ling, N., Bloom, F.E. and Guillemin, R. (1977). *Proc. Nat. Acad. Sci. (USA) 74,* 5162

Simantov, R. and Snyder, S. (1976). *Proc. Nat. Acad. Sci. (USA) 73,* 2515

Simantov, R., Kuhar, M.J., Uhl, G.R. and Snyder, S.H. (1977). *Proc. Nat. Acad. Sci. (USA) 74,* 2167

Smith, T.W., Hughes, J., Kosterlitz, H.W. and Sosa, R.P. (1976). In *Opiates and Endogenous Opioid Peptides* (Elsevier-North Holland, Amsterdam) p. 57

Terenius, L. and Wahlstrom, A. (1975). *Acta Physiol. Scand., 94,* 74

Tolis, G., Hickey, J. and Gupta, H. (1975). *J. Clin. Endocrinol. Metab., 41,* 797

Von Graffenried, G., del Pozo, E., Roubicek, J., *et al.* (1978). *Nature 272,* 729

Watson, S.J., Akil, H., Richard, C.W. and Barachas, J.D. (1978). *Nature 275,* 226

6　BEHAVIOURAL MODULATION BY ~
ADMINISTRATION OF ENKEPHAL~
ENDORPHINS

Richard D. Olson, Abba J. Kastin, Ga~
and David H. Coy

TABLE OF CONTENTS

I　INTRODUCTION　143

II　NATURAL BEHAVIOURS　144

III　LEARNED BEHAVIOURS　146

IV　CLINICAL BEHAVIOURS　147

V　SUMMARY AND CONCLUSIONS　148

REFERENCES　149

I. INTRODUCTION

After the discovery of opiate receptors in the brain by Goldstein *et al.* (1971), Pert and Snyder (1973), Simon *et al.* (1973) and Terenius (1973), and the subsequent identification of endogenous opiates by Hughes *et al.* (1975) and Terenius and Wahlstrom (1975a, b), several groups began to examine the role of these peptides in modulating behaviour. Initially, Jacquet and Marks (1976) and Bloom *et al.* (1976) noticed that the central administration of brain opiates resulted in profound analgesic and muscular effects. However, subsequent attempts to replicate these findings with systemic injections (Kosterlitz and Hughes, 1977) were not successful, thus restricting interest to effects of centrally administered opiates and their analogs.

A few investigators continued to explore the effects of peripherally administered opiates, for if endogenous opiates were ever to be clinically practical they obviously had to be effective when administered systemically. Accordingly, Szekely *et al.* (1977) demonstrated analgesia, Plotnikoff *et al.* (1976) showed increased activity in a DOPA potentiation test and Kastin *et al.* (1976b) reported that mice ran faster and made fewer errors in a maze, all after systemic injections of enkephalin analogs. Furthermore, as first proposed by Kastin *et al.* (1976b), it became clear that opiate peptides had both narcotic and non-narcotic effects and that the latter might ultimately prove to be the more important. This concept of dissociative effects has gained considerable support.

The success of these initial investigators in demonstrating that opiate peptides could have significant effects when administered peripherally, coupled with the development of new opiate analogs specifically designed to be more resistant to degradation and having greater ability to permeate the blood-brain barrier (Coy *et al.*, 1976), led to renewed interest by other groups in exploring the efficacy of systemic administration of opiate peptides. One problem still remained, however, before general acceptance of these findings was possible, and that was concrete evidence that opiate peptides were able to penetrate the blood-brain barrier. Behavioural and more direct evidence that this was the case has been reviewed elsewhere (Kastin *et al.*, 1976a, 1979a), but a very recent paper by Rapoport *et al.* (1980) now confirms that opioid peptides have a moderate cerebrovascular permeability that should allow penetration of the blood-brain barrier in three to eleven minutes, with a step increase in plasma concentration of unbound peptide. Hopefully, this report will dispel any lingering doubts about this issue.

The literature now contains numerous studies demonstrating behavioural effects in a wide range of organisms including lizards, chickens, goldfish, rats, monkeys and humans, all after systemic injection of opiate peptides and their analogs. The success of these experiments clearly indicates the clinical potential of opiate peptides.

II. NATURAL BEHAVIOUR

To survey the various forms of behavioural modulation that have occurred as a result of systemic administration of enkephalins and endorphins, it seemed appropriate to begin with natural behaviour such as activity, eating, drinking, etc. The very first study to report any behavioural effects after peripheral administration of an opiate peptide was by Plotnikoff *et al.* (1976). These investigators injected mice intraperitoneally (i.p.) with either met-enkephalin or its [D-ala^2]-analog and tested several paradigms of activity. Both forms of enkephalin increased activity in the DOPA potentiation test at all doses used (0.1, 1.0 and 10.0 mg/kg) after 30 minutes and, at the two larger doses, after 60 minutes. They also reported a slight but significant increase in the serotonin potentiation test in activity at both times for the 10-mg/kg dose. Finally, a decrease in footshock-induced fighting at lower doses and a slight reduction in audiogenic seizures was also described.

Two other studies have also focused primarily on activity. Olson *et al.* (1978c) studied the effects of intracranial (i.c.) and i.p. injections of 80 μg/kg of 21 substances on spontaneous activity in goldfish. All seven endorphins and two of four enkephalins tested reliably decreased activity, with very significant decreases being produced by a pentafluorinated enkephalin analog, [D-ala^2, F$_5$ phe^4]-met-enkephalin-NH$_2$. In regard to site of injection, effects after i.c. administration occurred, on average, after three minutes, while effects after i.p. administration averaged six minutes, the effects of the latter being weaker. In another study, Olson *et al.* (1978a) examined the effects of [D-ala^2]-β-endorphin on activity in squirrel monkeys and obtained variable biphasic results, with an increase after an intravenous (i.v.) injection of the smallest dose but a decrease after the largest dose. In all dose groups but the largest (800 μg/kg), they also found that the monkeys approached their food more rapidly after i.v. injections of the analog.

Another area where endogenous opioids seem to play an important role is in modulating food and water intake. Considerable evidence exists that elevated levels of endorphin are associated with obese rats (Margules *et al.*, 1978) and that intraventricular (i.v.t.) administration of either morphine or β-endorphin will cause marked increases in feeding (Belluzzi and Stein, 1977; Kenney *et al.*, 1978). King *et al.* (1979b) studied the effects of met-enkephalin, β-endorphin, and their [D-ala^2]-analogs on eating, drinking and activity measures in rats. Using 80-μg/kg i.p. injections, they found that activity was decreased by both forms of endorphin but that neither enkephalin had a significant effect on activity. They also showed that [D-ala^2]-met-enkephalin generated less eating and, therefore, lowered body weight. King *et al.* (1979a) administered naloxone in doses of 1.0, 2.0, 4.0 or 8.0 mg/kg i.p. to both normal and hypothalamic obese rats and showed that food intake was suppressed in both groups. Similar data were obtained using sweetened milk. They concluded that the opiate receptors located in the ventromedial hypothalamus were not essential for the effects of opiate agonists and antagonists on feeding behaviour. Kastin *et al.* (1979b) replicated these findings with intact rats as part of another experiment.

Using a swimming test to study passiveness in rats, Kastin *et al.* (1978) administered i.p. several opiate analogs and found that they reduced passiveness. Olson *et al.* (1979b) compared the effects of a potent opiate analog, [d-ala^2, F$_5$ phe^4]-met-enkephalin-NH$_2$, an opiate analog with virtually no narcotic action, [D-phe^4]-met-enkephalin, and a control solution of the diluent on tonic immobility in chickens. With doses ranging from 0.1 to 1000.0 µg/kg, all but the lowest doses extended the duration of tonic immobility. A biphasic relationship was obtained once again, 10.0 and 100.0 µg/kg being the most effective doses. It is significant to note that [D-phe^4]-met-enkephalin, the analog with negligible direct opiate action, could also be effective in extending latencies in this paradigm, clearly suggesting that the effect was not due to the opiate action of the peptide but rather to some dissociated action. Obviously the effects of opiate peptides vary with the tasks involved and probably as a function of the species being tested.

Several other papers have appeared documenting additional effects obtained after systemic injections of opiate peptides. Catlin *et al.* (1978) administered i.v. injections of either 10, 50, 100, 500 or 1000 µg/kg of β-endorphin to cats and found that doses above 100 µg/kg produced licking, some vomiting and relaxation of the nictitating membrane. They also emphasized that the effects of i.v.t. and i.v. injections were significantly different in magnitude. Tseng *et al.* (1978) were able to inhibit urine flow in rats with β-endorphin. Although it was far less effective when administered i.v. than i.v.t., it was still 24 times more potent than morphine when both were injected i.v.

Veith *et al.* (1978) administered 100-µg/kg i.p. injections of α-, γ- and β-endorphin and their [D-ala^2]-analogs to rats and found that β-endorphin increased grooming and [D-ala^2]-α-endorphin increased seminal discharge, while both γ-endorphin and [D-ala^2]-γ-endorphin increased several measures of emotionality. The specificity of these effects suggests that selective effects might result from differential actions on the same receptors. Weinberger *et al.* (1979) administered des-tyrosine[1]-γ-endorphin subcutaneously (s.c.) and found that it reduced rearings but had little effect on several measures characteristic of neuroleptic activity. They concluded that this opiate analog was not a neuroleptic and did not share the properties of haloperidol, thereby disagreeing with earlier reports by de Wied and his associates that the two substances were essentially equivalent. Panksepp *et al.* (1978) found that s.c. administration of β-endorphin reliably inhibited distress vocalization in chickens, but not nearly as effectively as when injected i.v.t. Furthermore, the frequency of distress vocalizations was lowered by [D-ala^2]-met-enkephalin. Finally, Sandman *et al.* (1979) injected rats s.c.

with 50 μg per rat of β-endorphin at two to seven days postnatally; when tested for analgesia at 90 days, there was a significant elevation in the threshold for thermal stimuli. They concluded that early exposure to β-endorphin might lead to permanent changes in behaviour as a result of alteration in the interaction of endogenous opiates with their binding sites during a critical period.

III. LEARNED BEHAVIOURS

The first study in the literature reporting learning effects from systemic administration of an opiate peptide was reported by Kastin *et al.* (1976b). Rats were administered i.p. injections of 80 or 800 μg/kg of met-enkephalin or [D-ala^2]-met-enkephalin-NH$_2$; it was found that experimental animals receiving either peptide ran faster and committed fewer errors in a twelve-choice Warden maze than did controls. Further, [D-phe^4]-met-enkephalin, an analog with virtually no direct opiate activity, was equally effective, thus suggesting that behavioural effects could be dissociated from opiate effects after i.p. administration.

Olson *et al.* (1978b) investigated the effects of endorphin and enkephalin analogs on fear habituation in goldfish and found that endorphin increased latencies, while enkephalin decreased latencies. However, conclusions about learning are difficult to draw because the locus of effect for both peptides appeared to be on the sensitization component rather than habituation, as evidenced by the immobilization of the fish following the presentation of the buzzer.

Several avoidance conditioning paradigms have been utilized to investigate the effects of opiate peptides on learning. Rigter *et al.* (1977) found that enkephalin injected s.c. before the retrieval test would attenuate carbon-dioxide-induced amnesia. Moreover, the effect was not naloxone-reversible, adding further support to the notion of dissociated behavioural and opiate effects. Rigter (1978) then evaluated the effects of a wide range of doses of met- and leu-enkephalin administered s.c. to rats, and found that no significant differences existed on the acquisition trial. However, met-enkephalin diminished amnesia in a dose-dependent fashion when administered before acquisition or retrieval, or at both times. Leu-enkephalin was not effective if injected before acquisition, but did reverse amnesia when given before retrieval or at both times. This work suggests that enkephalins may modulate memory processes, with the effect probably occurring during retrieval rather than storage.

Evaluating the effects of s.c. injections of doses of 0.1, 0.3 or 1.0 μg per rat of several opiate peptides, de Wied *et al.* (1978a) found that β-endorphin delayed extinction of a pole-jumping avoidance response, and that α- and β-endorphin were more potent than met-enkephalin in delaying pole-jumping extinction, although the changes caused by all three opiate peptides were significant. Furthermore, de Wied *et al.* (1978b) found that α-endorphin had some effects on open-field behaviour and on responsiveness to electric footshock. These findings were somewhat similar to those presented earlier by Veith *et al.* (1978) indicating increased emotionality. In addition, de Wied and his associates confirmed the earlier findings of Kastin *et al.* (1976b) indicating that behavioural results could be dissociated from those related to opiate-receptor sites and extended the idea to include extinction.

Utilizing the pole-jumping avoidance paradigm, de Wied *et al.* (1978b) examined the effects of [des-tyr[1]]-γ-endorphin and found that, while γ-endorphin facilitated extinction, DTγE was reliably more effective. Furthermore, in a passive avoidance paradigm, γ-endorphin facilitated acquisition, while DTγE attenuated performance. Based on these data, the authors concluded that DTγE or a closely related peptide was an endogenous neuroleptic. A third paper by the same group (de Wied *et al.*, 1978c) reported additional data in support of the idea that DTγE was an endogenous neuroleptic.

More recently, LeMoal *et al.* (1979) described the effects of s.c. injections of 10 μg/kg of α- and γ-endorphin on both appetitive and aversive tasks. Both forms of endorphin inhibited extinction of a runway task for water reward in water-deprived rats. Furthermore, in the aversive task, α-endorphin inhibited extinction of pole-jumping avoidance, while γ-endorphin facilitated extinction of the same response. The authors hypothesized that the effects were situation-dependent, contrary to the findings of de Wied's previous work.

Chipkin *et al.* (1978) injected an enkephalin analog i.p. 30 minutes before testing on lever pressing in an operant chamber and found a significant disruptive effect which was not naloxone-reversible. In related studies, i.v. injections of the potent enkephalin analog FK-33-824 proved to be an effective substitute for morphine in reinforcing operant responding in morphine-dependent rhesus monkeys (Mello and Mendelson, 1978). Similar results were reported by Roemer *et al.* (1977) using a single-dose suppression test.

Finally, Olson *et al.* (1979a) examined in rhesus monkeys the effects of s.c. injections of a very potent enkephalin analog, [D-ala[2], F$_5$phe[4]]-met-enkephalin-NH$_2$, at doses of 0.0, 0.1, 1.0, 10.0, 100.0 and 1000.0 μg/kg in both a discrimination-reversal task and a delayed-response task. No reliable effects were observed for peptide dose in the delayed response task, although a significant difference existed between delays of 0, 30 and 60 seconds, as had been expected. However, discrimination reversal was facilitated in a dose-dependent fashion, with all doses but the smallest producing significant effects.

IV. CLINICAL BEHAVIOUR

If one takes into account the studies presented earlier, as well as additional work with central injections, studies involving naloxone, and work with dialysis, it is apparent that considerable potential exists for the clinical use of brain opiates. Kline *et al.* (1977) published the first work with human patients, and demonstrated that injections of β-endorphin might produce beneficial changes in depressed and schizophrenic patients who knew what substances they were receiving. Although it certainly was not experimentally rigorous, the paper served to suggest a fantastic potential for the clinical use of endorphins. Kline and Lehmann (1978) subsequently extended their work to patients in categories other than depression or schizophrenia.

Verhoeven *et al.* (1978) injected schizophrenics with (des-tyr[1])-γ-endorphin and found improvement in all six patients, although it was reported to be short-lived in half of these individuals. Verhoeven *et al.* (1979) reported additional work with the same analog. In the first study, six patients were removed from all forms of neuroleptic medication and given 1 mg of DTγE i.m. daily, for seven days. In the second study, eight patients were employed in a double-blind procedure, the same dose of DTγE was administered for eight days in conjunction with standard neuroleptic medication. The results of both studies indicated some improvement of symptoms, but the more effective was the second study, in which the psychotic symptoms had virtually disappeared by the fourth day.

Furthermore, Krebs and Roubicek (1979) injected FK-33-824, a potent analog of met-enkephalin i.m. and created a brief improvement in the symptoms of four out of six psychotic patients. Nedopil and Ruther (1979) also used the same analog, and found improvement in five out of nine patients, although three dropped out early in the study.

These results look very promising, but a word of caution is necessary. Several studies were unable to prove that nalocone has any effect on altering schizophrenic symptoms (Lipinski *et al.*, 1979), suggesting at the very least that materials other than endogenous opiates must play a significant role in the etiology. This is further substantiated by the fact that Ross *et al.* (1979) found no difference in plasma β-endorphin-like immunoreactivity between 98 schizophrenic patients and 42 normal subjects, which was in direct contradiction to earlier reports (Wagemaker and Cade, 1977) of extremely high concentrations of [leu[5]]-β-endorphin in hemodialysates from schizophrenic patients. Additional clinical studies are eagerly awaited.

V. SUMMARY AND CONCLUSIONS

The first report of reliable modulation of behaviour by systemic administration of enkephalins was by the New Orleans group in 1976. At that time they also proposed that brain opiates had dissociative effects that could operate independently, and produce narcotic effects in some situations while producing non-narcotic effects in others. Subsequent investigations have now shown behavioural effects after systemic administration of brain opiates in numerous testing paradigms, with a wide variety of subjects ranging from goldfish to humans. Dose-response curves obtained in these studies for enkephalin and endorphin often assume an inverted U-shaped function with the apogee around 100 μg/kg. These studies, in conjunction with the limited clinical studies that have been reported, suggest that behavioural modulation by systemic administration of enkephalins and endorphins is realistic, and offers great potential for future clinical use.

ACKNOWLEDGEMENTS

This work was supported in part by The Medical Research Service of the Veterans Administration and NIH (NS 07664). The authors would like to thank Mrs Mary Berzas and Ms Rachel A. Loupe for their excellent clerical assistance with the preparation of this manuscript.

REFERENCES

Belluzzi, J. and Stein, L. (1977). *Nature 266*, 556
Bloom, F., Segal, D., Ling, N. and Guillemin, R. (1976). *Science 194*, 630
Catlin, D.H., George, R. and Li, C.H. (1978). *Life Sci., 23*, 2147
Chipkin, R.E., Stewart, J.M., Morris, D.H. and Crowley, T.J. (1978). *Pharmacol. Biochem. Behav., 9*, 129
Coy, D.H., Kastin, A.J., Schally, A.V., Morin, O., Caron, N.G., Labrie, F., Walker, J.M., Fertel, R., Berntson, G.G. and Sandman, C.A. (1976). *Biochem. Biophys. Res. Commun., 73*, 632
Goldstein, A., Lowney, L.I. and Pal, B.K. (1971). *Proc. Nat. Acad. USA, 68*, 1742
Hughes, J., Smith, T.W., Kosterlitz, H.W., Fothergill, L.A., Morgan, B.A. and Morris, H.R. (1975). *Nature 258*, 577
Jacquet, Y.F. and Marks, N. (1976). *Science 194*, 632
Kastin, A.J., Nissen, C., Schally, A.V. and Coy, D.H. (1976a). *Brain Res. Bull. 1*, 583
Kastin, A.J., Scollan, E.L., King, M.G., Schally, A.V. and Coy, D.H. (1976b). *Pharmacol. Biochem. Behav., 5*, 691
Kastin, A.J., Scollan, E.L., Ehrensing, R.H., Schally, A.V. and Coy, D.H. (1978). *Pharmacol. Biochem. Behav., 9*, 515
Kastin, A.J., Olson, R.D., Schally, A.V. and Coy, D.H. (1979a). *Life Sci. 25*, 401
Kastin, A.J., Olson, R.D., Ehrensing, R.H., Berzas, M.C., Schally, A.V. and Coy, D.H. (1979b). *Pharmacol. Biochem. Behav., 11*, 721
Kenney, N.J., McKay, L.D. and Woods, S.C. (1978). *Soc. Neurosci. Abstr., 4*, 176
King, B.M., Castellanos, F.X., Kastin, A.J., Berzas, M.C., Mauk, M.D., Olson, G.A. and Olson, R.D. (1979a). *Pharmacol. Biochem. Behav., 11*, 729
King, M.G., Kastin, A.J., Olson, R.D. and Coy, D.H. (1979b). *Pharmacol. Biochem. Behav., 11*, 407
Kline, N.S. and Lehmann, H.E. (1978). *Psychopharmacol. Bull. 14*, 12
Kline, N.S., Li, C.H., Lehmann, H.E., Lajtha, A., Laski, E. and Cooper, T. (1977). *Arch. Gen. Psychiatry 34*, 1111
Kosterlitz, H.W. and Hughes, J. (1977). *Brit. J. Psychiat., 130*, 298
Krebs, E. and Roubicek, J. (1979). *Pharmakopsychiat. Neuropsychopharmakol. 12*, 86
LeMoal, M., Koob, G.F. and Bloom, F.E. (1979). *Life Sci., 24*, 1631
Lipinski, J., Meyer, R., Kornetsky, C. and Cohen, B.M. (1979). *Lancet i (8129)*, 1292
Margules, D.L., Moisset, B., Lewis, M.J., Shibuya, H. and Pert, C.B. (1978). *Science 202*, 988
Mello, N.K. and Mendelson, J.H. (1978). *Pharmacol. Biochem. Behav., 9*, 579
Nedopil, N. and Ruther, E. (1979). *Pharmakopsychiat. Neuropsychopharmakol. 12*, 277
Olson, G.A., Olson, R.D., Kastin, A.J., Castellanos, F.X., Kneale, M.T., Coy, D.H. and Wolf, R.H. (1978a). *Pharmacol. Biochem. Behav., 9*, 687
Olson, R.D., Kastin, A.J., Michell, G.F., Olson, G.A., Coy, D.H. and Montalbano, D. (1978b). *Pharmacol. Biochem. Behav., 9*, 11
Olson, R.D., Kastin, A.J., Montalbano-Smith, D., Olson, G.A., Coy, D.H. and Michell, G.F.

(1978c). *Pharmacol. Biochem. Behav., 9*, 521

Olson, G.A., Olson, R.D., Kastin, A.J., Green, M.T., Roig-Smith, R., Hill, C.W. and Coy, D.H. (1979a). *Pharmacol. Biochem. Behav., 11*, 341

Olson, R.D., Kastin, A.J., LaHoste, G.J., Olson, G.A. and Coy, D.H. (1979b). *Pharmacol. Biochem. Behav., 11*, 705

Panksepp, J., Vilberg, T., Bean, N.J., Coy, D.H. and Kastin, A.J. (1978). *Brain Res. Bull. 3*, 633

Pert, C. and Snyder, S. (1973). *Science, 179*, 1011

Plotnikoff, N.P., Kastin, A.J., Coy, D.H., Christensen, C.W., Schally, A.V. and Spirtes, M.A. (1976). *Life Sci., 19*, 1283

Rapoport, S.I., Klee, W.A., Pettigrew, K.D. and Ohno, K. (1980). *Science 207*, 84

Rigter, H. (1978). *Science 200*, 83

Rigter, H., Greven, H. and Van Riezen, H. (1977). *Neuropharmacology 16*, 545

Roemer, D., Buescher, H.H., Hill, R.C., Pless, J., Bauer, W., Cardinaux, F., Closse, A., Hauser, D. and Huguenin, R. (1977). *Nature 268*, 547

Ross, M., Berger, P.A. and Goldstein, A. (1979). *Science 205*, 1163

Sandman, C.A., McGivern, R.F., Berka, C., Walker, J.M., Coy, D.H. and Kastin, A.J. (1979). *Life Sci., 25*, 1755

Simon, E.J., Hiller, J.M. and Edelman, I. (1973). *Proc. Nat. Acad. USA, 70*, 1947

Szekely, J.I., Ronai, A.Z., Dunai-Kovacs, Z., Miglecz, E., Berzetri, I., Bajusz, S. and Graf, L. (1977). *Europ. J. Pharmacol., 43*, 293

Terenius, L. (1973). *Acta. Pharmac. Tox., 32*, 317

Terenius, L. and Wahlstrom, A. (1975a). *Life Sciences, 16*, 1759

Terenius, L. and Wahlstrom, A. (1975b). *Acta. Physiol. Scand., 94*, 74

Tseng, L.F., Loh, H.H. and Li, C.H. (1978). *Int. J. Peptide Protein Res., 12*, 173

Veith, J.L., Sandman, C.A., Walker, J.M., Coy, D.H. and Kastin, A.J. (1978). *Pharmacol. Biochem. Behav., 20*, 539

Verhoeven, W.M., van Praag, H.M., Botter, P.A., Sunier, A., van Ree, J.M., and de Wied, D. (1978). *Lancet i (8072)*, 1046

Verhoeven, W.M., van Praag, H.M., van Ree, J.M. and de Wied, D. (1979). *Arch. Gen. Psychiat., 36*, 294

Wagemaker, H. and Cade, R. (1977). *Am. J. Psychiat., 134*, 684

Weinberger, S.B., Arnsten, A. and Segal, D.S. (1979). *Life Sci., 24*, 1637

de Wied, D., Bohus, B., van Ree, J.M., Kovacs, G.L. and Greven, H.M. (1978a). *The Lancet i (8072)*, 1046

de Wied, D., Bohus, B., van Ree, J.M. and Urban, I. (1978b). *J. Pharmacol. Exp. Ther., 204*, 507

de Wied, D., Kovacs, G.L., Bohus, B., van Ree, J.M. and Greven, H.M. (1978c). *Europ. J. Pharmacol., 49*, 427

7 EFFECT OF CANNABINOIDS AND NARCOTICS ON GONADAL FUNCTIONS

Alexander Jakubovic, Edith G. McGeer and Patrick L. McGeer

TABLE OF CONTENTS

I	INTRODUCTION	153
II	CANNABINOIDS	153
	A Effects in Humans	155
	B Effects in Animals	156
	C *In vitro* Effects	158
	D Possible Mechanisms of Action	160
III	NARCOTICS	161
	A Effects in Humans	161
	B Effects in Animals	163
	C *In vitro* Effects	164
IV	SUMMARY AND CONCLUSIONS	165
	REFERENCES	167

I. INTRODUCTION

Numerous studies have been reported on the pharmacological and biological activities of cannabinoids and narcotic derivatives. Some of these compounds have been shown to have complex biological effects on various systems, including the reproductive system. It is the purpose of this chapter to present a general overview of the reported effects of cannabinoids and narcotic analgesics on gonadal function. In order to do this we must consider some central as well as peripheral effects of these materials, since a normally functioning hypothalamic-pituitary gland system is essential for optimal gonadal function (Figure 7.1). Some pharmacological agents may temporarily or permanently affect male sexual function by virtue of a direct interaction on steroidogenic and/or spermatogenic tissue and/or by an indirect central influence on gonadal function due to interference with the hypothalamic-pituitary-gonadal axis. Moreover, atrophy of reproductive-organ tissue may occur if there is either chronic interference with local cellular metabolic processes or chronic deficiency in the supply of trophic, centrally produced hormones. Since the exact mechanisms of interactions between the CNS and the hypothalamic-pituitary-gonadal axis are not yet known, it is frequently impossible to define the site or sites of action of drugs which affect gonadal function, but it is necessary to remember that central as well as peripheral sites may be involved.

There is considerable controversy as to whether cannabis should be classified as a narcotic drug although there are a number of recent reports which indicate interactions between certain cannabinoids and the opioids, as well as a wide range of similar pharmacological actions (Kaymakcalan, 1979). In this review, however, the cannabinoids will be treated separately from the opioids, such as morphine, heroin, codeine and methadone.

II. CANNABINOIDS

Cannabinoids are C_{21} compounds unique to, and present in or derived from, the plant *Cannabis sativa* and its preparations (e.g., hashish, marihuana-cannabis). The formulas of the major natural cannabinoid, Δ^9-tetrahydrocannabinol (Δ^9-THC), and of Δ^8-tetrahydrocannabinol (Δ^8-THC) are given in Figure 7.2; at least 50 different cannabinoids have so far been identified.

Only some of these cannabinoids have the euphoric effects for which cannabis is widely known. Δ^9-THC is particularly active in this regard. Some other derivatives present in cannabis, such as cannabidiol (CBD) and cannabigerol (CBG),

Figure 7.1: A highly simplified diagrammatic representation of the hormonal regulation of female and male gonadal function by hypothalamic and pituitary hormones, and of the exerted feedback action (dotted lines) by gonadal sex hormones (estrogen, progesterone and testosterone).

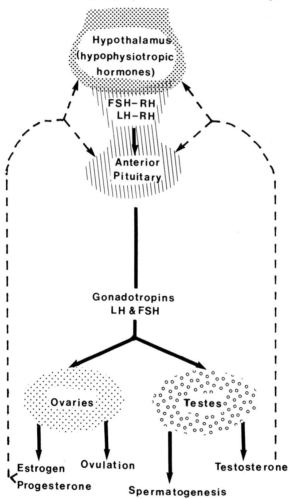

have little or no psychogenic activity. The mechanism of the euphoric effect is still unknown, but it has become clear that both psychogenically active and inactive cannabinoids may cause side effects due to interference with various metabolic processes, particularly those involving the synthesis of proteins and nucleic acids. One of the side effects, which has received considerable attention recently, is the action of cannabinoids on gonadal functions in humans as well as in experimental animals. Many of these reports are concerned with changes in

Figure 7.2: Structural formulas for some representative cannabinoids.

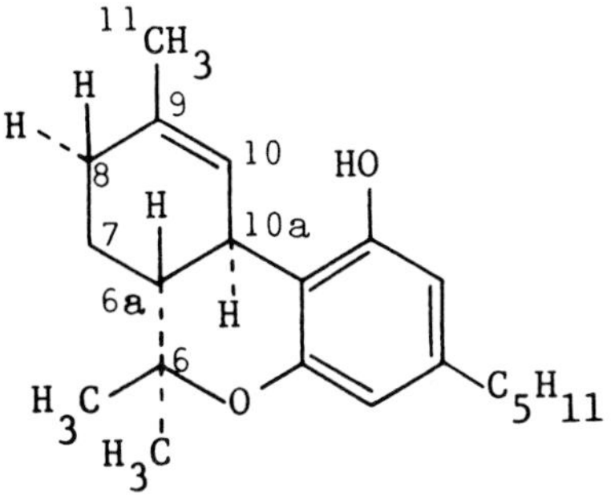

$$\Delta^9 - \text{ or } \Delta^1 \text{-THC} \qquad\qquad \Delta^8 - \text{ or } \Delta^{1(6)} \text{-THC}$$

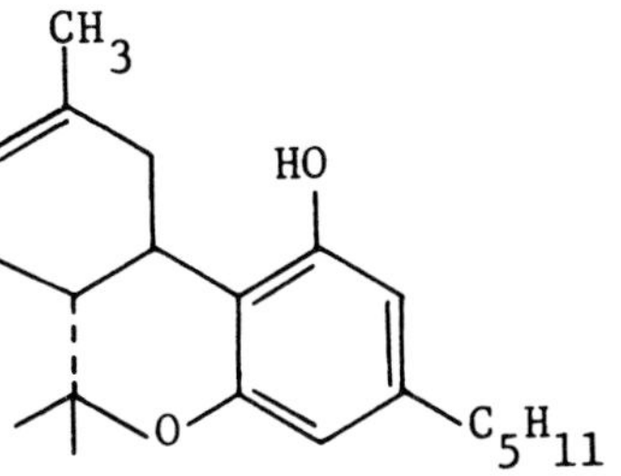

plasma levels of hormones which are important to the reproductive processes. As indicated in Figure 7.1, the pituitary hormones, LH and FSH, are important in both males and females. Other important hormones include: in males, testosterone produced by the testes; and, in females, estrogen and progesterone produced by the ovaries.

A. Effects in Humans

Although there are some negative findings, most of the available data indicate that chronic marihuana use in men causes reduced spermatogenesis and some increase in the percentage of abnormal sperm; these phenomena may or may not be accompanied by decreases in plasma hormonal levels. No data have been published yet on the effects of gonadal function in human females although there is one reference to unpublished experiments indicating that cannabinoids 'decreased luteal phase progesterone levels' in 30 women who were chronic marihuana users (Burstein *et al.*, 1979b).

A number of studies have reported a 30-44% decrease in plasma testosterone levels in young men after chronic (two to six months) use of marihuana. In addition, substantial decreases in plasma testosterone levels two to three hours after the smoking of one marihuana cigarette have been reported. In some of these studies, decreases in plasma LH levels were also reported, while in one study a high incidence (35%) of oligospermia and even impotence was found (Kolodny *et al.*, 1974, 1976; Cohen, 1976). In contrast to these observations, a number of studies have not found any significant change in testosterone or in LH plasma levels in man after chronic marihuana use for periods of up to 21 days (Hembree *et al.*, 1976; Mendelson *et al.*, 1978). Even some of the groups reporting significant decreases in plasma testosterone in marihuana users found that abstention from marihuana after chronic use produces a prompt increase in the hormonal level (Cohen, 1976; Kolodny *et al.*, 1976). Prolactin levels in human experiments were unchanged (Lemberger *et al.*, 1975; Cohen, 1976; Kolodny *et al.*, 1976), which seems to refute the suggestion that the gynecomastia reported in some cannabis users may be related to changes in prolactin secretion (Harmon and Aliapoulios, 1972).

There are reports which indicate that cannabis use may cause quantitative or qualitative changes in spermatogenesis in man, even in the absence of any change in serum hormonal levels, suggesting a possible direct action of cannabis on germinal epithelium. In these studies, four to six weeks of marihuana smoking by adult male subjects was found to cause a highly significant decrease in both the concentration of ejaculated and total sperm count; decreases in sperm motility as well as increases in the percentage of sperm with abnormal morphology were also noted (Hembree *et al.*, 1976, 1979). Frequent morphological abnormalities in sperm heads have also been found in other studies in chronic cannabis users. Acrosomal morphogenesis and synthesis of arginine-rich proteins appear to be greatly affected and there is frequently an incomplete condensation of the chromatin in sperm heads which are otherwise of normal size and shape (Issidorides, 1979).

B. Effects in Animals

Animal studies have yielded more consistent and conclusive evidence than available from human studies of alterations in gonadal function and reduction in androgen-dependent behavioural responses (aggression, copulatory behaviour) after marihuana or cannabinoid treatment.

Repeated injections of cannabis extracts to mice (Dixit *et al.*, 1974) or of Δ^9-THC to rats (Thompson *et al.*, 1973; Purohit *et al.*, 1979) have been shown to have adverse effects on the weights and morphologies of the testes and accessory sex organs. Other effects noted in some studies include increased percentage of abnormal sperm (Zimmerman *et al.*, 1979), arrest of spermatogenesis, regression of Leydig cells, significant decreases in testicular levels of RNA (Dixit *et al.*, 1974) in cytochrom P-450, and inhibition of the synthesis of testosterone by testis microsomes, but an increase in its metabolism by hepatic microsomes (Harclerode *et al.*, 1979). Since cytochrom P-450 is a marker for interstitial cells and is required for various enzyme activities which are important for testicular function, it has been suggested that the reduction in testicular cytochrom P-450 may be responsible, at least in part, for the decreased testosterone production. In addition, the defect in testosterone synthesis may be compounded by an increase in its metabolism (Harclerode *et al.*, 1979). Sharp (65%) decreases in serum levels of testosterone have also been found in male rhesus monkeys in the first 24 hours following acute doses of Δ^9-THC, with recovery during the subsequent 48 hours. The decreases in testosterone were accompanied by a 30% decrease in LH levels after six to twelve hours (R.G. Smith *et al.*, 1979).

In some experiments hypophysectomized and castrated as well as normal (sham-operated) rats have been used. Both Δ^9-THC and cannabinol significantly decreased testosterone and dihydrotestosterone levels, as well as the weights of accessory sex organs, in sham-operated rats, and blocked the stimulatory effects of hCG in hypophysectomized rats and of testosterone or dihydrotestosterone on the ventral prostate and seminal vesicles in castrated animals. These results suggested that cannabinoids may act both directly at the target organs and indirectly at the hypothalamic-hypophysial site (Purohit *et al.*, 1979).

The severity of the effects of cannabinoid treatment on male gonadal function presumably depends on the length of treatment, as well as the dosage used. Reduction of male reproductive-organ weights has been noted after long-term chronic treatment with cannabis extract or Δ^9-THC (Thompson *et al.*, 1973; Dixit *et al.*, 1974; Purohit *et al.*, 1979) but not after four days of Δ^9-THC administration (Ling *et al.*, 1973). Chronic treatment of rats with cannabis extract caused more pronounced decreases in peripheral hormonal levels than Δ^9-THC or a mixture of cannabinoids, indicating a possible important role of some other cannabis constituents (Maskarinec *et al.*, 1978). On the other hand, deterioration of sexual performance in male rats (Corcoran *et al.*, 1974) and mice (Dalterio *et al.*, 1978) has been reported even after acute exposure to cannabis or Δ^9-THC.

Only a few experiments have been done to assess the relative toxicities of various cannabinoids on testicular spermatogenesis *in vivo* (Dixit *et al.*, 1974; Zimmerman *et al.*, 1979). In mice treated with various cannabinoids for five days, an increased percentage of abnormal sperm was present and the relative toxicities were: Δ^9-THC > cannabinol > cannabidiol (Zimmerman *et al.*, 1979).

Less information exists on the effects of cannabinoids on the function of the reproductive system in female than in male animals. A very suitable model is the rhesus monkey, which has a menstrual cycle similar to that of women. In ovariectomized rhesus monkeys, various acute doses of Δ^9-THC (0.625-5 mg/kg) brought about a dose-dependent transient (six to 24 hour) diminution of LH and FSH plasma levels, which could be reversed by administration of LH-RH. This indicates that Δ^9-THC affects hypothalamic control of FSH and LH secretion by the pituitary. In such ovariectomized animals, the duration but not the extent of the decrease in LH levels was dose-dependent (Smith *et al.*, 1979a, b). In normal rhesus monkeys, however, both the levels of progesterone and the length of the luteal phase of the menstrual cycle were not affected by daily injections of Δ^9-THC during this phase. In rabbits, neither the hormonal secretion from the ovaries in response to hCG nor the number of ovulations were affected by treatment with various doses of Δ^9-THC (Asch *et al.*, 1979; Smith *et al.*, 1979b).

In similar experiments, chronic treatment with Δ^9-THC or with crude marihuana extracts resulted in a dose-related decrease in uterine and ovarian weights in rats, but with 90% recovery of the weights after 30 days abstention from treatment. This chronic treatment with cannabinoids also caused a marked reduction of hormonal stimulation of the uterus and ovary and thus prolongation of the diestrus stage of the uterine reproductive cycle (Fujimoto *et al.*, 1979).

Chronic administration of Δ^9-THC to proestrus rats (Chakravarty *et al.*, 1979) caused decreases in prolactin levels but not in FSH or LH levels in the pituitary or serum. The LH-RH content in the hypothalamus, however, was reduced in a dose-dependent manner, which suggests that Δ^9-THC affects gonadotropin secretion primarily by an action at the hypothalamic level. This is consistent with the findings in monkeys (Smith *et al.*, 1979a).

C. In vitro Effects

Spermatogenesis is a complex androgen-dependent process which is regulated indirectly by gonadotropins (LH and FSH) through their stimulation of testosterone production in the interstitial cells (Figure 7.1). Reduction of testicular steroidogenesis leads, in turn, to a decrease in spermatogenesis and in the functioning of androgen-dependent target organs. Impairment of testosterone synthesis by a drug may involve an indirect central effect and/or a direct adverse action on the biosynthetic pathways in the testes.

In order to study the possibility of direct effects on cannabinoids without any complicating *in vivo* hormonal relationships, the metabolism of rat testis was studied *in vitro*. Undisturbed metabolic activities of the seminiferous epithelium and Leydig cells are essential for the optimal functioning of both testis and androgen-dependent secondary sex organs.

Spermatogonial cells differentiate in a highly synchronous fashion into one type of differential cells, the spermatozoon. Spermatogenic cell populations, with their rapid biosynthetic processes, provide a suitable model for the study of direct drug actions.

In recent reports, the inhibitory effects of psychogenically active or inactive cannabinoids – Δ^9-THC, SP-111A (the water-soluble derivative of Δ^9-THC), 11-OH-Δ^9-THC, 8-β-OH-Δ^9-THC, cannabidiol (CBD), cannabinol (CBN) and cannabigerol (CBG) – on the biosynthesis of proteins, nucleic acids and lipids in rat-testis slices and spermatogonial-cell suspensions have been described (Jakubovic and McGeer, 1976, 1977). It was shown that the decreased protein synthesis caused by cannabinoids was general rather than specific, and was not caused by a lack of substrate in the intracellular amino-acid pool. The reduced RNA and/or DNA synthesis, however, correlated with a decreased synthesis of nucleotide precursors. Total ATP concentrations in the testicular cells also declined in the presence of cannabinoids (Jakubovic and McGeer, 1976, 1977). These results on protein and nucleic-acid synthesis are of particular interest, since it has been shown that stimulated testicular cell steroidogenesis is dependent on continuous synthesis of new RNA and proteins (Mendelson *et al.*, 1975a; Cooke *et al.*, 1979). These *in vitro* studies therefore suggested that the impaired testicular function seen in cannabinoid-treated animals and humans may depend, at least in part, on a direct action of cannabinoids on various metabolic processes in the testicular cells. This suggestion (Jakubovic and McGeer, 1976, 1977) is supported by recent reports of direct, *in vitro* inhibition by cannabinoids of testosterone production in mouse or rat testicular Leydig-cell suspensions

(Burstein *et al.*, 1979a; Jakubovic *et al.*, 1979a, b). The degree of inhibition by various cannabinoids is particularly pronounced in the latter type of preparation, where some cannabinoids, in less than micromolar concentrations, are effective in reducing both hCG (Figure 7.3) and dibutryl-cAMP stimulated testosterone synthesis (Jakubovic *et al.*, 1979a, b). These results indicate that the locus of interaction(s) is at some later stage than the cellular cAMP synthesis in the interstitial cells. In comparable *in vitro* experiments, cannabinoids have also been found to decrease protein synthesis (Table 7.1) in Leydig cells substantially. The effects of tested cannabinoids were found to be unrelated to their psychoactive effects. Their order of potency in inhibiting stimulated testosterone biosynthesis by Leydig cells *in vitro* was found to be: $8\text{-}\beta\text{-OH-}\Delta^9\text{-THC} \geqslant 11\text{-OH-}\Delta^9\text{-THC} > $ cannabinol (CBN) = cannabidiol (CBD) = cannabigerol (CBG) $> \Delta^9\text{-THC} = \Delta^8\text{-THC}$ (Jakubovic *et al.*, 1979a, b).

Figure 7.3: **Effect of cannabinoids upon testosterone production by rat Leydig cells in the presence of hCG. Results are expressed as percentages (mean ± SE) of testosterone produced in parallel controls incubated without cannabinoids. The result obtained in the absence of both hCG and cannabinoids is also indicated; p < 0.01.**

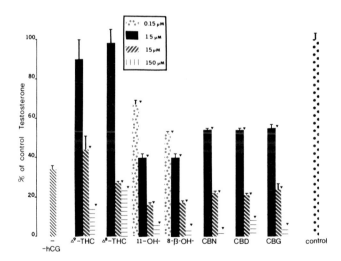

Table 7.1: Effect of Cannabinoids on L-[U-^{14}C]-Leucine, and dl-Methadone on [2-^{14}C]-Uridine Metabolism in Rat Leydig Cells in the Presence of Dibutyryl-cAMP*

A.	Percent of control dpm	
Cannabinoids 15 μM	Soluble fraction	Protein
Δ^9-THC	45 $\pm$ 5	32 $\pm$ 3
11-OH-Δ^9-THC	39 $\pm$ 5	27 $\pm$ 2
8-β-OH-Δ^9-THC	30 $\pm$ 2	8 $\pm$ 1
Cannabinol	34 $\pm$ 1	15 $\pm$ 1
B.		
dl-Methadone μM	Soluble fraction	RNA
10	74 $\pm$ 2	70 $\pm$ 3
50	67 $\pm$ 2	63 $\pm$ 5
100	47 $\pm$ 4	48 $\pm$ 4

*Cells were incubated with 1 mM dibutyryl-cAMP in the absence or presence of either cannabinoids (A) or dl-methadone (B) for 2 h at 37°C. L-[U-^{14}C]-leucine (A) or [2-^{14}C]-Uridine (B) was then added and incubation continued for 1 h. Results are expressed as percentage (mean $\pm$ SE) of d.p.m. in parallel control incubations; $p < 0.01$.

In preparations of mouse or rat Leydig cells, it has been shown that the inhibition of testosterone production is not caused by interference of Δ^9-THC with the binding of hCG to plasma membranes (Burstein *et al.*, 1979a). The cleavage of cholesterol palmitate by esterase activity from the cellular soluble fraction was significantly reduced by Δ^9-THC, but neither the uptake of cholesterol by Leydig-cell mitochondria nor the side-chain cleavage of cholesterol were affected. The inhibition of cholesterol esterase, which, in turn leads to reduced release of free precursor cholesterol from its ester storage form, may be partly responsible for the reduction of testosterone synthesis in Leydig cells treated with Δ^9-THC (Burstein *et al.*, 1979a). In similar experiments with isolated rat luteal cells, various cannabinoids inhibited both progesterone synthesis and cholesterol esterase activity (Burstein *et al.*, 1979b).

Still another possible site of action is suggested by the report that treatment of fresh bull sperm with Δ^9-THC led to motility and morphological changes as well as decreases in respiration, ATP content and glucose metabolism (Shahar and Bino, 1974).

D. Possible Mechanisms of Action

As outlined in the preceding sections, it is probable that the adverse effects of cannabinoids on gonadal function involve both central and peripheral actions — i.e., effects on the hypothalamus-pituitary system and on the gonads and accessory sex organs. The direct action at the gonadal level may depend on inhibition of essential biosynthetic processes including protein and nucleic-acid synthesis and cholesterol-esterase activity. After injection of radioactive Δ^9-THC, there was a long-lasting accumulation of relatively high amounts of cannabinoids in the gonadal fat tissue of both male and female mice as compared with other tissues examined (Rawitch *et al.*, 1979). The persistent presence of high concentrations of cannabinoids may increase the probability of a direct interference with gonadal cellular functions.

Even less is known about the possible mechanisms of action at the hypothalamic-pituitary level. Since estrogen can induce some of the effects found in humans and animals after use of cannabinoids (e.g., gynecomastia, oligospermia, atrophy of Leydig cells and reductions in testosterone levels, spermatogenesis, copulation and weights of testes and secondary sex organs), it has been suggested that cannabinoids may possess estrogenic-like activity and interact with receptor sites for these hormones on both the target organs and in the CNS. However, negative results with high doses of cannabis resin on rat uterine weight, as well as the lack of competition by Δ^9-THC for estrogen-receptor sites *in vitro*, indicate that cannabinoids are not estrogenic (Okey and Bondy, 1978; Smith *et al.*, 1979c). This attractive hypothesis therefore seems untenable, and leaves unanswered some questions concerning the complex mechanisms of action of the cannabinoids in both the gonads and the CNS.

III. NARCOTICS

Morphine and codeine (methylmorphine) are two well-known pharmacologically active alkaloids contained in opium, while heroin (diacetylmorphine), which is also a powerful narcotic analgesic, is prepared synthetically from morphine. Chemicals with pharmacological actions qualitatively identical to morphine, even though their chemical structures are quite different, include methadone and its long-acting congener 1-α-acetylmethadol (LAAM) (Figure 7.4). Both dl-methadone and LAAM have been in increasing use in maintenance programs for the treatment and rehabilitation of heroin- or morphine-dependent persons. The pharmacological activity of the dl-racemate of methadone is almost entirely related to the 1-isomer, which is about 50 times more active than the d-isomer.

A. Effects in Humans

There is evidence from a number of clinical studies indicating that the most prominent side effects in narcotic-dependent or methadone-treated persons are reduction or absence of libido and sexual potency (Martin *et al.*, 1973; Cicero *et al.*, 1975; Ling *et al.*, 1976; Kley *et al.*, 1977; Kreek, 1978; Wang *et al.*, 1978; Brambilla *et al.*, 1979). These effects are sometimes considered to be psychological in origin. There is considerable evidence, however, that these chemicals may induce physiological, pharmacological, biochemical and hormonal changes which may underlie the sexual disturbances. Almost all of the data are from

Figure 7.4: Structural formulas for some narcotic analgesics.

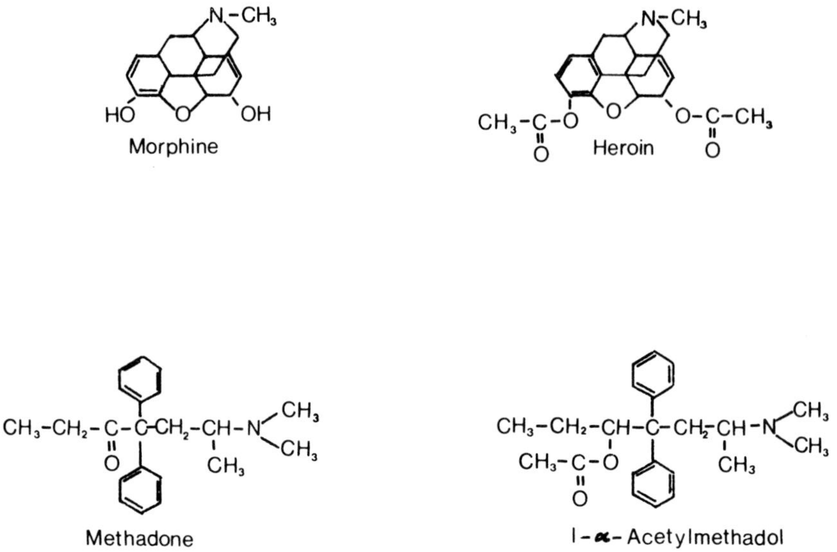

Morphine

Heroin

Methadone

l-α-Acetylmethadol

studies in males, and the most consistent finding is decreased plasma levels of testosterone. Changes have also been reported in blood levels of LH, FSH and prolactin, but these are less consistent.

There are some controversial reports on disturbances in sexual function as well as hormonal levels in addicts: much of the controversy may be related to differences in various factors such as age, dose, frequency and duration of drug use. The possibility of polydrug use also exists in some studies. More uniform results have been obtained under clinically controlled conditions.

In one recent study on 54 male heroin addicts, 92% reported decreased libido, 73% difficulties with ejaculation and 22% impotency. The same disturbances in gonadal and sexual functions were still found in some former addicts who had been abstinent for periods of 1.5-4.5 months (Wang *et al.*, 1978). Similar marked reductions in libido and sexual potency in a high percentage of individuals have been reported in other control studies on heroin addicts (Brambilla *et al.*, 1979) or persons on methadone- or LAAM-maintenance programs (Ling *et al.*, 1976; Mendelson *et al.*, 1976). Cicero *et al.* (1975) found a marked reduction in sexual activity and function, reduced secretory activity of seminal vesicles and prostate, and an increased sperm count but reduced sperm motility in subjects under methadone treatment. In heroin addicts studied by this group, however, the only significant abnormality was in sperm motility.

In all of the studies where a marked reduction in sexual activity was reported, a marked depletion was also found in plasma testosterone levels. In most such studies, a reduction in both total and free plasma testosterone levels is reported but the abnormality may be particularly evident in the free hormonal levels because of an increased capacity of the serum globulins to bind the sex hormones (Kley *et al.*, 1977; Wang *et al.*, 1978).

In some studies of persons on methadone-maintenance treatment, reductions in testosterone levels have been noted only at high doses (80-150 mg/day) and not at low doses (10-60 mg/day) (Mendelson *et al.*, 1975b). Another indication of a relation between dose and effect is the apparent existence of an inverse correlation between plasma methadone and plasma testosterone levels in adult males (Kreek, 1978).

There is still some doubt about the permanence of the effects. Martin *et al.* (1973) reported that some long-term methadone-treated subjects developed partial tolerance against the drug effects on sexual activity. In heroin addicts, Wang *et al.* (1978) found no correlation between the reduction in plasma testosterone and the duration of addiction and noted that normal levels of the hormone were quickly restored soon after withdrawal. Mendelson and Mello (1975) similarly found that the reduction of plasma testosterone was reversible in most addicts and normal levels were restored after one month's abstention. They suggest, however, that chronic heroin use at a critical age may cause longer-lasting detrimental effects.

Data on hormones other than testosterone are conflicting. Serum FSH and LH levels have been variously reported as normal (Kreek, 1978; Wang *et al.*, 1978), significantly diminished (Martin *et al.*, 1973; Brambilla *et al.*, 1979) or slightly increased (Kley *et al.*, 1977) in heroin addicts or persons on methadone maintenance. Changes in gonadal regulation and decreases in plasma testosterone levels have been observed after the withdrawal phase in male addicts following several years of addiction and methadone-maintenance treatment (Kley *et al.*, 1977). In these subjects there was no abnormality in adrenal or thyroid function or in serum LH levels but there were increased serum FSH levels, decreases in prolactin and changes in hormonal binding.

The few reports available on alterations of prolactin and estradiol levels due to narcotic analgesics are also inconsistent (Kley *et al.*, 1977; Kreek, 1978). In females addicted to heroin during their reproductive years, loss of libido, oligo- and amenorrhea, as well as infertility and spontaneous abortion have been observed (Kreek, 1978; Meites *et al.*, 1979). A few studies on females maintained on methadone indicate the occurrence of normal menstruation (Martin *et al.*, 1973; Kreek, 1978), ovulation, pregnancy and delivery, but an increased incidence of low birth weight and withdrawal syndromes in the neonates (Martin *et al.*, 1973).

B. Effects in Animals

Detrimental effects of narcotic analgesics on male gonadal functions and sexual activity have been well documented in a number of recent animal experiments.

Chronic administration of morphine (Tokunaga *et al.*, 1977) to male rats caused a markedly decreased copulation rate and fertilizing ability. In these experiments, as well as in others involving implantation of morphine pellets in rats or chronic administration of methadone to rats (Cicero *et al.,* 1976) or mice (Thomas and Dombrosky, 1975), there was atrophy of male accessory organs, with a decreased fluid content in seminal vesicles. Testicular weight has been variously reported as significantly reduced (Thomas and Dombrosky, 1975; Tokunaga *et al.,* 1977) or unchanged (Cicero *et al.,* 1976). Significant reductions in serum testosterone and LH levels were consistently found, but serum FSH levels were variously reported as increased (Tokunaga *et al.,* 1977) or unchanged (Cicero *et al.,* 1976). The effects of morphine were very rapid, and some were found as soon as six hours after pellet implantation into rats. They were reversed after seven days of withdrawal from morphine, and it was argued that the atrophy of the accessory sex organs was probably secondary to reduced gonadal steroidogenesis (Cicero *et al.,* 1976).

In similar work with various narcotics, Cicero *et al.* (1979) obtained data supporting their postulate that the neuroendocrine effects of narcotics are related to their opiate properties: according to these data, the neuroendocrine effects of various narcotics correlate with their opiate potency, are stereospecific and can be antagonized by naloxone or naltrexone.

Administration of morphine or methadone to either male or female rats prior to mating to untreated partners is reported to cause increased neonatal mortality and various detrimental effects in surviving progeny (Soyka *et al.,* 1978). It has been postulated that the presence of the narcotic in semen or in the female reproductive tract at a critical stage may be responsible for these effects. The concentration of narcotics in rat semen has not yet been measured, but in human subjects on methadone maintenance programs, methadone has been found in ejaculated semen, with a mean ratio of the semen to blood concentration of 1.8 (Gerber and Lynn, 1976).

Recent work with endorphins, peptides with opiate-like properties, has added significant new dimensions to the study of narcotic-gonadal function interactions. Both β-endorphin and the synthetic D-ala^2-met-enkephalinamide are reported to suppress completely copulatory behaviour in male rats in doses which do not affect motor activity. This inhibitory effect could be antagonized and prevented by the opiate antagonist, naloxone (Quarantotti *et al.,* 1978). Other reports in both animals and man also indicate that the opiate-like peptides appear to be intimately involved in gonadotropin secretion, as well as in the feedback regulation of neuroendocrine function by gonadal hormones. Some of these effects are probably mediated, at least in part, by specific opiate receptors in hypothalamic areas which influence the release of neurotransmitters from pathways projecting to the anterior pituitary (Blank *et al.,* 1979; Cicero *et al.,* 1979; Ieiri *et al.,* 1979; Meites *et al.,* 1979).

C. In vitro Effects

Although there is considerable evidence suggesting that most of the effects of morphine and endorphins on male sexual activity are mediated primarily by the action of these narcotics in the CNS (Martin *et al.*, 1973; Kley *et al.*, 1977; Wang *et al.*, 1978; Brambilla *et al.*, 1979; Cicero *et al.*, 1979; Meites *et al.*, 1979; Purohit *et al.*, 1979), the evidence so far presented has not ruled out a concomitant direct peripheral action of some narcotics on gonads. An important direct action of either morphine or heroin seems unlikely in view of *in vitro* studies indicating that these narcotics, even at 1 mM concentrations, have no effect on testosterone production in nonstimulated or hCG-stimulated whole rat testis (Cicero *et al.*, 1977; Jakubovic and McGeer, 1979). Under the same conditions, however, dl-methadone, its separate isomers and LAAM each significantly reduced testosterone production in either hCG- or dibutyryl-cAMP-stimulated rat testis (Jakubovic and McGeer, 1979). The effect of dl-methadone was even more marked in dibutyryl-cAMP-stimulated rat testicular Leydig-cell suspensions than in whole testis. Since these inhibitory effects on testosterone production could not be antagonized by naloxone, they presumably did not involve action at specific opiate receptors. Another possible mechanism was suggested by our data indicating that methadone and some related compounds have dose-related inhibitory effects on protein (Table 7.2), RNA and DNA biosynthesis in rat testicular-cell suspensions. Compounds active in this regard include dl-methadone, d-methadone, l-methadone, LAAM and various metabolites of methadone or LAAM. Nucleic-acid synthesis in Leydig-cell preparations was also significantly inhibited by dl-methadone (Table 7.1). The order of potency of various narcotics tested in inhibiting biosynthetic processes in the testicular-cell suspensions was: LAAM $\approx$ dl-methadone $\approx$ l-methadone $\geqslant$ d-methadone $\gg$ heroin $>$ codeine $>$ morphine (Jakubovic *et al.*, 1978; Jakubovic and McGeer, 1979). Some of the clinical findings on the severity of the side effects produced by methadone or heroin are consistent with the order of potency in inhibiting biosynthetic processes indicated by these experiments (Cicero *et al.*, 1975; Mendelson *et al.*, 1975b).

Since steroidogenesis in testicular Leydig cells is dependent on, and related to, continuous RNA and protein synthesis (Mendelson *et al.*, 1975a; Cooke *et al.*, 1979), we suggested that the possibility of combined direct and central effects on the gonadal function of some narcotics and/or their metabolites should be considered. Moreover, it seems important to note that any direct action of such materials on gonadal function may not be correlated with analgesic potency (Jakubovic and McGeer, 1979).

IV. SUMMARY AND CONCLUSIONS

The weight of evidence from clinical as well as experimental studies in humans and animals suggests that both the cannabinoids and narcotic analgesics may affect important processes involved in the regulation of sexual activity and function. Most of the available evidence is concerned with male sexual function, but the few data available from females suggest the possibility of a similar

Table 7.2: Effect of Narcotic Analgesics on L- [1-^{14}C] -Leucine Metabolism and Incorporation in Rat Testicular-cell Suspensions*

Additives	Concentration (mM)	O$_2$	^{14}CO$_2$	Percentage of control Soluble fraction (dmp/mg protein)	Protein
dl-Methadone	0.01	90 ± 5	85 ± 1**	91 ± 3	69 ± 9**
	0.05	90 ± 10	79 ± 5**	92 ± 2	46 ± 11**
	0.10	100 ± 10	70 ± 3**	112 ± 4	30 ± 3**
EMDP***	0.01	95 ± 2	91 ± 1	89 ± 2	74 ± 2**
	0.05	89 ± 8	89 ± 3	91 ± 3	53 ± 2**
	0.1	72 ± 10**	71 ± 3**	71 ± 1**	27 ± 1**
LAAM	0.01	106 ± 6	94 ± 2	103 ± 2	95 ± 2
	0.05	110 ± 11	96 ± 1	113 ± 2	81 ± 2**
	0.1	106 ± 3	85 ± 1	107 ± 3	55 ± 4**
Morphine	0.1	89 ± 3	84 ± 5	95 ± 3	92 ± 2
	0.5	92 ± 3	84 ± 3	90 ± 5	82 ± 4
	1.0	82 ± 3	86 ± 3	98 ± 4	71 ± 4**
Heroin	0.1	100 ± 4	93 ± 1	107 ± 2	84 ± 5
	0.5	100 ± 8	88 ± 2	133 ± 4**	73 ± 3**
	1.0	100 ± 4	84 ± 4	157 ± 4**	61 ± 2**

*Testicular cells were incubated in 3 ml Krebs-Ringer phosphate buffer with 10 mM glucose, under O$_2$, at 37°C for 60 minutes, with L- [1-^{14}C] -leucine as labeled substrate. Values are expressed as percentage (mean ± SE) of d.p.m. of control incubations.
**p < 0.02.
***A pharmacologically inactive methadone metabolite: 2-ethyl-5-methyl-3, 3-diphenyl-1-pyrroline.

involvement. The exact mechanisms of the interaction of these drugs with gonadal function are still unclear, but it seems probable that there are both central and peripheral sites of action. The central site of action would probably be in the hypothalamus and/or pituitary, but more likely in the hypothalamus where the drugs might alter the release of various neurotransmitters which, in turn, control the release of pituitary hormones. Such central action of the cannabinoids and narcotics may involve receptors similar or identical to those responsible for, respectively, their euphoric and analgesic properties. An additional mechanism of interference with gonadal function may involve direct action of the cannabinoids and narcotics at the gonadal level. *In vitro* experiments suggest that the activity of various cannabinoids or narcotic derivatives at this level is not related to their euphoric or analgesic potency and probably involves their inhibitory action on protein and nucleic-acid synthesis.

REFERENCES

Asch, R.H., Smith, C.G., Siler-Kohdr, T.M. and Pauerstein, C.J. (1979). *Fertil. Steril., 32*, 576

Blank, M.S., Panerai, A.E. and Friesen, H.G. (1979). *Science 203*, 1129

Brambilla, F., Resele, L., DeMaio, D. and Nobile, P. (1979). *Am. J. Psychiatry 136*, 314

Burstein, S., Hunter, S.A. and Scott Shoupe, T. (1979a). *Mol. Pharmacol., 15*, 633

Burstein, S., Hunter, S.A. and Scott Shoupe, T. (1979b). *Res. Commun. Chem. Pathol. Pharmacol., 24*, 413

Chakravarty, I., Shah, P.G., Sheth, A.R. and Glosh, J.J. (1979). *J. Reprod. Fert., 57*, 113

Cicero, T.J., Bell, R.D., Wiest, W.G., Allison, J.H., Polakoski, K. and Robins, E. (1975). *N. Engl. J. Med., 292*, 882

Cicero, T.J., Meyer, E.R., Bell, R.D. and Koch, G.A. (1976). *Endocrinology 98*, 367

Cicero, T.J., Bell, R.D., Meyer, E.R. and Schweitzer, J. (1977). *J. Pharmacol. Exp. Ther., 200*, 76

Cicero, T.J., Schainker, B.A. and Meyer E.R. (1979). *Endocrinology 104*, 1286

Cohen, S. (1976). *Ann. NY Acad. Sci., 282*, 211

Cooke, B.A., Janszen, F.H.A., van Driel, M.J.A. and van der Molen, H.J. (1979). *Mol. Cell. Endocrinol., 14*, 181

Corcoran, M.E., Amit, Z., Malsbury, C.W. and Daykin, S. (1974). *Res. Commun. Chem. Pathol. Pharmacol., 7*, 779

Dalterio, S., Bartke, A., Roberson, C., Watson, D. and Burstein, S. (1978). *Pharmacol. Biochem. Behav., 8*, 673

Dixit, V.P., Sharma, V.N. and Lohiya, N.K. (1974). *Eur. J. Pharmacol., 26*, 111

Fujimoto, G.T., Kostellow, A.B., Rosenbaum, R., Morrill, G.A. and Bloch, E. (1979). In *Marihuana: Biological Effects. Analysis, Metabolism, Cellular Responses. Reproduction and Brain*, Vol. 22/23, ed. G.G. Nahas and W.D.M. Paton (Pergamon Press, Oxford-New York) p. 441

Gerber, N. and Lynn, R.K. (1976). *Life Sci., 19*, 787

Harclerode, J., Nyquist, S.E., Nazar, B. and Lowe, D. (1979). In *Marihuana: Biological Effects. Analysis, Metabolism, Cellular Responses, Reproduction and Brain*, Vol. 22/23, ed. G.G. Nahas and W.D.M. Paton (Pergamon Press, Oxford-New York) p. 395

Harmon, J. and Aliapoulios, M.A. (1972). *N. Engl. J. Med., 287*, 936

Hembree, W.C. III, Zeidenberg, P. and Nahas, G. (1976). In *Marihuana: Chemistry, Biochemistry and Cellular Effects*, ed. G.G. Nahas (Springer-Verlag, New York) p. 521

Hembree, W.C., III, Nahas, G.G., Zeidenberg, P. and Huang, H.F.S. (1979). In *Marihuana: Biological Effects. Analysis, Metabolism, Cellular Responses, Reproduction and Brain*, Vol. 22/23 ed. G.G. Nahas and W.D.M. Paton (Pergamon Press, Oxford, New York) p. 429

Ieiri, T., Chen, H.T. and Meites, J. (1979). *Neuroendocrinology 29*, 288

Issidorides, M.R. (1979). In *Marihuana: Biological Effects. Analysis, Metabolism, Cellular Responses, Reproduction and Brain*, Vol. 22/23 ed. G.G. Nahas and W.D.M. Paton (Pergamon Press, Oxford, New York) p. 377

Jakubovic, A. and McGeer, P.L. (1976). In *Marihuana: Chemistry, Biochemistry and Cellular Effects*, ed. G.G. Nahas (Springer-Verlag, New York) p. 223

Jakubovic, A. and McGeer, P.L. (1977). *Toxicol. Appl. Pharmacol., 41*, 473

Jakubovic, A. and McGeer, E.G. (1979). *Mol. Pharmacol., 16*, 970

Jakubovic, A., McGeer, E.G. and McGeer, P.L. (1978). *Biochem. Pharmacol., 24*, 123

Jakubovic, A., McGeer, E.G. and McGeer, P.L. (1979a). *Mol. Cell. Endocrinol., 15*, 41

Jakubovic, A., McGeer, E.G. and McGeer, P.L. (1979b). In *Marihuana: Biological Effects. Analysis, Metabolism, Cellular Responses, Reproduction and Brain*, Vol. 22/23, ed. G.G. Nahas and W.D.M. Paton (Pergamon Press, Oxford, New York) p. 251

Kaymakcalan, S. (1979). In *Marihuana: Biological Effects. Analysis, Metabolism, Cellular Responses, Reproduction and Brain*, Vol. 22/23 ed. G.G. Nahas and W.D.M. Paton (Pergamon Press, Oxford, New York) p. 591

Kley, H.K., Oellerich, M., Wiegelmann, W., Herrmann, J., Rudorff, K.H., Nieschlag, E. and Krüskemper, H.L. (1977). *Horm. Metab. Res., 9*, 484

Kolodny, R.C., Masters, W.H., Kolodner, R.M. and Toro, G. (1974). *N. Engl. J. Med., 290*, 872

Kolodny, R.C., Lessin, P., Toro, G., Masters, W.H. and Cohen, S. (1976). In *The Phama-cology of Marihuana*, ed. M.C. Braude and S. Szara (Raven Press, New York) p. 217

Kreek, M.J. (1978). *Ann. NY Acad. Sci., 311*, 110

Lemberger, L., Crabtree, R., Rowe, H. and Clemens, J. (1975). *Life Sci., 16*, 1339

Ling, G.M. Thomas, J.A., Usher, D.R. and Singhal, R.L. (1973). *Int. J. Clin. Pharmacol., 7*, 1

Ling, W., Charuvastra, V.C., Kaim, S.C. and Klett, C.J. (1976). *Arch. Gen. Psychiatry 33*, 709

Martin, W.R., Jasinski, D.R., Haertzen, C.A., Kay, D.C., Jones, B.E., Mansky, P.A. and Carpenter, R.W. (1973). *Arch. Gen. Psychiatry 28*, 286

Maskarinec, M.P., Shipley, G., Novotny, M., Brown, D.J. and Forney, R.B. (1978). *Separatum Experentia 34*, 88

Meites, J., Bruni, J.F., Van Vugt, D.A. and Smith, A.F. (1979). *Life Sci., 24*, 1325

Mendelson, C., Dufau, M.L. and Catt, K.J. (1975a). *Biochem. Biophys. Acta 411*, 222

Mendelson, J.H. and Mello, N.K. (1975). *Clin. Pharmacol. Ther., 17*, 529

Mendelson, J.H., Mendelson, J.E. and Patch, V.D. (1975b). *J. Pharmacol. Exp. Ther., 192*, 211

Mendelson, J.H., Inturrisi, C.E., Renault, P. and Senay, E.C. (1976). *Clin. Pharmacol. Ther., 19*, 371

Mendelson, J.H., Ellingboe, J., Kuehnle, J. and Mello, N.K. (1978). *J. Pharmacol. Exp. Ther., 207*, 611

Okey, A.B. and Bondy, G.P. (1978). *Science 200*, 312

Purohit, V., Singh, H.H. and Ahluwalia, B.S. (1979). *Biol. Reprod., 20*, 1039

Quarantotti, B.P., Corda, M.G., Paglietti, E., Biggio, G,and Gessa, G.L. (1978). *Life Sci., 23*, 673

Rawitch, A.B., Rohrer, R. and Vandaris, R.M. (1979). *Gen Pharmacol., 10*, 525

Shahar, A. and Bino, T. (1974). *Biochem. Pharmacol., 23*, 1341

Smith, C.G., Besch, N.F., Smith, R.G. and Besch, P.K. (1979a). *Fertil. Steril., 31*, 335

Smith, C.G., Smith, M.T., Besch, N.F., Smith, R.G. and Asch, R.H. (1979b). In *Marihuana: Biological Effects. Analysis, Metabolism, Cellular Responses, Reproduction and Brain*, Vol. 22/23, ed. G.G. Nahas and W.D.M. Paton (Pergamon Press, Oxford, New York) p. 449

Smith, R.G., Besch, N.F., Besch, P.K. and Smith, C.G. (1979). *Science 204*, 325

Soyka, L.F., Joffe, J.M., Peterson, J.M. and Smith, S.M. (1978). *Pharmacol. Biochem. Behav., 9*, 405

Thomas, J.A. and Dombrosky, J.T. (1975). *Arch. Internat. Pharmacodyn. Ther., 215*, 215

Thompson, G.R., Mason, M.M., Rosenkrantz, H. and Braude, M.C. (1973). *Toxicol. Appl. Pharmacol., 25*, 373

Tokunaga, Y., Muraki, T. and Hosoya, E. (1977). *Japan. J. Pharmacol., 27*, 65

Wang, C., Chan, V. and Yeung, R.T.T. (1978). *Clin. Endocrinol., 9*, 455

Zimmerman, A.M., Bruce, W.R. and Zimmerman, S. (1979). *Pharmacology 18*, 143

8 EFFECT OF NEUROLEPTICS ON PITUITARY FUNCTION IN MAN

Samarthji Lal and Ram B. Rastogi

TABLE OF CONTENTS

I	INTRODUCTION	171
II	HYPOTHALAMIC-PITUITARY AXIS	171
III	NEUROTRANSMITTER REGULATION OF PITUITARY SECRETION	173
IV	EFFECT OF NEUROLEPTICS ON NEUROTRANSMITTER FUNCTION	173
V	EFFECT OF NEUROLEPTICS ON PROLACTIN SECRETION	174
VI	EFFECT OF NEUROLEPTICS ON GROWTH-HORMONE SECRETION	176
VII	EFFECT OF NEUROLEPTICS ON LUTEINIZING-HORMONE AND FOLLICLE-STIMULATING-HORMONE SECRETION	178
VIII	EFFECT OF NEUROLEPTICS ON THYROID-STIMULATING-HORMONE AND ADRENOCORTICOTROPIC-HORMONE SECRETION	179
IX	EFFECT OF NEUROLEPTICS ON POSTERIOR PITUITARY SECRETION	179
X	SUMMARY AND CONCLUSIONS	179
	REFERENCES	180

INTRODUCTION

Soon after the introduction of neuroleptics (major tranquillizers; antischizo-phrenic agents) into the treatment of psychiatric disorders there appeared reports of side effects consisting of galactorrhea (Gäde and Heinrich, 1955; Winnik and Tennenbaum, 1955) and menstrual disorders (Polishuk and Kulcsar, 1956). It was natural, then, that interest should be stimulated in the study of the effects of neuroleptics on endocrine function. One of the early monographs on this topic in animals and man appeared in 1967 (de Wied, 1967). Since that time it has become established that neuroleptics block dopamine receptors and in this way increase prolactin (PRL) secretion. The effect of neuroleptics on other pituitary hormones is less clearly defined, and the precise way in which they induce various endocrine pathologies remains unclear.

In recent years attention has focused on the experimental use of neuroleptics to elucidate the role of dopaminergic mechanisms in pituitary hormone secretion (Lal et al., 1977). Also, the PRL-elevating property of neuroleptics has been used to assess dopamine-receptor function in psychiatric (Gruen et al., 1978) and neurological disorders and to evaluate the mode of action of psychotropic drugs in man (Lal and Nair, 1979).

It is evident that considerable species differences exist in neurotransmitter control of pituitary secretion (Weiner and Ganong, 1978). The present chapter provides a review of the effect of neuroleptics on pituitary hormone secretion. Much of the background information in Sections II and IV is based on animal data, but the remaining sections are devoted exclusively to findings in man.

II. HYPOTHALAMIC-PITUITARY AXIS

The hypothalamus contains several discrete neuronal groups (nuclei) which contain neurosecretory cells. These cells synthesize hypothalamic hormones. Part of the hypothalamus, the median eminence, lies outside the blood-brain barrier where it is in close proximity to capillaries of the pituitary portal vasculature. No barrier exists between the portal blood vessels and the substance of the median eminence.

The pituitary, which lies outside the blood-brain barrier, consists of the neurohypophysis (posterior lobe) and the adenohypophysis. The posterior lobe is derived from an outgrowth of the developing brain and maintains a direct neural connection with the hypothalamus. Cell bodies in the supraoptic and paraventricular nuclei synthesize oxytocin (OX) and antidiuretic hormone (ADH) which are transported down the axons to the posterior lobe, where they are stored in nerve endings until released into the general circulation. ADH is also contained in axon terminals on the portal capillary bed in the median eminence, and is secreted into the portal blood.

The adenohypophysis is composed of the pars distalis (anterior lobe) and the pars tuberalis, which overlies the median eminence. In man, there is no distinct intermediate lobe. Growth hormone (GH), PRL, thyroid-stimulating hormone (TSH), adrenocorticotropic hormone (ACTH), luteinizing hormone (LH) and follicle-stimulating hormone (FSH) are synthesized, stored and secreted by specific cell populations of the anterior lobe. The major functional connection between the anterior lobe and the hypothalamus is the portal pituitary system. Within this vascular system, blood flows from the hypothalamus to the anterior pituitary as well as in the reverse direction (Oliver *et al.*, 1977). There are few nerve fibres in the anterior lobe, and none are involved in the control of anterior pituitary secretion.

The existence of several hypothalamic factors which have an inhibitory or releasing effect on pituitary hormones has been proposed, but to date only three have been identified: thyrotropin-releasing hormone (TRH), luteinizing-hormone-releasing hormone (LHRH) and growth-hormone-release-inhibiting hormone (somatostatin). These peptides are released into the portal system and regulate anterior pituitary function by changing the activity of a membrane-bound adenylate cyclase in the pituitary (Labrie *et al.*, 1976).

The hypothalamus contains several neurotransmitter substances: dopamine, noradrenaline, serotonin, histamine, adrenaline, acetylcholine and γ-aminobutyric acid (Brownstein, 1977). These substances modulate hypothalamic peptidergic function, or are themselves released into the portal system to act on pituitary receptors and hence hormone release. It is now evident that dopamine is itself the major physiological prolactin-inhibitory factor (Gibbs and Neill, 1978).

Neurotransmitter-sensitive adenylate cyclases have been identified in the hypothalamus (Sato *et al.*, 1974; Ahn and Makman, 1977), as well as in the anterior and posterior pituitary (Ahn *et al.*, 1979a, b), of various animal species. Dopamine-receptor sites have been characterized on pituitary-membrane fractions from several species including man (Goldsmith *et al.*, 1979). Such binding sites have been noted in both anterior and posterior pituitary, as well as in the median eminence of animals (Cronin *et al.*, 1978). Recently, it has been shown that the β-rotameric form is the preferred conformation for activation of the dopamine receptor in rat anterior pituitary that regulates PRL secretion (Rick *et al.*, 1979). In bovine hypothalamus, norepinephrine binding sites have also been identified (Deupree, 1977).

In addition to biogenic amines and hypothalamic factors, prostaglandins, opioid and other peptides may influence pituitary secretion in animals (Morley *et al.*, 1979). Pituitary hormones themselves, as well as target-organ hormones, may also affect neuroendocrine function at the level of the hypothalamus and/or pituitary either directly or indirectly by altering catecholamine mechanisms. Finally, extrahypothalamic brain sites may also influence hypothalamic control of pituitary function. Thus, the potential sites of action of drugs, both ana-tomical and biochemical, direct and indirect, are many.

III. NEUROTRANSMITTER REGULATION OF PITUITARY SECRETION

Most studies on neurotransmitter regulation of pituitary secretion have focused on GH and PRL secretion. Much less information is available for other pituitary hormones. Different physiological and experimental stimuli of pituitary secretion may be subserved by completely different regulatory mechanisms, and this may explain some of the discrepant conclusions in the literature.

GH secretion is modulated by stimulatory α-adrenergic (Imura *et al.*, 1971; Lal *et al.*, 1975; Lancranjan and Marbach, 1977), stimulatory dopaminergic (Lal *et al.*, 1972, 1973, 1977) and inhibitory β-adrenergic mechanisms (Imura *et al.*, 1968, 1971; Camanni *et al.*, 1974). The role of a stimulatory serotonergic mechanism (Nakai *et al.*, 1974) remains controversial. Recent evidence points to a cholinergic role in sleep-induced GH secretion (Mendelson *et al.*, 1979).

Predominant regulation of PRL secretion in man, as in other species, is via a tonic inhibitory dopaminergic mechanism (Martin *et al.*, 1974; Langer *et al.*, 1979). Inhibitory H_2-histaminergic (Carlson and Ippoliti, 1977), inhibitory cholinergic (Lal *et al.*, 1979) and stimulatory serotonergic (Wirtz-Justice *et al.*, 1976) control systems have also been proposed but are less well established.

Some evidence points to an inhibitory dopaminergic mechanism on basal LH secretion (Leblanc *et al.*, 1976; Lachelin *et al.*, 1977) but a stimulatory dopaminergic and noradrenergic mechanism controlling the mid-cycle LH surge (Leppäluoto *et al.*, 1976). In contrast, catecholamine systems play little role in FSH secretion. Serotonin exerts a stimulatory role on ACTH secretion (Imura *et al.*, 1973; Cavagnini *et al.*, 1975; Plonk and Feldman, 1976); an inhibitory α-adrenergic system has also been described (Lancranjan *et al.*, 1979). There is little evidence for a physiological role for catecholamines in TSH regulation, though some authors have postulated an inhibitory role for dopamine (Besses *et al.*, 1975; Delitala *et al.*, 1979).

Little is known about neurotransmitter regulation of ADH or OX. Some data are compatible with an inhibitory noradrenergic (Shimamoto and Miyahara, 1976) and a stimulatory nicotinic cholinergic modulation of ADH secretion (Husain *et al.*, 1975).

IV. EFFECT OF NEUROLEPTICS ON NEUROTRANSMITTER FUNCTION

A characteristic of neuroleptics, including clozapine, is their ability to block dopamine receptors. There is evidence for at least two different types of dopamine receptors: those linked to adenylate cyclase and those that are not linked to adenylate cyclase (Kebabian, 1978). Clinically effective antischizophrenic agents block the stereospecific binding of ^{3}H-haloperidol to dopamine receptors at concentrations which correlate directly with their clinical potency (Seeman *et al.*, 1976). A similar correlation also exists with respect to inhibition of dopamine-sensitive adenylate cyclase but exceptions exist particularly in the case of the butyrophenones, which are much more clinically potent than would be indicated by their inhibition of dopamine-sensitive adenylate cyclase (Iversen, 1975). Also, benzamide derivatives, such as sulpiride (a weak neuroleptic) and metoclopramide (a predicted antischizophrenic agent; Stanley and Wilk, 1979) have no effect on dopamine-sensitive adenylate cyclase (Kebabian, 1978).

Neuroleptics differ in their ability to block norepinephrine receptors. Whereas chlorpromazine blocks dopamine and noradrenaline receptors equally, haloperidol has a greater effect on dopamine receptors, and pimozide and fluspirilene affect only dopamine receptors (Andén *et al.*, 1970). The selectivity of pimozide has also been demonstrated in human cerebral-cortex slices (Tsang and Lal, 1977). However, selectivity of action is a dose-dependent phenomenon, so that beyond certain dose ranges selectivity may be lost.

Neuroleptics show wide differences in intrinsic anticholinergic activity. Clozapine and thioridazine are much more potent antimuscarinic than anti-dopaminergic drugs, whereas butyrophenones have only weak anticholinergic properties (Iversen, 1975).

In addition, neuroleptics block serotonin receptors (Horn, 1975) and alter the turnover of both serotonin (Shanack and Hornykiewicz, 1980; Rastogi *et al.*, unpublished observation) and GABA (Mao *et al.*, 1978). Thus, neuroleptics have multiple biochemical sites of action, which must be taken into account in the interpretation of studies on hormone secretion. Pharmacokinetic characteristics of drugs, their ability to cross the blood-brain barrier and to concentrate preferentially in different brain regions must also be considered in data interpretation. Normal fluctuations in serum hormonal concentrations make conclusions on drug action based on single or infrequent sampling difficult to accept uncritically.

V. EFFECT OF NEUROLEPTICS ON PROLACTIN SECRETION

Neuroleptics of diverse clinical structure including rauwolfia alkaloids (Lee *et al.*, 1976), phenothiazines, thioxanthenes, butyrophenones, diphenylbutylpiperidines, dihydroindolones, dibenzoxapines (Gruen *et al.*, 1978) and benzamide derivatives (McCallum *et al.*, 1976; L'Hermite *et al.*, 1978a) increase serum PRL in man. Several classes of neuroleptics also augment PRL in cerebrospinal fluid (Sedvall, 1979) and, in the case of benzamide derivatives, have been shown to enhance the already elevated levels of PRL in each trimester of pregnancy (Guitelman *et al.*, 1978). With the exception of reserpine, which depletes dopamine, these drugs increase PRL secretion by blocking dopamine receptors. Phenothiazines such as promazine and promethazine which have little or no neuroleptic activity neither increase PRL secretion nor block dopamine receptors (Gruen *et al.*, 1978). The thioxanthene, cis-flupenthixol, is an effective antischizophrenic agent, blocks dopamine receptors and increases PRL secretion in man, whereas the trans-isomer is devoid of these properties (Crow *et al.*, 1978).

Antagonism exists between the dopaminergic agents L-DOPA, CB 154, lergotrile, lisuride and ET 495, which decrease PRL secretion, and neuroleptics (Lal and Nair, 1980). Neuroleptic-induced PRL secretion is antagonized by an infusion of dopamine (Langer *et al.*, 1979), which suggests a pituitary site of action for neuroleptics. In stalk-sectioned patients, in whom the inhibitory influence of the hypothalamus on PRL secretion is removed, PRL secretion is increased by TRH, which acts on the pituitary, but not chlorpromazine (Lister *et al.*, 1974; Woolf, *et al.*, 1974). The latter data also indicate a hypothalamic site of action for neuroleptics.

For several neuroleptics, clinical antipsychotic potency correlates with potency in increasing PRL secretion (Gruen *et al.*, 1978). This has led to the suggestion that the capacity of a drug to increase PRL in man may be a good screening technique for potential antischizophrenic agents (Gruen *et al.*, 1978). It has also been postulated that the antischizophrenic effect of neuroleptics may be mediated by PRL itself (Horrobin, 1977). However, the activity of certain drugs points to limitations to the former supposition and questions the latter. Thus, the benzamide derivatives, sulpiride and metoclopramide, are potent stimulators of PRL secretion but weak neuroleptics. Clozapine, a dibenzodiazepine which is a clinically effective antischizophrenic agent, has little or no effect on PRL secretion (Meltzer *et al.*, 1979; Nair *et al.*, 1979) at doses which inhibit apomorphine-induced growth-hormone secretion (Nair *et al.*, 1979). It is unlikely that the anticholinergic properties of clozapine account for its surprisingly weak effect on PRL secretion. Thus, the anticholinergic agent, benztropine, enhances PRL secretion induced by haloperidol (Lal *et al.*, 1979), a butyrophenone with weak intrinsic anticholinergic properties. The differential effect of clozapine on GH and PRL secretion and the failure of clozapine to induce parkinsonism or galactorrhea may indicate that there are differences in dopamine receptors or dopaminergic mechanisms which modulate PRL secretion and extrapyramidal function on the one hand and GH secretion and antischizophrenic activity on the other (Lal and Nair, 1980).

The potent ability of clozapine to stimulate PRL secretion in the rat (Meltzer *et al.*, 1975), in contrast to its effect in man, points to important species differences.

Elevated serum PRL concentrations are maintained with continued neuroleptic treatment (Lal and Nair, 1980) and return to normal within a few days of neuroleptic discontinuation (Meltzer and Fang, 1976). The magnitude of the increase in the clinical-treatment setting is affected by many variables, which include gender, menopausal status, co-administration of antiparkinsonian anti-cholinergic drugs and chronicity of treatment in premenopausal women (DeRivera *et al.*, 1975). The effect of gender and menopausal status is in keeping with the observation that estrogens enhance neuroleptic-induced PRL secretion (Buckman *et al.*, 1976). In patients treated beyond six months, basal concentrations in men do not exceed 50 ng/ml and in women 100 ng/ml (Lal and Nair, 1980). PRL concentrations may also be within the normal range, but decrease when neuroleptics are discontinued (Lal and Nair, 1980).

The capacity of neuroleptics, including metoclopramide, to maintain elevated PRL concentrations with continued use is believed to play a key role in the development of menstrual irregularities, amenorrhea, galactorrhea (Beumont *et al.*, 1974a; de Wied, 1967; Aono *et al.*, 1978), gynecomastia, decreased libido and impotence (Shader and DiMascio, 1970; Falaschi *et al.*, 1978). Because of differing criteria for the definition of galactorrhea, the fact that many normal women can express a milk-like fluid from the breast and the many clinical variables that surround neuroleptic use, the reported incidence of galactorrhea varies from 5% to 57% (de Wied 1967; Beumont *et al.*, 1974a). A high proportion of psychiatric patients, particularly schizophrenics and manic-depressives, have menstrual irregularities and sexual dysfunction associated with the mental illness itself, so that this must be taken into consideration in evaluating the frequency of neuroleptic-induced endocrine dysfunction.

Elevated serum PRL plays a role in mammary cancer in mice. This has raised the question as to whether chronic neuroleptic treatment may predispose patients to breast cancer. However, epidemiological studies have failed to establish an increased risk (Ettigi *et al.*, 1973; Kodlin and McCarthy, 1978; Schyve *et al.*, 1978).

In unmedicated schizophrenics, the PRL response to chlorpromazine is similar to controls (Gruen *et al.*, 1978). It has been pointed out that chronic neuroleptic treatment increases dopamine-receptor sensitivity. However, in chronic schizophrenics (with and without tardive dyskinesia) who are withdrawn from chronic neuroleptic therapy for two to 15 weeks, apomorphine has no effect on basal PRL concentrations (Ettigi *et al.*, 1976). Also, in patients with tardive dyskinesia who are withdrawn from neuroleptics, the PRL response to haloperidol remains unchanged (Asnis *et al.*, 1979). In contrast, the PRL response to TRH is enhanced (Brambilla *et al.*, 1976). These data suggest that hypothalamic-pituitary dopamine receptors are not rendered supersensitive by chronic neuroleptic therapy, and their sensitivity may in fact be decreased.

Study of the PRL response to a single injection of haloperidol provides a means of illuminating the effect of psychoactive drugs on at least one dopaminergic system. In such investigations neither lithium (Lal *et al.*, 1978) nor the GABA-ergic agents, baclofen and sodium valproate (Nair *et al.*, 1980), altered the PRL response to haloperidol. These results suggest that none of the agents alter dopamine-receptor function, at least in the hypothalamic-pituitary axis. Whether one can generalize the findings to other dopamine systems in brain is unclear.

VI. EFFECT OF NEUROLEPTICS ON GROWTH-HORMONE SECRETION

The effect of neuroleptics on GH secretion appears to depend on the nature of the physiological or experimental stimulus on GH secretion (Lal and Nair, 1980). Decreases in basal GH secretion have been described with some neuroleptics but not others. Unfortunately, in most reports conclusions are not easy to accept, because normal basal GH concentrations are low to begin with and often approach the lower level of assay sensitivity (Lal and Nair, 1980). The early studies which suggested a suppressive effect of neuroleptics on GH secretion led to the use of such agents in the treatment of disorders of GH secretion. Kolodny *et al.* (1971) found that chlorpromazine decreased GH levels in a patient with acromegaly and Costin *et al.* (1973) reported normalization of GH responses in a patient with hypothalamic gigantism following chlorpromazine therapy; however, this patient was also receiving radiation treatment. Schaison *et al.* (1974) noted a greater suppression of GH levels in acromegalics following a glucose load while on sulpiride. However, subsequent studies have failed to confirm a therapeutic role for either chlorpromazine (AvRuskin *et al.*, 1973; Dimond *et al.*, 1973; Alford *et al.*, 1974) or sulpiride (Berthezene *et al.*, 1975) or to demonstrate changes in GH secretory patterns following a variety of stimuli in such patients.

Sleep-induced GH secretion is unaffected by acute (Takahashi *et al.*, 1968; Clarenbach *et al.*, 1978) or chronic neuroleptic treatment (Syvälahti and Pekkarinen, 1977). Pimozide decreases exercise-induced GH secretion but enhances episodic GH release (Schwinn *et al.*, 1976). Some data suggest that the rebound increase in GH secretion that follows glucose suppression is decreased by chlorpromazine (Benjamin *et al.*, 1969). Reserpine decreases the GH response to insulin-hypoglycemia (Cavagnini and Peracchi, 1971). Of 13 studies which investigated the effect of other neuroleptics on insulin-hypoglycemia-induced GH secretion, five reported a blunting effect on the GH response and in the remainder no decrease was observed (Lal and Nair, 1980). Pimozide antagonizes GH secretion induced by arginine (Schwinn *et al.*, 1976) and diazepam (Koulu *et al.*, 1979), but has no effect on glucagon-induced GH secretion (Masala *et al.*, 1978). The arginine response is unaffected by reserpine (Cavagnini and Peracchi, 1971). Neuroleptics antagonize GH secretion stimulated by dopamine infusion (Langer *et al.*, 1979), L-DOPA (Mims *et al.*, 1975; Mori *et al.*, 1977), CB 154 (Dammacco *et al.*, 1976) and methylphenidate (Janowsky *et al.*, 1978). Also, the increase in GH secretion induced by apomorphine is antagonized by chlorpromazine (Lal *et al.*, 1973), pimozide (Lal *et al.*, 1977) and clozapine (Nair *et al.*, 1979). The antagonism of neuroleptics towards GH secretion that follows dopamine infusion indicates that neuroleptics, in part at least, antagonize GH secretion at a site outside the blood-brain barrier.

In chronic schizophrenic patients (with and without tardive dyskinesia) withdrawn from chronic neuroleptic therapy (five years or more) for two to 15 weeks, the GH response to apomorphine is diminished compared with controls (Ettigi *et al.*, 1976). Ettigi *et al.* (1976) found that the diminished GH response was more marked in patients with tardive dyskinesia than in those without tardive dyskinesia. The peak GH response correlated inversely with the duration of neuroleptic therapy but not with the duration of psychosis. These data point to a diminished dopamine-receptor sensitivity, which is probably a sequel of neuroleptic therapy rather than a consequence of the schizophrenic psychosis. Recently, Brambilla *et al.* (1979) have shown that, following neuroleptic withdrawal, two groups of patients can be identified: those who show no change in GH response and those who show a hyperresponsivity to L-DOPA as a result of prior neuroleptic treatment. The latter were characterized by the absence of CRAG-HLA-1-type antigens. Adverse sequelae arising from neuroleptic effects on GH secretion have not been observed.

VII. EFFECT OF NEUROLEPTICS ON LUTEINIZING-HORMONE AND FOLLICLE-STIMULATING-HORMONE SECRETION

The effect of neuroleptics on LH secretion appears dependent on the physiological state of the subject and, possibly, the class of neuroleptic.

A variety of benzamide derivatives has been shown to be without effect on basal LH and FSH secretion (Falaschi *et al.*, 1978; L'Hermite *et al.*, 1978a, 1979; Tamagna *et al.*, 1979). However, in postmenopausal women a decrease in both hormones follows sulpiride administration (Mancini *et al.*, 1975). Using single samples, Collu *et al.* (1975) reported a decrease in LH with pimozide in men. Women on a variety of neuroleptics but with regular menses showed no change in basal LH concentration (Beumont *et al.*, 1974a).

Chlorpromazine has no effect on pulsatile LH secretion (Santen and Bardin, 1973) but, in women with sulpiride-induced amenorrhea-galactorrhea, the pulsatile fluctuations disappear (Maneschi and Martorana, 1978).

The gonadotropin response to LHRH is unaffected by sulpiride (L'Hermite *et al.*, 1978b), metoclopramide (Falaschi *et al.*, 1978) or pimozide (Leppäluoto *et al.*, 1976). In women with galactorrhea and/or menstrual disorders induced by sulpiride or metoclopramide, an exaggerated LH response to LHRH has been noted (Aono *et al.*, 1978). In patients given a variety of long-acting neuroleptics, a normal but delayed LH response to LHRH has been described (Huws and Groom, 1977). Neither phenothiazines nor haloperidol affect the LH or FSH release following clomiphene citrate (Brambilla *et al.*, 1977). Sulpiride is reported to block the LH peak induced by estrogens (L'Hermite *et al.*, 1978b), whereas a variety of other neuroleptics was found to have no such effect (Weiss *et al.*, 1977).

In amenorrheic subjects on neuroleptics, the mid-cycle peak of LH is absent. Following discontinuation of neuroleptics, the LH peak returns, as does menstruation (Beumont *et al.*, 1974a). Leppäluoto *et al.* (1976) found that pimozide blocks the mid-cycle LH but not the FSH surge. The rise in basal temperature associated with ovulation was unaffected by pimozide, which suggested that ovulation was not blocked.

VIII. EFFECT OF NEUROLEPTICS ON THYROID-STIMULATING-HORMONE AND ADRENOCORTICOTROPIC-HORMONE SECRETION

Few data are available on the effect of neuroleptics in TSH or ACTH secretion. A single dose of metoclopramide or sulpiride increases TSH secretion (Healy and Burger, 1977; Massara *et al.*, 1978), whereas chronic administration of metoclopramide has no effect (Tamagna *et al.*, 1979). A decrease in this hormone was noted following short-term pimozide administration using single samples (Collu *et al.*, 1975). Chronic administration of chlorpromazine or thioridazine inhibits the TSH response to TRH, whereas pimozide (Lamberg *et al.*, 1977; Linnoila *et al.*, 1978), sulpiride or metoclopramide have no effect (Aono *et al.*, 1978).

The circadian rhythm of corticosteroid secretion is unchanged by chlorpromazine (Saldanha *et al.*, 1972). Little (Saldanha *et al.*, 1972) or no change (Beumont *et al.*, 1974b) in cortisol response to insulin-hypoglycemia is observed following neuroleptic treatment. Chronic administration of metoclopramide has no effect on morning cortisol concentrations (Falaschi *et al.*, 1978).

IX. EFFECT OF NEUROLEPTICS ON POSTERIOR PITUITARY SECRETION

Little is known about the effect of neuroleptics on posterior pituitary secretion. Data for OX are lacking. Evidence gathered by de Wied (1967) pointing to a diuretic effect of chlorpromazine in man is compatible with an inhibition of ADH release. However, Shah *et al.* (1973), who measured plasma ADH in patients with schizophrenia and anxiety neurosis before and during chlorpromazine therapy using a bioassay (ethanol-hydrated rat), found that chlorpromazine increased ADH secretion. Also, induction of the syndrome of inappropriate ADH secretion has been described in isolated case reports following thiothixine (Ajlouni *et al.*, 1974), fluphenazine (DeRivera *et al.*, 1975), haloperidol (Matuk and Kalyanaraman, 1977) and thioridazine (Matuk and Kalyanaraman, 1977; Vincent and Emery, 1978), which also suggest a stimulatory effect on ADH secretion. In contrast, Kendler *et al.* (1978), in normal volunteers, found no effect of a single injection of haloperidol (1 mg) on plasma ADH levels using a radioimmunoassay over a relatively short sampling period (80 minutes). The dose of haloperidol used was sufficient to increase PRL concentration. The discrepancy with the findings of Shah *et al.* (1973) may indicate that it is the antimuscarinic effect of chlorpromazine that affects ADH secretion rather than its dopamine- and noradrenergic-receptor blocking properties.

X. SUMMARY AND CONCLUSIONS

Neuroleptics have many biochemical and anatomical sites of action. Differences in pharmacological spectrum exist between different classes of neuroleptics, though one common property is their ability to block dopamine receptors. Neuroleptics induce a variety of endocrine disorders. All except clozapine increase prolactin secretion, and this increase is sustained with chronic administration, though the magnitude is affected by several clinical variables. Though hyperprolactinemia has been implicated as a factor in mammary cancer in animals, no increased risk in neuroleptic-treated subjects has been substantiated. Neuroleptics, including clozapine, antagonize dopaminergic-mediated GH secretion. The effect on other stimuli of GH secretion is dependent on the nature of the physiological or experimental stimulus. Dopamine-receptor sensitivity in the hypothalamic-pituitary axis is not increased by chronic neuroleptic treatment. The effect of neuroleptics on LH secretion appears dependent on the physiological state of the subject and, possibly, on the class of neuroleptic. Neuroleptics block the midcycle surge of LH but not FSH. Few data are available on the effect of neuroleptics on TSH, ACTH or posterior pituitary secretion.

REFERENCES

Ahn, H.S. and Makman, M.H. (1977). *Brain Res., 138,* 125

Ahn, H.S., Gardner, E. and Makman, M.H. (1979a). *Europ. J. Pharmacol., 53,* 313

Ahn, H.S., Feldman, S.C. and Makman, M.H. (1979b). *Brain Res., 166,* 422

Ajlouni, K., Kern, M.W., Tures, J.F., Theil, G.B. and Hagen, T.C. (1974). *Arch. Int. Med., 134,* 1103

Alford, F.P., Baker, H.W.G., Burger, H.G., Cameron, D.P. and Keogh, E.J. (1974). *J. Clin. Endocr. Metab., 38,* 309

Andén, N.-E., Butcher, S.G., Corrodi, H., Fuxe, K. and Ungerstedt, U. (1970). *Europ. J. Pharmacol., 11,* 303

Aono, T., Shiogi, T., Kinugasa, T., Onishi, T. and Kurachi, K. (1978). *J. Clin. Endocr. Metab., 47,* 675

Asnis, G.M., Sachar, E.J., Langer, G., Halpern, F.S. and Fink, M. (1979). *Psychopharmacology 66,* 247

AvRuskin, T.W., Sau, K., Tang, S., Juan, C. (1973). *J. Clin. Endocr. Metab., 37,* 380

Benjamin, F., Casper, D.J. and Kolodny, H.H. (1969). *Obstet. Gynec., 34,* 34

Berthezene, F., Hugues, B. and Mornex, R. (1975). *Lyon Med., 13,* 57

Beumont, P.J.V., Gelder, M.G., Friesen, H.G., Harris, G.W., Mackinnon, P.C.B., Mandelbrote, B.M. and Wiles, D.H. (1974a). *Brit. J. Psychiat., 124,* 413

Beumont, P.J.V., Corker, C.S., Friesen, H.G., Kolakowska, T., Mandelbrote, M., Marshall, J., Murray, M.A.F. and Wiles, D.H. (1974b). *Brit. J. Psychiat., 124,* 420

Besses, G.S., Burrow, G.N., Spaulding, S.W., Donabedian, R.K. and Pechinski, T. (1975). *J. Clin. Endocr. Metab., 41,* 985

Brambilla, F., Guastalla, A., Guerrini, A., Rovere, C., Legnani, G., Sarno, M. and Riggi, F. (1976). *Acta Psychiat. Scand., 54,* 275

Brambilla, F., Rovere, C., Guastalla, A., Guerrini, A., Riggi, F. and Burbati, G. (1977). *Acta Psychiat. Scand., 66,* 399

Brambilla, F., Bellodi, L., Negri, F., Smeraldi, E. and Malagoli, G. (1979). *Psychoneuroendocrinology 4,* 329

Brownstein, M. (1977). *Fed. Proc., 36,* 1960

Buckman, M.T., Peake, G.T. and Srivastava, L.S. (1976). *J. Clin. Endocr. Metab., 43,* 901

Camanni, F., Massara, F. and Molinatti, G.M. (1974). *Biomedicine 21,* 241

Carlson, H.E. and Ippoliti, A.F. (1977). *J. Clin. Endocr. Metab., 45,* 367

Cavagnini, F. and Peracchi, M. (1971). *J. Endocrinol., 51,* 651

Cavagnini, F., Panerai, A.E., Valentini, F., Bulgheroni, P., Peracchi, M. and Pinto, M. (1975).

J. Clin. Endocr. Metab., 41, 143

Clarenbach, P., Prunkl, R., Riegler, M. and Cramer, H. (1978). *Neuroscience 3,* 345

Collu, R., Jéquier, J.-C., Leboeuf, G., Letarte, J. and Ducharme, J.R. (1975). *J. Clin. Endocr. Metab., 41,* 981

Costin, G., Fefferman, R.A. and Kogut, M.D. (1973). *J. Pediat., 88,* 419

Cronin, M.J., Roberts, J.M. and Weiner, R.I. (1978). *Endocrinology 103,* 302

Crow, T.J., Johnstone, E.C., Longden, A. and Owne, F. (1978). *Adv. Biochem. Psychopharmacol., 19,* 301

Dammacco, F., Rigillo, N., Tafaro, E., Gagliardi, F., Chetri, G. and Dammacco, A. (1976). *Horm. Metab. Res., 8,* 247

Delitala, G., Wass, J.A.H., Stubbs, W.A., Jones, A., Williams, S. and Besser, G.M. (1979). *Clin. Endocr., 11,* 1

DeRivera, J.L.G. (1975). *Ann. Int. Med., 82,* 811

DeRivera, J.L., Lal, S., Ettigi, P., Hontela, S., Muller, H.F., and Friesen, H.G. (1975). *Clin. Endocrinol., 5,* 273

Deupree, J.D. (1977). *Neuropharmacology 16,* 557

Dimond, R.C., Brammer, S.R., Atkinson, R.I., Howard, W.J. and Earll, J.M. (1973). *J. Clin. Endocr. Metab., 36,* 1189

Ettigi, P., Lal, S. and Friesen, H.G. (1973). *Lancet 2,* 266

Ettigi, P., Nair, N.P.V., Lal, S., Cervantes, P. and Guyda, H. (1976). *J. Neurol. Neurosurg. Psychiat., 39,* 870

Falaschi, P., Frajese, G., Sciarra, F., Rocco, A. and Conti, C. (1978). *Clin. Endocrinol., 8,* 427

Gäde, E.B. and Heinrich, K. (1955). *Nervenarzt 26,* 49

Gibbs, D.M. and Neill, J.D. (1978). *Endocrinology 102,* 1895

Goldsmith, P.C., Cronin, M.J. and Weiner, R.I. (1979). *J. Histochem. Cytochem., 27,* 1205

Gruen, P.H., Sachar, E.J., Langer, G., Altman, N., Leifer, M., Frantz, A. and Halpern, F.S. (1978). *Arch. Gen. Psychiat., 35,* 108

Guitelman, A., Aparicio, N.J., Mancini, A. and Debeljuk, L. (1978). *Fertil. Steril., 30,* 42

Healy, D.L. and Burger, H.G. (1977). *Clin. Endocrinol., 7,* 195

Horn, A.S. (1975). In *Handbook of Psychopharmacology,* Vol. 2, ed. L.L. Iversen and S.H. Snyder (Plenum Press, New York) p. 179

Horrobin, D.F. (1977). *Postgrad. Med. J., 53* (suppl 4), 160

Husain, M.K., Frantz, A.G., Ciarochi, F. and Robinson, A.G. (1975). *J. Clin. Endocr. Metab., 41,* 1113

Huws, D. and Groom, G.V. (1977). *Postgrad. Med. J., 53* (suppl 4), 175

Imura, H., Kato, Y., Ikeda, M., Morimoto, M., Yawata, M. and Fukase, M. (1968). *J. Clin. Endocr. Metab., 28,* 1079

Imura, H., Kato, Y., Ikeda, M., Morimoto, M. and Yawata, M. (1971). *J. Clin. Endocr. Invest., 50,* 1069

Imura, H., Nakai, Y. and Yoshimi, T. (1973). *J. Clin. Endocr. Metab., 36,* 204

Iversen, L.L. (1975). *Science 188,* 1084

Kebabian, J.W. (1978). *Life Sci., 23,* 479

Janowsky, D.S., Leichner, P., Parker, D., Judd, L.L., Huey, L. and Clopton, P. (1978). *Arch. Gen. Psychiat., 35,* 1384

Kendler, K.S., Weitzman, R.E. and Rubin, R.T. (1978). *J. Clin. Endocr. Metab., 47,* 204

Kodlin, D. and McCarthy, N. (1978). *Cancer 41,* 761

Kolodny, H.D., Sherman, L., Singh, A., Kim, S. and Benjamin, F. (1971). *New Engl. J. Med., 284,* 819

Koulu, M., Lammintausta, R., Kangas, L. and Dahlström, S. (1979). *J. Clin. Endocr. Metab., 48,* 119

Labrie, F., Pelletier, G., Borgeat, P., Drouin, J., Ferland, L. and Belanger, A. (1976). In *Frontiers of Neuroendocrinology,* Vol. 4, ed. L. Martini and W.F. Ganong (Raven Press, New York) p. 63

Lachelin, G.C.L., Leblanc, H. and Yen, S.S.C. (1977). *J. Clin. Endocr. Metab., 44,* 728

Lal, S. and Nair, N.P.V. (1979). In *Neuroendocrine Correlates in Neurology and Psychiatry,* ed. E.E. Müller and A. Agnoli (Elsevier/North Holland Biomedical Press, Amsterdam) p. 179

Lal, S. and Nair, N.P.V. (1980). In *Neuroactive Drugs in Endocrinology,* ed. E.E. Müller (Elsevier/North Holland Biomedical Press, Amsterdam) p. 223

Lal, S., de la Vega, C.E., Sourkes, T.L. and Friesen, H.G. (1972). *Lancet 2*, 661
Lal, S., de la Vega, C.E., Sourkes, T.L. and Friesen, H.G. (1973). *J. Clin. Endocr. Metab.*, *37*, 719
Lal, S., Tolis, G., Martin, J.B., Brown, G.M. and Guyda, H. (1975). *J. Clin. Endocr. Metab.*, *41*, 827
Lal, S., Guyda, H. and Bikadoroff, S. (1977). *J. Clin. Endocr. Metab.*, *44*, 766
Lal, S., Nair, N.P.V. and Guyda, H. (1978). *Acta Psychiat. Scand.*, *57*, 91
Lal, S., Mendis, T., Cervantes, P., Guyda, H. and DeRivera, J.L. (1979). *Neuropsychobiology 5*, 327
Lamberg, B.-A., Linnoila, M., Fogelholm, R., Olkinuora, M., Kotilainen, P. and Saarinen, P. (1977). *Neuroendocrinology 24*, 90
Lancranjan, I. and Marbach, P. (1977). *Metabolism 26*, 1225
Lancranjan, I., Ohnhaus, E. and Girard, J. (1979). *J. Clin. Endocr. Metab.*, *49*, 227
Langer, G., Sachar, E.J., Nathan, R.S., Tabrizi, M.A., Perel, J.M. and Halpern, F.S. (1979). *Psychopharmacology 65*, 161
Leblanc, H., Lachelin, G.C.G., Abu-Fadil, S. and Yen, S.S.C. (1976). *J. Clin. Endocr. Metab.*, *43*, 668
Lee, P.A., Kelly, M.R. and Wallin, J.D. (1976). *J. Amer. Med. Assoc.*, *235*, 2316
Leppäluoto, J., Mannisto, P., Ranta, T. and Linnoila, M. (1976). *Acta Endocrinol.*, *81*, 455
L'Hermite, M., Denayer, P., Golstein, J., Virasoro, E., Vanhaelst, L., Copinschi, G. and Robyn, C. (1978a). *Clin. Endocr.*, *9*, 195
L'Hermite, M., Delogne-Desnoeck, J., Michaux-Duchene, A. and Robyn, C. (1978b). *J. Clin. Endocr. Metab.*, *47*, 1132
L'Hermite, M., MacLeod, R.M. and Robyn, C. (1979). *J. Clin. Endocr. Metab.*, *49*, 317
Linnoila, M., Lamberg, B-A., Seppala, T. and Karonen, S-L. (1978). *Acta Parmacol. Toxicol.*, *43*, 72
Lister, R.C., Underwood, L.E., Marshall, R.N., Friesen, H.G. and Vanwyk, J.J. (1974). *J. Clin. Endocr. Metab.*, *39*, 1148
McCallum, R.W., Sowers, J.R., Hershman, J.M. and Sturdevant, R.A.L. (1976). *J. Clin. Endocr. Metab.*, *42*, 1148
Mancini, A.M., Guitelman, A., Debeljuk, L. and Vargas, C.A. (1975). *J. Endocr.*, *67*, 127
Maneschi, M. and Martorana, A. (1978). In *Clinical Psychoneuroendocrinology in Reproduction*, ed. L. Carenza, P. Pancheri and L. Zichella (Academic Press, London)p. 139
Mao, C.L., Marco, E., Revuilla, A. and Costa, E. (1978). In *Interactions between Putative Neurotransmitters*, ed. S. Garattini, J.F. Pujol and R. Samanin (Raven Press, New York) p. 151
Martin, J.B., Lal, S., Tolis, G. and Friesen, H.G. (1974). *J. Clin. Endocr. Metab.*, *39*, 180
Masala, A., Delitala, G., Alagna, S., Devilla, L., Rovasio, P.P. and Lotti, G. (1978). *Metabolism 27*, 921
Massara, F., Camanni, F., Belforte, L., Vergano, V. and Molinatti, G.M. (1978). *Clin. Endocrinol.*, *9*, 419
Matuk, F. and Kalyanaraman, K. (1977). *Arch. Neurol.*, *34*, 374
Meltzer, H.Y. and Fang, V.S. (1976). *Arch. Gen. Psychiat.*, *33*, 279
Meltzer, H.Y., Daniels, S. and Fang, V.S. (1975). *Life Sci.*, *17*, 339
Meltzer, H.Y., Goode, D.J., Schyve, P.M., Young, M. and Fang, V.S. (1979). *Am. J. Psychiat.*, *136*, 1550
Mendelson, W.B., Jacobs, L.S., Gillin, J.C. and Wyatt, R.J. (1979). *Psychoneuroendocrinology 4*, 341
Mims, R.B., Scott, C.L., Modebe, O. and Bethune, J.E. (1975). *J. Clin. Endocr. Metab.*, *40*, 256
Mori, M., Kobayashi, I., Shimoyama, S., Uehara, T., Nemoto, T. Fukuda, H. and Kamio, N. (1977). *Endocr. Jap.*, *24*, 149
Morley, J.E., Melmed, S., Briggs, J., Carlson, H.E., Hershman, J.M., Solomon, T.E., Lamers, C. and Damassa, D.A. (1979). *Life Sci.*, *25*, 1201
Nair, N.P.V., Lal, S., Cervantes, P., Yassa, R. and Guyda, H. (1979). *Neuropsychobiology 5*, 136
Nair, N.P.V., Lal, S., Schwartz, G. and Thavundayil, J.X. (1980). *Adv. Biochem. Psychopharmacol, 24*, 437

Nakai, Y., Imura, H., Sakurai, H., Kurahachi, H. and Yoshimi, T. (1974). *J. Clin. Endocr. Metab., 38,* 446
Oliver, C., Mical, R.S. and Porter, J.C. (1977). *Endocrinology 101,* 598
Plonk, J. and Feldman, J.M. (1976). *J. Clin. Endocr. Metab., 42,* 291
Polishuk, W.Z. and Kulcsar, S. (1956). *J. Clin. Endocr., 16,* 292
Rick, J., Szabo, M., Payne, P., Kovanthana, N., Cannon, J.G. and Frohman, L.A. (1979). *Endocrinology 104,* 1234
Saldanha, V.F., Havard, C.W.H., Bird, R. and Gardner, R. (1972). *Clin. Endocrinol., 1,* 173
Santen, R.J. and Bardin, C.W. (1973). *J. Clin. Invest., 52,* 2617
Sato, A., Onaya, T., Kotani, M., Harada, A. and Yamada, T. (1974). *Endocrinology 94,* 1311
Schaison, G., Croisier, J.C., Nathan, C. and Gilbert-Dreyfus (1974). *Ann. d'Endocr., 35,* 103
Schwinn, G., Schwarck, H., McIntosh, C., Milstrey, H.-R., Willms, B. and Köbberling, J. (1976). *J. Clin. Endocr. Metab., 43,* 1183
Schyve, P.M., Smithline, F. and Meltzer, H.Y. (1978). *Arch. Gen. Psychiat., 35,* 1291
Sedvall, G. (1979). In *Neuroendocrine Correlates in Neurology and Psychiatry,* ed. E.E. Müller and A. Agnoli (Elsevier/North Holland Biomedical Press, Amsterdam) p. 195
Seeman, P., Lee, T., Chau-Wong, M. and Wong, K. (1976). *Nature 261,* 717
Shader, R.J. and DiMascio, A. (1970). *Psychotropic Drug Side Effects* (Williams and Wilkins Co., Baltimore)
Shah, D.K., Wig, N.N. and Chaudhury, R.R. (1973). *Indian J. Med. Res., 61,* 771
Shannack, K.S. and Hornykiewicz, O. (1980). In *Long-term Effects of Neuroleptics,* ed. F. Cattabeni and E. Costa (Raven Press, New York) p. 323
Shimamoto, K. and Miyahara, M. (1976). *J. Clin. Endocr. Metab., 43,* 201
Stanley, M. and Wilk, S. (1979). *Life Sci., 24,* 1907
Syvälahti, E. and Pekkarinen, A. (1977). *J. Neural Transm., 40,* 221
Takahashi, Y., Kipnis, D.M. and Daughaday, W.H. (1968). *J. Clin. Invest., 47,* 2079
Tamagna, E.I., Lane, W., Hershman, J.M., Carlson, H.E., Sturdevant, R.A.L., Poland, R.E. and Rubin, R.T. (1979). *Horm. Res., 11,* 161
Tsang, D. and Lal, S. (1977). *Can. J. Physiol. Pharmacol., 55,* 1263
Vincent, F.M. and Emery, S. (1978). *Ann. Int. Med., 89,* 147
Weiner, R.I. and Ganong, W.F. (1978). *Physiol. Rev., 58,* 905
Weiss, G., Schmidt, C., Kleinberg, D.L. and Ganguly, M. (1977). *Clin. Endocrinol., 7,* 423
de Wied, D. (1967). *Pharmacol. Rev., 19,* 251
Winnik, H.Z. and Tennenbaum, L. (1955). *La Presse Medicale 63,* 1092
Wirtz-Justice, A., Puhringer, W., Lacoste, V., Graw, P. and Gastpar, H. (1976). *Pharmakopsychiat. Neuropsychopharm., 9,* 277
Woolf, P.D., Jacobs, L.S., Danofrio, R., Burday, S.Z. and Schalch, D.S. (1974). *J. Clin. Endocr. Metab., 38,* 71

PART THREE

NEUROENDOCRINE REGULATION AND
PATHOGENESIS OF MENTAL ILLNESS

9 NEUROENDOCRINE STUDIES ON THE PATHOGENESIS OF DEPRESSION

Noboru Hatotani, Junichi Nomura and Isao Kitayama

TABLE OF CONTENTS

I   INTRODUCTION   189

II   NEUROENDOCRINE STUDIES ON DEPRESSION   189
   A.   Hypothalamo-pituitary-adrenal axis   189
   B.   Hypothalamo-pituitary-thyroid axis   190
   C.   Growth hormone and prolactin   191
   D.   Hypothalamo-pituitary-gonadal axis   192
   E.   Biogenic amines   192

III   STRESS-INDUCED MODEL OF DEPRESSION IN ANIMALS   193
   A.   Psychoendocrine model of depression   193
   B.   Method of producing a model of a stress-induced depression in female rats   194
   C.   Change of brain monoamines in depression-model rats   196
      a. Histochemical fluorescence investigations   196
      b. Studies on the turnover rate of catecholamines   197
      c. Other preliminary findings   197
   D.   Discussion   198

IV   SUMMARY AND CONCLUSIONS   201

REFERENCES   202

I. INTRODUCTION

Depression is the most fundamental psychosomatic reaction caused by various etiological factors of which some are known to be genetic, some are environmental and some are of physical nature. The essence of depression is a reduction of vital potency which is directly manifested by symptoms such as depressed mood, retardation of the psychic process, decrease of individual drives and general activity and impairment of basic biological rhythms. Various kinds of individual personality reactions develop on the basis of these fundamental symptoms and make up diverse clinical pictures. Despite the variety of its clinical manifestations, depression is the most therapeutically accessible state by such biological treatments as antidepressants or electroconvulsive therapy. It is indisputable that the basic symptom of depression is brought about through a final common biological pathway irrespective of etiology. As a matter of fact, depression is the principal field upon which biological studies have recently been focused, and neuro-endocrine findings, in particular, are accumulating. In this chapter, the significant results of clinical psychoneuroendocrine studies on depression (Ettigi and Brown, 1977; Hatotani et al., 1978) will be briefly reviewed and then an animal model of depression will be described in an attempt to explore further the role of neurotransmitters and the hypothalamo-pituitary system in the pathogenesis of depression (Hatotani et al., 1977, 1979).

II. NEUROENDOCRINE STUDIES ON DEPRESSION

A. Hypothalamo-pituitary-adrenal Axis

The hypothalamo-pituitary-adrenal (HPA) axis has the following characteristics (Carroll, 1972): (1) acute physical and psychological stresses of various types cause an increase in the HPA activity, probably by altering the inhibiting control of the central nervous system (CNS); (2) the HPA function is regulated by a circadian program organized by the CNS and is characterized by episodic bursts of corticotropin-releasing factor (CRF), adrenocorticotropic hormone (ACTH) and cortisol; (3) the HPA function receives a negative-feedback control by the circulating glucocorticoid levels. The function of the HPA axis in depresssion is the most thoroughly studied aspect of clinical psychoneuroendocrinology.

From many findings on corticosterone levels in depressed patients, it has been concluded that the gross output of glucocorticoids is elevated in depression. The elevation could be a result of such nonspecific factors as emotional distress, anxiety, ego disorganization or environmental changes (Sachar, 1967; Sachar et al., 1970). However, taking these factors into account, it seems evident that the hypersecretion of cortisol in certain depressive illnesses is not simply a stress response, but rather a reflection of some fundamental neuroendocrine disturbance. The importance of a detailed sampling schedule of plasma cortisol was recently emphasized and it was reported that depressives, while ill, secreted substantially more cortisol, had more secretory episodes and more minutes of active secretion. Moreover, increased cortisol secretion was observed during the night, at which time cessation of cortisol production should normally have occurred (Sachar *et al.*, 1973a). As a result, the diurnal rhythm of plasma cortisol is frequently disturbed in depressed patients. These findings suggest an abnormal disinhibition of the neuroendocrine centers regulating the release of ACTH.

The HPA function has been further investigated by means of the dexamethasone suppression test. The administration of 1-2 mg of dexamethasone at midnight suppresses the plasma cortisol and 17-OHCS levels for at least 24 hours by negative feedback. Carroll (1977) summarized the results of eight investigations and found that 70 of 152 depressive patients (46%) had abnormal suppression responses. In these studies, a single plasma cortisol level was obtained in the morning after dexamethasone. However, when blood samples were more frequently obtained during 24 hours, one could often find cases who showed normal suppression in the morning but resumed cortisol secretion within a short time (Carroll *et al.*, 1976). The inadequate response to dexamethasone is interpreted as a failure of the normal CNS inhibiting mechanism for ACTH-cortisol secretion. Hypersecretion of cortisol, the disturbance of the circadian rhythm of cortisol secretion and the failure of response to dexamethasone are common findings confirmed by many investigators, and seem to indicate a central disinhibition of the HPA axis during the depressive episodes.

B. Hypothalamo-pituitary-thyroid Axis

It has been clinically observed that patients with thyroid diseases are apt to be mentally ill. The hypothalamo-pituitary-thyroid axis may be an important integral part of the psychoendocrine system for maintenance of normal mental function. Protein-bound iodine, basal metabolic rate (BMR) and rate of uptake of radioactive iodine have been reported to be within the normal range in depressed patients. However, the response of thyroid-stimulating hormone (TSH) to infused synthetic thyrotropin-releasing hormone (TRH) was shown to be significantly lower in a large proportion of depressed patients when compared with normal subjects (Ehrensing *et al.*, 1974; Takahashi *et al.*, 1974). Hatotani and coworkers (Hatotani *et al.*, 1977; Yamaguchi *et al.*, 1977) found that 19 out of 51 depressed patients (37%) showed abnormal TSH responses to TRH in terms of exaggerated, diminished and delayed responses. The basal level of triiodothyronine (T_3) and its response to TRH were significantly lower in patients with delayed or diminished TSH responses than in normal subjects. These findings suggest that the central regulating mechanism of the hypothalamo-pituitary-thyroid system is disturbed in some groups of depressed patients.

Prange *et al.* (1969) reported that depressed patients recovered more quickly if one added a small amount of T_3 to a usual regimen of imipramine. The observation has been confirmed by many workers also for other tricyclic anti-depressants. The possible explanation of this effect may be that T_3 enhances the sensitivity of the central adrenergic receptor and thus has an additive or synergistic effect with imipramine. It is also proposed that slight shifts in the thyroid state toward hypothyroidism, although within the normal range, affect the CNS in such a fashion as to produce a predisposition to retarded depression. Yamaguchi *et al.* (1975) observed that cases of persistent depression refractory to anti-depressants tended to show relatively low BMR, while other indices of thyroid function were within the normal range and they responded well to additional thyroid medication. The latent hypothyroidism, probably due to hypothalamo-pituitary dysfunction, might make these patients resistant to antidepressants.

C. Growth Hormone and Prolactin

Basal growth-hormone (GH) levels have been reported to be normal in depressed patients, despite the fact that GH is considered to be one of the stress-responsive hormones. GH responses to various stimuli have been studied in depression as a test of hypothalamic function. Among these stimuli, GH responses to L-dihydroxyphenylalanine (L-DOPA) and to insulin-induced hypoglycemia are believed to be catecholaminergically mediated. Sachar *et al.* (1972) reported impaired GH responses to L-DOPA in depressed patients. However, when their age and sex were taken into consideration, GH responses were the same in con-trol subjects and depressed patients (Sachar *et al.*, 1975). Impaired GH responses to insulin hypoglycemia in depressed patients were reported by many workers (Müller *et al.*, 1969; Sachar *et al.*, 1971; Endo *et al.*, 1974). When ten post-menopausal depressed women were matched with normal postmenopausal women for age, body weight, baseline GH levels and blood-sugar response to insulin, a significant reduction of GH response was observed in depressed women (Gruen *et al.*, 1975). The discrepancy between GH responses to hypoglycemia and L-DOPA in depression remains to be elucidated. It was also reported that depressed patients showed inadequate GH responses to 5-hydroxytryptophan (5-HTP) (Takahashi *et al.*, 1973), and that TRH, which does not affect GH levels in normal subjects, caused an abnormal increase in GH in depressed patients (Maeda *et al.*, 1975).

An elevation of basal prolactin plasma levels was also observed in depressed patients compared with normal subjects (Sachar *et al.*, 1973b). However, when L-DOPA was given to the two groups, no difference was noticed in prolactin suppression. As for prolactin responses to TRH, some investigators (Maeda *et al.*, 1975) reported them as being enhanced, and others (Ehrensing *et al.*, 1974) as being diminished in depressed patients. GH- and prolactin-response tests to various stimuli may provide further support for the concept of hypothalamic dysfunction and of functional defects of neurotransmitters in depression.

D. Hypothalamo-pituitary-gonadal Axis

Affective disorders associated with the menstrual cycle, pregnancy and puerperium have recently been reviewed by Hatotani *et al.* (1978). The secretion of gonadotropin in depression has not been extensively studied. Altman *et al.* (1975) reported reduced plasma luteinizing-hormone (LH) levels in depressed postmenopausal women compared with normal postmenopausal women. This finding might be a reflection of the diminished hypothalamic catecholaminergic activity in depression. As for plasma LH responses to LH-releasing hormone (LHRH), a disturbance was preliminarily reported in male secondary depressed patients (Ettigi *et al.*, 1979). Sachar *et al.* (1973c) determined plasma testosterone concentration in depressed men with no meaningful findings.

Several investigators pointed out abnormal fraction patterns of urinary 17-ketosteroids and estrogens in depression or in acute psychoses (Hatotani *et al.*, 1962; Coppen *et al.*, 1967; Mendels, 1969). Hatotani and his group regarded these disturbances as a result of hepatic dysfunction, which could be a manifestation of breakdown of the cerebro-hepatic homeostasis. The control of hepatic steroid metabolism by the CNS was demonstrated experimentally (Nomura *et al.*, 1979). Disturbances of the hypothalamo-hepatic axis may play a role in the pathogenesis of some groups of affective disorders.

E. Biogenic Amines

It has been generally accepted that functional defects of biogenic amines in the CNS may underlie the pathogenesis of affective disorders. Neuroendocrine abnormalities in patients with affective disorders as described above may well be the result of such altered neurotransmitter function, since neurotransmitters act on the hypophysiotropic area in the hypothalamus. Aminergic pathways in the CNS have been extensively studied (Ungerstedt, 1971). Biogenic amines are believed to modulate the activity of the central neurons which are involved in the regulation of mood and behaviour, as well as sleep and autonomic functions. Furthermore, various drugs bring about a change in mood and behaviour by altering the activity of neurotransmitters. Reserpine-induced sedation, mood elevation by monoamine-oxidase inhibitors and the mode of effect of tricyclic antidepressants have all provided support for the biogenic-amine hypothesis of depression. The simplest hypothesis suggests that some types of depression are associated with an altered availability of one or another biogenic amine at functionally important receptor sites in the CNS.

Many authors have studied the urinary and cerebrospinal-fluid (CSF) metabolites of catecholamines and indoleamines in depressed patients. Among urinary norepinephrine metabolites, the major part of 3-methoxy-4-hydroxyphenylglycol (MHPG) has its origin in the metabolism of norepinephrine in brain (Maas *et al.*, 1973). In the depressed phase of primary affective disorder, the bipolar group excreted approximately 30% less MHPG than the age- and sex-matched normal volunteers (Maas *et al.*, 1968). The results of studies of urinary and CSF MHPG in depression have recently been summarized by Wehr and Goodwin (1977). Various aspects of indoleamine metabolism in depression were reviewed by van Praag (1977). Several authors have reported a decrease in the accumulation of 5-hydroxy-indoleacetic acid (5-HIAA) in lumbar CSF after probenecid in depressed patients, this is considered to be an indication of a reduced 5-hydroxytryptamine (5-HT) turnover in the CNS. However, the above finding is by no means universal in depression, and Van Praag (1977) states that the suspected 5-HT defect is probably a factor predisposing to, rather than causing, a pathological depression of mood.

The various biochemical findings have been used to identify biochemical subgroups of depression and to predict selective responses to specific anti-depressant drugs. Schildkraut *et al.* (1978), measuring urinary excretion of MHPG and other metabolites of norepinephrine, classified the group of manic-depressive or schizoaffective depressions and the group of schizophrenia-related depressions. Depressed patients with low pretreatment levels of MHPG may be expected to respond well to imipramine and desipramine (Maas *et al.*, 1972; Beckman and Goodwin, 1975). On the other hand, patients with low levels of 5-HIAA in the CSF may benefit from an antidepressant which chiefly potentiates 5-HT or from the combination of a 5-HT precursor with an antidepressant (van Praag, 1977). The effects of 5-HT precursors (L-tryptophan and 5-hydroxytryp-tophan) and a catecholamine precursor (L-DOPA) on depression were reviewed by Mendels *et al.* (1975). The results, though not necessarily consistent with the biogenic-amine hypothesis of affective disorders, do not allow abnormal adrenergic or serotonergic activity in depressed patients to be ruled out. Further studies are required to test the biogenic-amine hypothesis.

III. STRESS-INDUCED MODEL OF DEPRESSION IN ANIMALS

A. Psychoendocrine Model of Depression

From the results of neuroendocrine studies discussed above, it appears that a dysfunction of the hypothalamo-pituitary system and a metabolic disturbance of biogenic amines are the most significant findings in depressive illness. The function of the hypothalamo-pituitary system, which is controlled by monoaminergic systems from the limbic system and the lower brain stem, involves the integration of mood and behaviour, consciousness, instinctive drives, autonomic nervous and endocrine functions and biological rhythms. Therefore, the basic symptoms of depression appear to be underpinned by the disturbance of all these functions. A final common biological pathway in the pathogenesis of depression is supposed to be a dysfunction of monoaminergic systems in the CNS. It is a common experience that the onset of depression, in most instances, is preceded by a stressful situation of either psychological or physical nature. When the loading of stress exceeds an individual's ability to cope with it, it is probable that the resultant metabolic decompensation of brain monoamines would lead to persistent disturbance of the regulatory mechanism of the CNS, reflected in the symptomatology of depression. According to such a conceptual model of depression, it is conceivable that stress plays an important part in the pathogenesis of depression, in which a metabolic disturbance of brain monoamines is essential. This stress model of depression is pertinent not only to reactive depression but also to endogenous depression. An individual who is genetically predisposed to depressive psychosis would be more prone to develop depression precipitated by stress than would an individual with no such predisposition. It is postulated that depression is the commonest pattern of decompensation of the CNS function.

In order to verify the relevance of such a conceptual model of depression, it would be helpful to produce an animal model of depression by means of stress. It is well known that adult female rats show a regular cycle of spontaneous running acitivity corresponding to the estrous cycle (Richter, 1960). We took advantage of this species-specific baseline behaviour, which is tied to an endocrine function, in order to derive an animal model of depression (Hatotani *et al.*, 1977, 1979).

B. Method of Producing a Model of Stress-induced Depression in Female Rats

Adult female Wistar rats were raised in revolving cages of 1-m circumference and their spontaneous running activity, recorded as the number of revolutions of the drum, was measured at 10.00 a.m. every day. Water and food intake, rectal temperature, body weight, motor responsiveness (running activity in a 10-s period following a sound stimulus) and the percentage of cornified cells in a vaginal smear were recorded at the same time. Only rats which showed a regular cycle of spontaneous running activity corresponding to the estrous cycle were selected for the experiment. The selected rats were moved into drums revolving automatically at five revolutions per minute, so that they were exposed to stress in terms of forced running. Food and water were accessible during the stress. When the fatigue reached its maximum and the rectal temperature came down to 33°C or less, the rats were relieved from the stress and given a rest for 24 hours. This sequence of stress and rest was repeated three times. The duration of each stress was two to five days, varying from rat to rat. The animals were then brought back to the revolving cages, and their running activity, vaginal smear, food and water intake, rectal temperature and motor responsiveness were again recorded daily.

The rats showed no spontaneous running activity for several days after the stress. In about half of the rats, the spontaneous running activity started to reappear in about two weeks, and resumed regular cyclicity. These animals were termed *recovery rats*. In the other half of the rats, as shown in Figure 9.1, the spontaneous running activity continued to remain very low for more than two weeks after the stress, and showed no cyclic regularity. The vaginal smear showed a pattern of constant diestrus or irregular reappearance of cornified cells. The intake of food and water was almost the same as before the stress. Motor responsiveness and rectal temperature returned to normal levels within several days after the stress. Body weight also recovered within ten days and continued

Figure 9.1: **Effect of stress on running activity, body weight, rectal temperature, motor responsiveness, food and water intake and percentage of cornified cells in a depression-model rat. Note that, after stress, the running activity and its cyclicity almost disappeared for six weeks; the vaginal smear showed the pattern of constant diestrus or irregular reappearance of cornified cells. In contrast, food and water intake was almost the same as before the stress. Motor responsiveness and rectal temperature returned to normal levels within several days after the stress. Body weight recovered within ten days and continued to increase thereafter.**

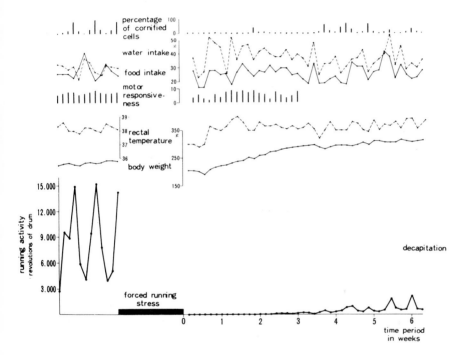

to increase thereafter. Therefore, these animals seemed to have completely recovered from physical exhaustion. Occasionally, they showed aptosis, sleep disturbances and pronounced aggression, though in general they were inactive and seclusive. This stress-induced, prolonged inactive state accompanied by the abolition of hormone-dependent cyclic behaviour, a state different from simple exhaustion, was supposed to be a depression analog in rats. These animals were termed *depression-model rats*. Reversal of symptoms by clinically effective treatment techniques is proposed as one of the criteria which might validate an animal model of depression (McKinney, 1977). In our preliminary study, imipramine injection in these inactive rats seemed to enhance the resumption of normal cyclic activity, as illustrated in Figure 9.2.

Figure 9.2: Effect of imipramine on the running activity of a depression-model rat. Imipramine (i.p.) was injected daily from two weeks after the stress. Note that the first series of imipramine injections (4 mg/kg for ten days) was not very effective; however, the rat quickly recovered its normal cyclic running activity during the second series of imipramine injections (8 mg/kg for ten days).

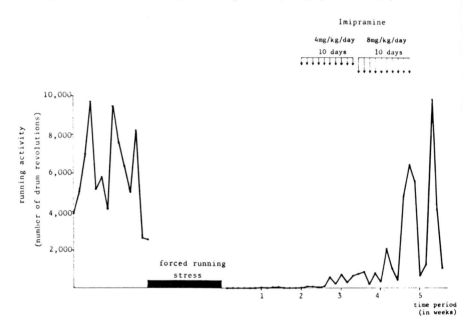

C. Change of Brain Monoamines in Depression-model Rats

a. Histochemical Fluorescence Investigations. Experimental animals, paired with age-matched controls on the diestrus day 1, were decapitated, and brain tissues were treated according to the histochemical fluorescence method of Falck *et al.* (1962). Twelve regions selected from the main monoaminergic neuronal systems, such as the ascending noradrenaline (NA) system, the nigrostriatal dopamine (DA) system and the ascending serotonin (5-HT) system, were examined by means of fluorescence microscopy. In some experiments, the fluorescence intensity of catecholamines was measured semiquantitatively according to a modification of the technique described by Lichtensteiger (1970). The following results were obtained consistently.

(1) Fluorescence intensity of catecholamines in depression-model rats increased in the cell groups of the ascending NA systems: nucleus reticularis lateralis (A1), ventral part of nucleus commisuralis (A2), tractus rubrospinalis (A5), locus coeruleus (A6) and nucleus subcoeruleus (A7), which are situated in the medulla oblongata and the pons.

(2) In contrast, fluorescence intensity of catecholamines in depression-model rats decreased in cell bodies in the nucleus arcuatus (A12) and in their nerve terminals in the external layer of the median eminence, which belong to the tuberoinfundibular DA system.

(3) The above described findings were also seen in rats which were examined immediately after the stress, but not in recovery rats. No remarkable change was observed in the fluorescence intensity of the nigrostriatal DA system and of the ascending 5-HT system.

Representative examples are shown in Figures 9.3 and 9.4. Figure 9.3 contains fluorescence microphotographs of NA-containing cell bodies in the locus coeruleus. The depression-model rat (B) showed much stronger fluorescence than the control (A), but the recovery rat (C) displayed almost the same fluorescence intensity as the control (A'). Figure 9.4 shows DA-containing cell bodies in the nucleus arcuatus. Fluorescence seen in the control rat (A) almost vanished in the rat killed immediately after the stress (B) and recovered only about half the intensity of its original level in the despression-model rat (C).

b. *Studies on the Turnover Rate of Catecholamines.* In this case, α-methyl-p-tyrosine methylester (250 mg/kg) was injected intraperitoneally three hours before the decapitation in order to inhibit the synthesis of catecholamines. The turnover rate of catecholamines was estimated by comparing the disappearance rate of fluorescence in control and experimental rats. Figure 9.5 shows fluorescence microphotographs of the nucleus supraopticus. The fluorescence intensity of catecholamines in the steady state (without α-methyl-p-tyrosine) was approximately the same in the control rat (A), the depression-model rat (B) and the recovery rat (C). The disappearance rate of fluorescence after the injection of α-methyl-p-tyrosine was reduced in the depression-model rat (B') as compared with the control rat (A'). The disappearance rate in the recovery rat (C') was almost the same as in the control rat (A'). This finding suggests that the turnover rate of catecholamines decreased in the depression-model rat. The same result was obtained in other regions in which nerve terminals of the ascending NA system are concentrated, such as the nucleus motorius dorsalis nervi vagi, the internal layer of the median eminence, the retrochiasmatic area, the nucleus dorsomedialis, the nucleus paraventricularis and the nucleus interstitialis striae terminalis. In contrast, the turnover rate of catecholamines in these brain areas of a rat exposed to acute stress (forced running for 24 hours) was observed to be increased.

c. *Other Preliminary Findings.* In biochemical studies of the depression-model rat, norepinephrine, dopamine and serotonin were measured in the cerebral cortex, midbrain, hypothalamus, medulla oblongata and pons, striatum, hippocampus and cerebellum (Nomura *et al.*, 1978). Immediately after the forced-running stress, norepinephrine levels in the hypothalamus decreased significantly,

Figure 9.3: Fluorescence microphotographs of noradrenaline-containing cell bodies in the locus coeruleus. Note that the depression-model rat (B) showed much stronger fluorescence than the control (A). The recovery rat (C) resumed almost the same fluorescence intensity as the control (A'). Magnification: ×50.

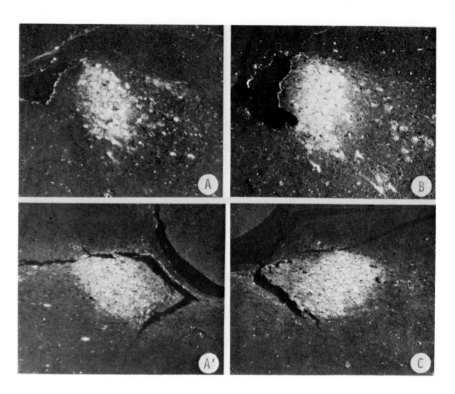

and remained at a lower level in the depression-model rat. The recovery rat, in contrast, showed no difference from the controls. Biochemical investigation of the metabolism of catecholamines in each nucleus is now in progress in order to verify histochemical findings and to obtain a detailed kinetic analysis.

D. Discussion

The above-mentioned animal model of depression, which was produced as a possible research strategy, has the following advantages: (a) the experimental animals have a definite baseline behaviour with regular cyclicity corresponding to the estrus cycle; (b) animals were exposed to long-lasting, uncontrollable stress, which may be an important factor preceding the onset of depression; (c) the stress-induced inactive state of animals accompanied by impairment of the biological rhythm can be differentiated from simple physical exhaustion; (d) taking account of the lifespan of the rat, the duration of this inactive state is comparable in length to human depression; (e) the changes in animal behaviour

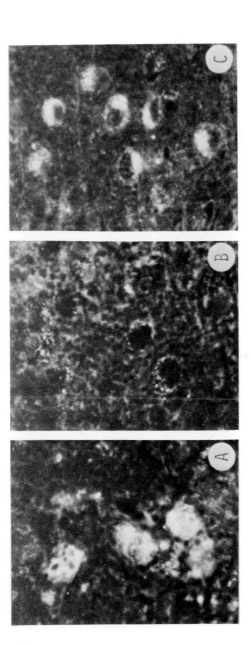

Figure 9.4: Fluorescence microphotographs of dopamine-containing cell bodies in the nucleus arcuatus. Note that the fluorescence seen in the control (A) almost vanished in the rat killed immediately after the stress (B). The fluorescence recovered only half the intensity of its original level in the depression-model rat (C). Magnification: ×400.

Figure 9.5: Fluorescence microphotographs of the nucleus supraopticus. Note that the fluorescence intensity of catecholamines in the steady state (without α-methyl-p-tyrosine) was approximately the same in the control rat (A), the depression-model rat (B) and the recovery rat (C). The disappearance rate of fluorescence three hours after the intraperitoneal injection of α-methyl-p-tyrosine (250 mg/kg) was reduced in the depression-model rat (B') as compared with the control (A'). The disappearance rate in the recovery rat (C') was almost the same as in the control (A'). Magnification: ×100.

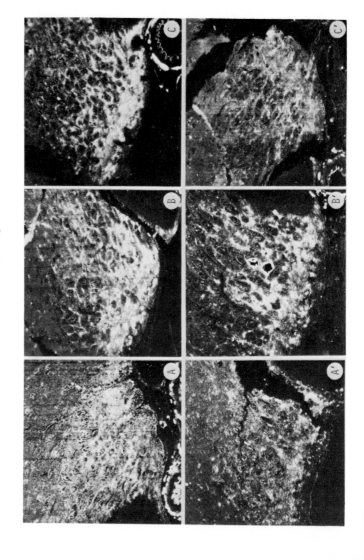

can be evaluated objectively in terms of running activity and its cyclicity; (f) the recovery rat shows almost the same findings in brain monoamines as the control rat; (g) the antidepressant drug normalizes the animal behaviour. Taking all this into consideration, it would be reasonable to regard this animal model as a depression analog.

The most significant findings concerning brain monoamines in the depression-model animals were: (a) an increase in fluorescence intensity of cell groups in the ascending NA system; (b) a decrease in fluorescence intensity of cell bodies and nerve terminals in the tuberoinfundibular DA system; (c) a decrease in turn-over rate of catecholamines in nerve terminals of the ascending NA system.

It is well known that the ventral pathway of the ascending NA system terminates mainly in the hypothalamus (Ungerstedt, 1971) and that the tuberoinfundibular system has a close relation to the regulation of the estrus cycle (Fuxe and Hökfelt, 1969). The function of these brain regions is obviously related to emotion, spontaneity and the sleep mechanism, as well as to endocrine and autonomic nervous functions. Dysfunction of these areas may be related to the development of depressive symptoms. It is also interesting to note that the turn-over rate of catecholamines in nerve terminals of the ascending NA system increased during the acute stress and decreased after the long-lasting stress. One may speculate that the turnover rate of catecholamines, which adaptively increases during acute stress, changes to a decrease at a certain critical point in the course of long-lasting stress, and that this alteration of catecholamine metab-olism remains unrestored in depression-model animals. The relationship between the increase in catecholamine levels in cell bodies and the decrease in cate-cholamine turnover in nerve terminals of the ascending NA system is also a suggestive finding, and should be studied further. Thus, a metabolic disorder of brain catecholamines might be a final common pathway in the pathogenesis of depression and an appropriate animal model would greatly help in detailed experimental studies of this problem.

IV. SUMMARY AND CONCLUSIONS

The main points of the present review may be summarized as follows: (1) hypothalamo-pituitary dysfunction appears to be the most significant finding of neuroendocrine studies of depression; (2) the underlying pathology of this dysfunction is probably a metabolic disturbance of brain monoamines; (3) brain monoamines were examined by the fluorescence histochemical method in a stress-induced animal model of depression; (4) an increase in fluorescence intens-ity was found in the cell groups of the ascending NA system; (5) cell bodies and nerve terminals of the tuberoinfundibular DA system showed a decrease of fluorescence intensity; (6) the turnover rate of catecholamines in nerve terminals of the ascending NA system decreased.

REFERENCES

Altman, N., Sachar, E.J., Gruen, P.H., Halpern, F. and Eto, S. (1975). *Psychosom. Med., 37,* 274

Beckman, H. and Goodwin, F.K. (1975). *Arch. Gen. Psychiat., 32,* 17

Carroll, B.J. (1972). In *Depressive Illness: Some Research Studies,* ed. B. Davies, B.J. Carroll and R.M. Mobray (C.C. Thomas, Springfield) p. 23

Carroll, B.J. (1977). In *Handbook of Studies on Depression,* ed. G.D. Burrows (Excerpta Medica, Amsterdam) p. 325

Carroll, B.J., Curtis, G.C. and Mendels, J. (1976). *Arch. Gen. Psychiat., 33,* 1039, 1051

Coppen, A., Julian, T., Fry, D.E. and Marks, V. (1967). *Brit. J. Psychiat., 113,* 269

Ehrensing, R.H., Kastin, A., Schalch, D.S., Friesen, H.G., Vargas, J.R. and Schally, A.V. (1974). *Am. J. Psychiat., 131,* 714

Endo, M., Endo, J., Nishikubo, M., Yamaguchi, T. and Hatotani, N. (1974). In *Psychoneuroendocrinology,* ed. N. Hatotani (S. Karger, Basel) p. 22

Ettigi, P.G. and Brown, G.M. (1977). *Am. J. Psychiat., 134,* 493

Ettigi, P.G., Brown, G.M. and Seggie, J.A. (1979). *Psychosom. Med., 41,* 203

Falck, B., Hillarp, N.A., Thieme, G. and Torp, A. (1962), *J. Histochem. Cytochem., 10,* 348

Fuxe, K. and Hökfelt, T. (1969). In *Frontiers in Neuroendocrinology,* ed. W.F. Ganong and L. Martini (Oxford University Press, New York) p. 47

Gruen, P.H., Sachar, E.J. Altman, N. and Sassin, J. (1975). *Arch. Gen. Psychiat., 32,* 31

Hatotani, N., Ishida, C., Yura, R., Kato, Y., Nomura, J., Wakoh, T., Takekoshi, A., Yoshimoto, S., Yoshimoto, K. and Hiramoto, K. (1962). *Folia Psychiat. Neurol. Jap., 16,* 248

Hatotani, N., Nomura, J., Yamaguchi, T. and Kitayama, I. (1977). *Psychoneuroendocrinology 2,* 115

Hatotani, N., Nomura, J., Wakoh, T. (1978). In *Perspectives in Endocrine Psychobiology,* ed. F. Brambilla, P.K. Bridges, E. Endröczi and G. Heuser (Akadémiai Kiadó, Budapest) p. 423

Hatotani, N., Nomura, J., Inoue, K. and Kitayama, I. (1979). *Psychoneuroendocrinology 4,* 155

Lichtensteiger, W. (1970). *Prog. Histochem. Cytochem., 1,* 185

Maas, J.W., Fawcett, J.A. and Dekirmenjian, H. (1968). *Arch. Gen. Psychiat., 19,* 129

Maas, J.W., Fawcett, J.A. and Dekirmenjian, H. (1972). *Arch. Gen. Psychiat., 26,* 252

Maas, J.E., Dekirmenjian, H., Garver, D., Redmond, D.E. and Landis, D.H. (1973). *Eur. J. Pharmacol., 23,* 121

McKinney, W.T. (1977). In *Animal Models in Psychiatry and Neurology,* ed. J. Hanin and E. Usdin (Pergamon Press, Oxford) p. 117

Maeda, K., Kato, Y., Ohgo, S., Chihara, K., Yoshimoto, Y., Yamaguchi, N., Kuromaru, S. and Imura, H. (1975). *J. Clin. Endocr. Metab., 40,* 501

Mendels, J. (1969). *Brit. J. Psychiat., 115,* 581

Mendels, J., Stinnett, J.L., Burns, D. and Frazer, A. (1975). *Arch. Gen. Psychiat., 32,* 22

Müller, P.S., Heninger, G.R. and McDonald, R.K. (1969). *Arch. Gen. Psychiat., 21,* 587

Nomura, J., Kitayama, I. and Hatotani, N. (1978). *Folia Psychiat. Neurol. Jap., 32,* 159

Nomura, J., Hisamatsu, K., Hatotani, N., Kamiya, S., Higashimura, T. and Hattori, H. (1979). *Psychoneuroendocrinology 4,* 47

Prange, A.J., Wilson, I.C., Rabon, A.M. and Lipton, M.A. (1969). *Am. J. Psychiat., 126,* 457

Richter, C.P. (1960). *Proc. Nat. Acad. Sci., 46,* 1506

Sachar, E.J. (1967). *Arch. Gen. Psychiat., 17,* 544

Sachar, E.J., Kanter, S.S., Buie, D., Engle, R. and Mehlman, R. (1970). *Am. J. Psychiat., 126,* 1067

Sachar, E.J., Finkelstein, J. and Hellman, L. (1971). *Arch. Gen. Psychiat., 25,* 263

Sachar, E.J., Mushrush, G., Perlow, M., Weitzman, E.D. and Sassin, J. (1972). *Science 178,* 1304

Sachar, E.J., Hellman, L., Roffwarg, H.P., Halpern, F.S., Fukushima, D.K. and Gallagher, T.F. (1973a). *Arch. Gen. Psychiat., 28,* 19

Sachar, E.J., Frantz, A.G., Altman, N. and Sassin, J. (1973b). *Am. J. Psychiat., 130,* 1362

Sachar, E.J., Halpern, F., Rosenfeld, R.S., Gallagher, T.F. and Hellman, L. (1973c). *Arch. Gen. Psychiat., 28,* 15

Sachar, E.J., Altman, N., Gruen, P.H., Glassman, A., Halpern, F.S. and Sassin, J. (1975). *Arch. Gen. Psychiat., 32,* 502

Schildkraut, J.J., Orsulak, P.J., Schatzberg, A.F., Gudeman, J.E., Cole, J.O., Rohde, W.A. and LaBrie, R.A. (1978). *Arch. Gen. Psychiat., 35,* 1427, 1436

Takahashi, S., Kondo, H., Yoshimura, M., Ochi, Y. and Yoshimi, T. (1973). *Folia Psychiat. Neurol. Jap., 27,* 197

Takahashi, S., Kondo, H., Yoshimura, M. and Ochi, Y. (1974). *Folia Psychiat. Neurol. Jap., 28,* 355

Ungerstedt, U. (1971). *Acta Physiol. Scand. Suppl., 367,* 1

Van Praag, H.M. (1977). In *Handbook of Studies on Depression,* ed. G.D. Burrows (Excerpta Medica, Amsterdam) p. 303

Wehr, T. and Goodwin, F.K. (1977). In *Handbook of Studies on Depression,* ed. G.D. Burrows (Excerpta Medica, Amsterdam) p. 283

Yamaguchi, T., Nomura, J., Nishikubo, M., Tsujimura, R. and Hatotani, N. (1975). *Folia Psychiat. Neurol. Jap., 29,* 230

Yamaguchi, T., Hatotani, N., Nomura, J. and Ushijima, Y. (1977). *Folia Psychiat. Neurol. Jap., 31,* 173

10 THYROID HORMONE IN THE REGULATION OF NEUROTRANSMITTER FUNCTION AND BEHAVIOUR

Radhey L. Singhal and Ram B. Rastogi

TABLE OF CONTENTS

I   INTRODUCTION   207

II   THYROID HORMONE: ONTOGENESIS OF MONOAMINERGIC
       SYSTEMS IN THE BRAIN   208

III   EFFECT OF APOMORPHINE ON BEHAVIOURAL ACTIVITY
       IN HYPERTHYROID RATS: EVIDENCE FOR DA-RECEPTOR
       SUPERSENSITIVITY   210

IV   MENTAL ILLNESSES RELATED TO ABNORMAL THYROID
       FUNCTION: RELEVANCE OF AN ANIMAL MODEL OF
       HYPERTHYROIDISM   212

V   EFFECT OF DIAZEPAM ON BEHAVIOURAL ACTIVITY
       AND BRAIN NOREPINEPHRINE METABOLISM IN
       HYPERTHYROID RATS   213
       A. Effects on serotonin metabolism: a possible mechanism
          of antianxiety action   214

VI   HYPERTHYROID ANIMALS AS A MODEL OF
       HYPERACTIVITY: STUDIES ON THE MODE OF
       ANTIMANIC ACTION OF LITHIUM   215
       A. Effects of lithium on NE and DA metabolism in discrete
          brain regions of hyperthyroid rats   215

VII   POTENTIATION OF ANTIDEPRESSANT ACTION OF
       TRICYCLICS BY L-TRIIODOTHYRONINE AND
       THYROTROPIN-RELEASING HORMONE   217

VIII   SUMMARY AND CONCLUSIONS   219

REFERENCES   219

I. INTRODUCTION

Recent developments in the field of psychoneuroendocrinology have led to the suggestion that neuroendocrine strategies represent a useful approach for evaluating brain function during psychiatric illnesses. Studies have revealed that there are significant alterations in the levels of circulating hormones during affective disorders and psychoses. Since evidence also indicates that hormones affect the metabolism of putative neurotransmitters, it is likely that changes in the mental state during certain endocrine disorders might be the sequelae of alterations in the central monoamine metabolism produced by altered levels of hormones. Additionally, it has been shown that neurotransmitters such as catecholamines and indoleamine play a major role in the regulation of neuroendocrine function. Hence, it is important to determine which of the two abnormalities (i.e., changes in hormonal secretion induced by altered levels of central amines or aberrant metabolism of putative neurotransmitters induced by altered hormonal levels) has the prime role in the pathophysiology of these psychiatric illnesses.

An impressive body of evidence has also emerged emphasizing the importance of various external and internal stimuli in the structural and biochemical ontogeny of the central nervous system as determinants of adult physiological and behavioural processes. It is now believed that, among the internal stimuli, thyroid hormones play a vital role in the maturation of brain. Using developing rat cerebellum as a model (in newborn rat, the development of cerebellum is less advanced compared with that of most other parts of the brain: e.g., the cell number in cerebellum is only about 3% of the adult value, in contrast with about 50% in the forebrain; Patel *et al.*, 1979), thyroxine treatment was found to cause increased cell proliferation during the first week of life (Weichsel, 1974; Patel *et al.*, 1979). In contrast, radio- or chemical-thyroidectomy at birth delayed the cerebellar cell acquisition (Balazs, 1971; Nicholson and Altman, 1972). Gelber *et al.* (1964) observed that administration of L-thyroxine significantly enhanced the leucine-^{14}C incorporation into protein of the developing brain. Thyroid hormone has also been shown to stimulate RNA polymerase activity and the synthesis of ribosomal and, perhaps, messenger RNA. This chain of reactions ultimately enhances the cellular content of functional ribosomes, which subsequently leads to increased synthesis of brain enzymes involved in several important pathways (Tata *et al.*, 1963; Garcia-Argiz *et al.*, 1967).

Studies have shown that the levels of neurotransmitters and the activity of their synthesizing and degrading enzymes are relatively low at birth, and increase progressively during early neonatal life. Since the optimal levels of thyroid hormones are required particularly during the critical period of growth and development (which is the first three weeks of life in rats and six months in children), any deviation in thyroid functioning during neonatal life could lead to abnormal levels of neurotransmitters and related behaviour. The aims of this chapter are: (1) to focus attention on studies concerning the influence of altered thyroid status on certain neurotransmitter systems; (2) to correlate behavioural abnormalities seen during hyperthyroidism with the altered metabolism of norepinephrine (NE), dopamine (DA) and 5-hydroxytryptamine (5-HT) in discrete brain regions; and (3) to examine the validity of hyperthyroid animals as a model for the study of over-functioning in catecholaminergic and serotonergic neurons in the brain.

II. THYROID HORMONE: ONTOGENESIS OF MONOAMINERGIC SYSTEMS IN THE BRAIN

As summarized in Table 10.1, neonatal hypothyroidism led to marked interference with ontogenic increases of a variety of enzymes involved in the synthesis and metabolism of NE, DA, 5-HT and acetylcholine (ACh) in the brain. The endogenous levels of these biogenic amines were also lowered in brains of hypothyroid rats; however, the level of 5-hydroxyindoleacetic acid (5-HIAA) was increased, which is consistent with previous data (Toth and Csaba, 1966). It is assumed that higher levels of this indoleamine metabolite might perhaps be the result of morphological and biochemical alterations in the brain which could have impaired its efflux mechanism(s). The emergence of spontaneous alternation behaviour was delayed (Sobrian *et al.*, 1976) and hypothyroidism greatly decreased behavioural activity in developing animals. Since spontaneous alternation involves intact hippocampal circuitry, it is likely that neonatal thyroidectomy delays the maturation of hippocampal as well as cerebral and cerebellar tissues. More recently, Schalock *et al.* (1979) found that neonatal hypothyroidism induced by propylthiouracil resulted in decreased performance on avoidance and escape learning.

Hyperthyroidism, on the other hand, significantly accelerated the motor performance, as well as the rate of catecholamine synthesis and turnover in the brain (Rastogi and Singhal, 1976a). The synthesis of ACh was also increased following neonatal administration of L-triiodothyronine, as evidenced by enhanced choline acetyltransferase activity and higher levels of brain stem ACh (Rastogi *et al.*, 1977b). In contrast to hypothyroidism, where little or no change in brain amine metabolism could be detected if the radiothyroidectomy was induced after three weeks of neonatal life (Rastogi and Singhal, 1974a, b), the changes in catecholamine metabolism of hyperthyroid animals did not seem to be age-related, since thyroxine administration to adult mice and rats enhanced the turnover of both NE and DA (Beley *et al.*, 1975; Svensson, 1976). However, these data are at variance with the earlier studies of Prange *et al.* (1970) and Emlen *et al.* (1972), who reported reduced catecholamine synthesis in adult rats

Table 10.1: Summary of Behavioural and Neurochemical Alterations Seen in Brains of Neonatal Hypo- and Hyperthyroid Rats

Parameter	Hypothyroidism* (^{131}I or methimazole)	Hyperthyroidism* (T$_3$ treatment)
Tyrosine	No change	Increased
Tyrosine hydroxylase	Decreased	Increased
Norepinephrine	Decreased	No change
4-Hydroxy-3-methoxyphenyl glycol	–	Increased (Keller *et al.*, 1974)
Dopamine	Decreased	Increased
Homovanillic acid	–	Increased
3, 4-Dihydroxyphenylacetic acid	–	Increased
Rate of catecholamine synthesis	–	Increased (Beley *et al.*, 1975; Svensson, 1976)
^{3}H-NE uptake in P$_2$ pellet of brain	–	No change
Responsiveness of NE receptor sites	–	Increased (Strombom *et al.*, 1977)
Responsiveness of DA receptor sites	–	Increased
Tryptophan	No change	Increased
Tryptophan hydroxylase	Decreased	Increased
5-Hydroxytryptophan decarboxylase	No change	–
5-Hydroxytryptamine	Decreased	Unchanged
5-Hydroxyindoleacetic acid	Increased (Toth and Csaba, 1966)	Increased
Rate of 5-Hydroxytryptamine synthesis	–	Increased
^{3}H-5-HT uptake in P$_2$ pellet of brain	–	No change
Choline	–	–
Acetylcholine	^{131}I increased; methimazole unchanged (Hrdina *et al.*, 1975)	Increased
Choline acetyltransferase	Decreased	Increased
Acetylcholine esterase	No change	Increased
Monoamine oxidase	No change	No change
Catechol-O-methyltransferase	Increased	Decreased
Spontaneous locomotor activity	Decreased	Increased
Spontaneous alternation behaviour	Delayed (Sobrian *et al.*, 1976)	–

* A dash indicates that no data are available.

Source: Except for the references cited in the table, all other findings are based on the work published from authors' laboratories. The following are the major references: Rastogi and Singhal (1974a, b; 1976a, b; 1978; 1979); Rastogi *et al.* (1975; 1976a, b; 1977a, b; 1979; 1980b).

treated with thyroxine. These latter investigators suggested that thyroxine treatment in adult animals decreased the turnover of catecholamines, probably by enhancing the sensitivity of pre- and/or postsynaptic receptor sites. It is interesting that repeated treatment with thyroxine in adult mice has been shown to increase NE turnover as well as the responsiveness of NE receptor sites to clonidine, a NE-receptor agonist (Strombom *et al.*, 1977).

III. EFFECT OF APOMORPHINE ON BEHAVIOURAL ACTIVITY IN HYPERTHYROID RATS: EVIDENCE FOR DA-RECEPTOR SUPERSENSITIVITY

In an effort to examine the status of DA-ergic neurons in hyperthyroid animals, neonatally L-triiodothyronine-treated and age-matched rats were challenged to apomorphine, and the consequential changes in behavioural activity and brain DA metabolism were studied.

As shown in Figure 10.1, daily injection of L-triiodothyronine (T_3) for 30 days in developing rats elevated locomotor activity to 218% of control values taken as 100%. T_3 treatment *per se* elicited no stereotyped behaviour, except for occasional circling movement followed by seizures that were seen in some animals immediately after injection. The hyperthyroid animals were more aggressive to handling than normal rats of the same age group. Apomorphine injection in normal rats led to stereotyped behaviour consisting of hypermobility, sniffing, gnawing and occasional rearing on hind legs. As expected, the stereotyped behaviour and hypermobility in normal rats persisted only for less than an hour after each injection of apomorphine. No change in spontaneous locomotor activity could therefore be recorded 24 hours after the 15th injection of apomorphine in normal animals. By contrast, administration of apomorphine further enhanced the locomotor performance in hyperthyroid animals by 242%; this potentiative effect of apomorphine lasted 24 hours or longer after the last injection. Furthermore, the stereotyped behaviour was much more pronounced in apomorphine-treated hyperthyroid animals, as compared to age-matched normal rats receiving apomorphine alone. Hyperthyroid rats treated with apomorphine marched in a row with Straub tail, and displayed 'bizarre social behaviour' consisting of a 'wrestling' posture without fighting (i.e., 'mock fighting') when left in pairs after apomorphine injection. Similar behavioural changes are seen after the administration of amphetamine or L-DOPA together with a peripherally acting decarboxylase inhibitor (Evetts *et al.*, 1970). It is possible that thyroid hormone amplifies the action of apomorphine on the DA-ergic system of the brain by one or more of the following three mechanisms: (1) treatment of neonatal rats with thyroid hormone alters the sensitivity and/or increases the number of DA receptors in the brain; (2) neonatal hyperthyroidism is accompanied by enhanced turnover of DA, thus making more of it available at the receptor site which is being stimulated by apomorphine; and (3) as in adult rats (Engstrom *et al.*, 1975), hyperthyroidism in developing animals increases both the responsiveness of DA-receptor sites and the turnover of catecholamines

Figure 10.1: Effect of chronic apomorphine treatment on spontaneous locomotor activity of normal and neonatally hyperthyroid rats. Each bar represents the mean ± SEM of six rats in a group. One-day-old rats were injected s.c. with T_3 (10 μg/100 g/day) for 30 days to induce hyperthyroidism. Groups of rats pretreated with the vehicle (0.02 N NaOH) or T_3 for 15 days since birth were subsequently given apomorphine (1 mg/kg/day, s.c.) alone or in combination with T_3, respectively, for the remaining 15 days. Then, 24 hours after the last injection of the drug or vehicle, animals were placed individually in a plastic cage resting on a selective-activity meter, to record their spontaneous locomotor activity over a period of 25 minutes, preceded by five minutes of exploration. Data in parentheses express results in percentages, taking the values for control or T_3-treated rats as 100%. An asterisk denotes a statistically significant difference when compared with the values of control rats ($p < 0.05$), while a dagger denotes a statistically significant difference when compared with the values of T_3-treated rats ($p < 0.05$).

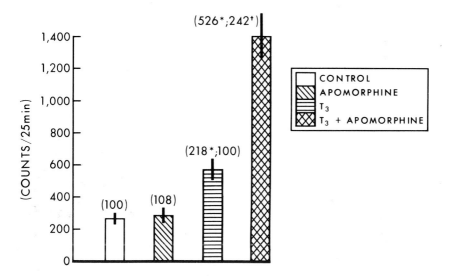

in the brain. Nigrostriatal dopaminergic neurons are critical components of the motor system and are assumed to be activated by muscular movements (Bartholini *et al.*, 1969). Hence, it is likely that the hypermobility and exaggerated responses to environmental stimuli seen in hyperthyroid subjects could partly be due to increased synthesis and turnover of striatal DA and/or hypersensitivity of DA-receptor sites. Animal data that injection of minute amounts of NE induces hyperactivity (Broitman and Donoso, 1971) suggest that hypermobility seen in hyperthyroid rats may, in part, be related to increased turnover of NE in the brain.

Additionally, neonatal hyperthyroidism significantly increased the rate of synthesis and turnover of 5-HT in mid-brain and several other brain regions examined (Rastogi and Singhal, 1976a). Post *et al.* (1973) observed significantly higher levels of 5-HIAA in the cerebrospinal fluid of man subjected to increased psychomotor activity. However, with the existing data, it is difficult to say whether the increased functioning of serotonergic neurons indeed has a role in hyperactivity or else is responsible for some other mental disturbance such as sleep disorders and anxiety (Wheatley, 1972) frequently seen in hyperthyroid patients.

IV. MENTAL ILLNESSES RELATED TO ABNORMAL THYROID FUNCTION: RELEVANCE OF AN ANIMAL MODEL OF HYPERTHYROIDISM

Several studies in human and subhuman primate subjects have demonstrated that the pituitary-thyroidal system can be stimulated under psychologically stressful conditions. Significantly elevated plasma protein-bound-iodine (PBI) levels have been found in a group of army officers subjected to a simulated military stress situation. Mason *et al.* (1973) reported increases in both plasma thyrotropin (TSH) and thyroxine levels in all of the eight normal young men during the 20-minute anticipatory period immediately before their first experimental session involving exercise to the point of exhaustion on a bicycle ergometer. More recently, elevation of plasma TSH levels has also been found in relation to both the chair-restraint and conditioned-avoidance situation in the monkey (Mason, 1975).

Clinical studies have revealed that excessive thyroid secretion produces psychological symptoms including emotional lability, restlessness, irritability, over-reactiveness with predominant anxiety and tension (Eayrs, 1960; Whybrow and Ferrell, 1974). In fact, Wheatley (1972) has reported the presence of anxiety in hyperthyroid subjects. More severe mental disorders may also be seen in hyperthyroid patients if left untreated with antithyroid drugs. Psychosis of the acute organic type has frequently been encountered in severe cases and also during 'thyroid crisis'. The incidence of other psychoses in hyperthyroid individuals has been under some dispute (Bursten, 1961). Lidz and Whitehorn (1949) detected evidence of psychosis in 20% of thyrotoxic patients attending an outpatient clinic. Furthermore, Kleinschmidt *et al.* (1956) considered that 20% of their 84 thyrotoxics were schizophrenic or borderline psychotics. More recently, Rinieris *et al.* (1980) found significantly low levels of serum thyroxine and free-thyroxine index in 24 schizophrenics after six weeks' treatment with chlorpromazine, trifluoroperazine or clozapine. Evidence also exists suggesting some biochemical as well as behavioural similarities between hyperthyroidism and mania, such as increased synthesis and turnover of catecholamines, hyper-mobility, sleeplessness and exaggerated responses to environmental stimuli, etc. (Maletzky and Blachley, 1971). All these observations led us to hypothesize that neonatally hyperthyroid animals could serve as an appropriate model of anxiety and mania. Even though considerable research has been carried out to delineate the neuronal basis of anxiety and mania, the exact nature of the underlying disturbances in the functioning of monoaminergic neurons is not clearly understood. We have therefore employed neonatally hyperthyroid animals as an experimental model to elucidate the mode of action of certain anxiolytic as well as antimanic drugs.

V. EFFECTS OF DIAZEPAM ON BEHAVIOURAL ACTIVITY AND BRAIN NOREPINEPHRINE METABOLISM IN HYPERTHYROID RATS

As shown in Table 10.2, diazepam (DZP) treatment in normal rats produced a tranquilizing effect, as evidenced by depressed locomotor activity. Furthermore, this benzodiazepine suppressed the T_3-stimulated increase in locomotor activity to values which were statistically indistinguishable from those for control animals.

DZP treatment for 15 days in normal rats increased the accumulation of NE without altering the rate of synthesis, synaptosomal uptake of ^{3}H-NE in P_2 pellet (Rastogi *et al.*, 1979) or the activity of the catabolizing enzyme monoamine oxidase (MAO) (unpublished data). These findings suggest that DZP interferes with the neuronal release of this catecholamine, and antagonizes the T_3-induced rise in NE turnover in the brain. This gains support from the data

Table 10.2: Influence of Fifteen-day Treatment with DZP on Spontaneous Locomotor Activity and MOPEG Levels in Whole-Brain (Minus Striatum) of Neonatally Hyperthyroid Rats*

Treatment	Percentage change	
	Spontaneous locomotor activity	MOPEG
Control	(100)	(100)
DZP	(49)**	(72)**
T_3	(212**; 100)	(193**; 100)
T_3 + DZP	(95; 45***)	(117; 61***)

* One-day-old rats were injected daily with T_3 (10 μg/100 g, s.c.) for 30 days to induce hyperthyroidism. A group of 15-day-old rats was injected with DZP (10 mg/kg, s.c.) for 15 days. Furthermore, a group of T_3-treated rats was injected daily with DZP (10 mg/kg, s.c.) along with T_3 for 15 days, beginning from 15 days of age. Controls received daily an equal volume of vehicle. The spontaneous locomotor activity was recorded for 25 minutes, 18 hours after the last injection of T_3 or DZP. The spontaneous locomotor counts in control animals were 228 ± 20 per five minutes in 25 minutes whereas the MOPEG value in whole brain (minus striatum) was 0.47 ± 0.03 μg/g of wet weight of tissue.

** Statistically significant difference when compared with the values of control rats (p <0.05).

*** Statistically significant difference when compared with the values of T_3-treated rats (p <0.05).

Source: Modified from Rastogi *et al.* (1979).

that the levels of 4-hydroxy-3-methoxyphenylglycol (MOPEG) (the main meta-bolite of NE) are significantly lowered in hyperthyroid rats receiving DZP (Table 10.2). In line with our findings, Corrodi *et al.* (1971) had earlier reported that benzodiazepines antagonize the stress-induced rise in catecholamine turnover.

A. Effects on Serotonin Metabolism: A Possible Mechanism of Antianxiety Action

Like NE, serotonin levels were markedly elevated following DZP treatment in brains of normal as well as hyperthyroid rats (Rastogi *et al.*, 1979) suggesting that the turnover of this indoleamine was reduced after benzodiazepine admini-stration. An increased retention of ^{14}C-5-HT injected intraventricularly was observed in rats pretreated with oxazepam for six days (Wise *et al.*, 1972). Using the conflict test and a variety of drugs which specifically affect either NE or 5-HT turnover, these workers later demonstrated that, whereas reduced turn-over of NE may be associated with the antidepressant action, decreased turnover of 5-HT in the brain is responsible for the anxiolytic action of benzodiazepines (Stein *et al.*, 1975). If benzodiazepines reduce the turnover of 5-HT, the concen-tration of its metabolite 5-HIAA should be lowered. However, higher levels of 5-HIAA were reported in several areas of brains of normal rats treated with DZP as well as bromazepam, fenobam and clobazam (Chase *et al.*, 1970; Rastogi *et al.*, 1976b, 1977a, 1980a; Lapierre *et al.*, 1981). A similar rise in 5-HIAA levels of hyperthyroid rats treated with DZP was reported by Rastogi *et al.* (1979). Chase *et al.* (1970) demonstrated that benzodiazepine treatment impedes the efflux of 5-HIAA from the brain; hence it is likely that higher levels of 5-HIAA in several brain regions of normal and hyperthyroid rats seen after DZP treatment could be due to their poor clearance from the brain.

Studies have also implicated serotonergic neurons in anxiety (Stein *et al.*, 1975). It has been suggested that anxiety is associated with increased functioning of the serotonergic neuronal system. Since evidence suggests some similarity between hyperthyroidism and anxiety (Wheatley, 1972), it seems probable that the enhanced turnover of 5-HT might be associated with anxiety, whereas increased turnover of NE with agitation is seen in anxious patients. Clinical studies measuring the levels of these monoamines and their metabolites in biological fluids of anxious patients treated with benzodiazepines are currently underway in our laboratory to substantiate this hypothesis. Our animal data available thus far suggest that hyperthyroid animals may be used as an experi-mental model for investigating the underlying mechansim of action of benzo-diazepines; however, further research involving the measurement of several other psychophysiological parameters of anxiety in hyperthyroid rats is desirable. Furthermore, the role of GABA-ergic neurons in mediating some of the be-havioural changes seen during hyperthyroidism needs to be examined. It has been shown that DZP enhances the functioning of GABA-ergic neurons in the brain (Costa *et al.*, 1975); hence the effect of DZP and GABA should be investi-gated in this animal model. GABA is a powerful inhibitory neurotransmitter in the brain, and its level may be lowered during hyperthyroidism. Furthermore, long-term DZP treatment should antagonize T_3-induced decreases in brain GABA levels. Such experiments are currently in progress.

VI. HYPERTHYROID ANIMALS AS A MODEL OF HYPERACTIVITY: STUDIES ON THE MODE OF ANTIMANIC ACTION OF LITHIUM

Lithium is known to show avidity for thyroid gland; it interferes with the production of thyroxine, resulting in a deficiency of the circulating hormone, which, in turn, may lead to enhanced levels of thyroid-stimulating hormone (Schou *et al.*, 1968; Perrild *et al.*, 1978; Transbol *et al.*, 1978). Chronic treatment with T_3 also has been found to elicit a protective action against the toxic effects of lithium. Despite the suggestive association, only a handful of attempts have been made to relate thyroid function to manic illness (Maletzky and Blachley, 1971). The neurochemical and behavioural studies have shown that hyperthyroidism has a number of features in common with mania (Beley *et al.*, 1975; Goodwin and Sack, 1973; Messiha *et al.*, 1970; Singhal and Rastogi, 1978). We therefore employed neonatally hyperthyroid animals as a model to elucidate further the mode of antimanic action of lithium. Our data demonstrate that, whereas administration of lithium carbonate (60 mg/kg/day) for ten days beginning from 20 days of age failed to alter locomotor activity in normal animals, it antagonized the T_3-stimulated rise in mobility (Rastogi and Singhal, 1977a). It is interesting that lithium has been shown to decrease the amphetamine-(Segal *et al.*, 1975) and L-DOPA-stimulated (Smith, 1976) increases in locomotor activity; both of these pharmacological agents, like T_3, are known to increase DA levels in the brain.

A. Effects of Lithium on NE and DA Metabolism in Discrete Brain Regions of Hyperthyroid Rats

As shown in Table 10.3, long-term administration of lithium in normal animals enhanced the synthesis and turnover of DA, as evidenced by increased TH activity and higher levels of the metabolite, 3, 4-dihydroxyphenylacetic acid (DOPAC) in striatum. The endogenous levels of DA in striatum were slightly decreased (23%). In contrast, NE levels remained unaltered in hypothalamus (as were DA levels) in lithium-treated rats, suggesting that increased utilization of this catecholamine probably kept pace with increased synthesis (Rastogi and Singhal, 1977b). Segal *et al.* (1975) have reported a similar rise in TH activity of substantia nigra and caudate-putamen. An increased turnover of NE has also been found in rats exposed acutely or chronically to this alkali metal (Schildkraut *et al.*, 1969; Greenspan *et al.*, 1970; Poitou and Bohuon, 1975). The enhanced turnover of NE and DA seen after lithium treatment is somewhat surprising, since the classical hypothesis is that excess of brain NE (Goodwin and Sack, 1973) and possibly DA (Messiha *et al.*, 1970) accompany mania. If this is true, then lithium should reduce the turnover of these brain amines rather than

Table 10.3: Effect of Lithium Treatment on TH Activity, DA, DOPAC and NE Levels in Certain Brain Areas of Neonatally Hyperthyroid Rats*

Treatment	Striatum			Hypothalamus		Midbrain
	TH (nmole DOPA/g/h)	DA (μg/g)	DOPAC (μg/g)	DA (μg/g)	NE (μg/g)	COMT (nmole/g/h)
Control	78.31 ± 5.27 (100)	6.89 ± 0.55 (100)	1.93 ± 0.14 (100)	0.67 ± 0.04 (100)	1.79 ± 0.09 (100)	389.4 ± 18.9 (100)
Li	101.02 ± 6.1 (129)**	5.31 ± 0.41 (77)**	2.59 ± 0.19 (134)**	0.65 ± 0.03 (97)	1.68 ± 0.12 (94)	319.3 ± 12.6 (82)**
T$_3$	108.9 ± 6.3 (139**;100)	11.22 ± 0.67 (163**;100)	5.08 ± 0.36 (263**;100)	0.85 ± 0.05 (127**;100)	1.75 ± 0.08 (98;100)	311.14 ± 9.7 (80**;100)
T$_3$ + Li	76.1 ± 6.1 (97;70***)	7.86 ± 0.54 (114;70***)	2.28 ± 0.23 (118;45***)	0.65 ± 0.04 (97;76***)	3.05 ± 0.23 (175**;175***)	401.41 ± 18.3 (103;129***)

* Each value is the mean ±SEM of seven animals in the group. One-day-old rats were injected daily with T$_3$ (10 μg/100 g; s.c.) for 30 days. Groups of rats pretreated with 0.02 N NaOH (vehicle) or T$_3$ for 20 days since birth were subsequently injected with lithium (60 mg/kg; i.p.) alone or in combination with T$_3$, respectively, for the remaining ten days. Animals were killed 24 hours after the last injection of T$_3$ or lithium. Data in parentheses express results in percentages, taking the values for control or T$_3$-treated animals as 100%.
** Statistically significant difference when compared with the values of control rats (p < 0.05).
*** Statistically significant difference when compared with the values of T$_3$-treated rats (p < 0.05).

Source: Modified from Rastogi and Singhal (1977b).

increase it, as seen in normal animals. Our data on neonatally hyperthyroid animals (Table 10.3), however, show that lithium antagonized the T_3-stimulated increase in both synthesis and turnover of DA and NE in brain. The elevated levels of NE in hypothalamus and other brain regions of lithium-treated hyperthyroid rats (Rastogi and Singhal, 1977b) might be due to accumulation of NE, the evoked liberation of which is presumed to be decreased *in situ* after lithium administration (Katz *et al.*, 1968; Bindler *et al.*, 1971). Our more recent finding that lithium treatment decreased brain levels of MOPEG, an extraneuronal metabolite of NE, suggests that this antimanic drug lowers the concentration of NE within the synaptic cleft. The data that lithium antagonized amphetamine-, L-DOPA- and T_3-stimulated increases in locomotor activity, while producing no change in normal animals, would suggest that lithium exerts antiphasic or damping effects upon mood swings, which might be mediated via catecholaminergic systems in the brain. Hence, the mode of antimanic action of lithium should not be derived only from data obtained in normal animals with little or no change in neurotransmitter systems.

How does lithium exert behavioural-suppressant effects in hyperthyroid rats? Does lithium antagonize the T_3-stimulated rise in brain NE and DA turnover directly, or does it elicit its effects primarily by interfering with thyroid function? Studies indicate that lithium decreases the secretion of thyroxine (Fieve and Platman, 1968; Shopsin *et al.*, 1969). Such an action of lithium has, in fact, made this alkali metal clinically useful in the treatment of thyrotoxicosis (Lazarus *et al.*, 1974). Hence, the question as to whether lithium alters NE and DA metabolism in hyperthyroid animals primarily by influencing the functioning of catecholaminergic neurons in the brain remains unanswered at present. However, our studies suggest that hyperthyroid animals in which lithium decreased catecholamine synthesis and turnover (the property that an antimanic drug should ideally possess), may be employed as a tool to investigate the mode of action of new antimanic agents.

VII. POTENTIATION OF ANTIDEPRESSANT ACTION OF TRICYCLICS BY L-TRIIODOTHYRONINE AND THYROTROPIN-RELEASING HORMONE

Prange *et al.* (1968, 1969) reported that administration of T_3 together with imipramine potentiated the antidepressant action of the latter compound in both retarded and nonretarded depressed patients. In subsequent years, these authors found that TRH also elicited considerable antidepressant properties (Prange and Wilson, 1972), although controversy seems to surround this issue (Prange *et al.*, 1978). Loosen *et al.* (1974), in a double-blind placebo-controlled study, reported that TRH produced beneficial effects in depressed alcoholics three hours after its injection, with only mild side effects. It was also found that the antidepressant action of TRH was associated with increased central dopaminergic activity, as evidenced by elevated baseline levels of growth hormone and low baseline levels of prolactin. Animal studies have also shown that TRH enhances DA as well as NE turnover in the brain (Keller *et al.,* 1974; Constantinidis *et al.,* 1974; Agarwal *et al.,* 1977). Neurophysiologic studies of Koranyi *et al.* (1976) revealed that, following a few systemic injections of TRH, changes in multiple-unit activity (MUA) were similar to those seen after a single injection of imipramine.

Unlike tricyclics, TRH possesses a more rapid onset of antidepressant action (Prange and Wilson, 1972) and is relatively devoid of side effects on the cardiovascular system and CNS, and does not suppress rapid-eye-movement (REM) sleep (Koranyi *et al.,* 1976). We therefore examined the influence of concurrent treatment of TRH and imipramine on NE and DA turnover in rat brain. Our data demonstrate that this tripeptide markedly potentiated the effects of imipramine on NE and DA turnover. The levels of HVA and MOPEG were markedly elevated in rats treated daily with TRH and imipramine for ten days, as compared to those given either of these two drugs singly (Table 10.4). Thus, by including TRH in the therapeutic regimen, the doses of imipramine and the inherent side effects seem to be minimized (Rastogi *et al.,* 1980b).

Table 10.4: Effects of Imipramine Alone and in Combination with TRH on Striatal HVA and Whole-brain MOPEG Levels*

Treatment	HVA (μg/g)	MOPEG (μg/g)
Control	0.61 ± 0.02 (100)	0.39 ± 0.02 (100)
TRH	0.72 ± 0.03 (118)**	0.57 ± 0.03 (148)**
Imipramine	0.81 ± 0.05 (133**; 100)	0.50 ± 0.04 (128**; 100)
Imipramine + TRH	0.97 ± 0.05 (159**; 120)	0.71 ± 0.06 (182**; 142)

* Each value represents the mean ± SEM of six rats in each group (except the control group, which consisted of twelve animals). TRH was injected daily in a dose of 20 mg/kg (in two equally divided doses at 8.0 a.m. and 5.0 p.m.), while imipramine was given at the dose level of 10 mg/kg for ten days by the i.p. route. Appropriate controls received equal volumes of physiological saline. Animals were killed 18-20 hours after the last injection. Data in parentheses express results in percentage, taking the values of control and imipramine-treated rats as 100%.

** Statistically significant difference when compared with the values of control rats (p $<$0.05).

*** Statistically significant difference when compared with the values of imipramine-treated animals (p $<$0.05).

Source: Modified from Rastogi *et al.* (1980b).

VIII. SUMMARY AND CONCLUSIONS

Results presented in this chapter demonstrate that thyroid hormone exerts an important regulatory influence on the ontogenic pattern of a variety of neurotransmitter systems in rat brain. Behavioral changes including anxiety and elation seen in hyperthyroid subjects may be associated with increased turnover of brain serotonin and catecholamines. However, the role of other neurotransmitters (e.g., GABA) and peptides (substance P, endorphins) needs to be studied, in order to gain deeper insight into the neuronal mechanisms underlying the behavioral changes seen during thyroid disorders. The results support the contention that hyperthyroid animals present a model for investigation of the etiology and treatment strategies of psychiatric disorders such as anxiety and mania. Lastly, it is shown that TRH enhances the turnover of brain NE and DA and potentiates the effects of imipramine on these monoamines. However, in the light of the clinical data that TRH does not improve all depressive illnesses, there is an urgent need to synthesize chemically related hypothalamic peptides which would have a longer biological halflife and an ability to penetrate the blood-brain barrier faster than TRH.

ACKNOWLEDGEMENTS

This work was supported by a grant from the Ontario Mental Health Foundation.

REFERENCES

Agarwal, R.A., Rastogi, R.B. and Singhal, R.L. (1977). *Neuroendocrinology 23*, 236

Balazs, R. (1971). *Proc. Conf., 1969*, 273

Bartholini, G., Blum, J.E. and Pletscher, A. (1969). *J. Pharm. Pharmacol., 21*, 297

Beley, A., Beley, P. and Bralet, J. (1975). *Arch. Int. Physiol. Biochem., 83*, 471

Bindler, E.H., Wallach, M.B. and Gershon, S. (1971). *Arch. Int. Pharmacodyn. Ther., 190*, 150

Broitman, S.T. and Donoso, A.O. (1971). *Experientia 27*, 1380

Bursten, B. (1961). *Arch. Gen. Psychiat., 4*, 267

Chase, T.N., Katz, R.I. and Kopin, I.J. (1970). *Neuropharmacology 9*, 103

Constantinidis, J., Gaillard, J.M., Hovaguimiam, T. and Tissot, R. (1974). *Experientia 30*, 1182

Corrodi, H., Fuxe, K., Lidbrink, P. and Olson, L. (1971). *Brain Res., 29*, 1

Costa, E., Guidotti, A., Mao, C.C. and Suria, A. (1975). *Life Sci., 17*, 167

Eayrs, J.T. (1960). *Br. Med. Bull., 16*, 122

Emlen, W., Segal, D.S. and Mandell, A.J. (1972). *Science 175*, 79

Engstrom, G., Strombom, U. Svensson, T.H. and Waldeck, B. (1975). *J. Neural. Trans., 37*, 1

Evetts, K.D., Uretsky, N.J., Iversen, L.L. and Iversen, S.D. (1970). *Nature 225*, 961

Fieve, R.R. and Platman, S.R. (1968). *Am. J. Psychiat., 125*, 527

Garcia-Argiz, C.A., Pasquini, J.M., Kaplun, B. and Gomez, C.J. (1967). *Brain. Res., 6*, 635

Gelber, S., Campbell, P.L., Deibler, G.E. and Sokoloff, L. (1964). *J. Neurochem., 11*, 221

Goodwin, F.K. and Sack, R.L. (1973). In *Frontiers in Catecholamine Research*, ed. E. Usdin and S. Snyder (Pergamon Press, Oxford) p. 1157

Greenspan, K., Aronoff, M.A. and Bojdanski, D.F. (1970). *Pharmacology 3*, 129

220   *Thyroid Hormone and Regulation*

Hrdina, P.D., Gosh, P.K., Rastogi, R.B. and Singhal, R.L. (1975). *Can. J. Physiol. Pharmacol.,*
53, 709
Katz, R.I., Chase, T.N. and Kopin, I.J. (1968). *Science 162,* 466
Keller, H.H., Bartholini, G. and Pletscher, A. (1974). *Nature 248,* 529
Kleinschmidt, H.J., Waxenberg, S.E. and Cukor, R. (1956). *J. Mt. Sinai Hosp. NY, 23,* 131
Koranyi, L., Tamasy, V., Lissak, K., Kiraly, I. and Borsy, J. (1976). *Psychopharmacology*
49, 197
Lapierre, Y.D., Rastogi, R.B. and Singhal, R.L. (1981). *Gen. Pharmacol.* (in press)
Lazarus, J.H., Richards, A.R. and Addison, G.M. (1974). *Lancet ii,* 1160
Lidz, T. and Whitehorn, J. (1949). *J. Am. Med. Assoc., 139,* 698
Loosen, P.T., Prange, A.J., Jr. and Wilson, I.C. (1974). *Arch. Gen. Psychiat., 36,* 540
Maletzky, B. and Blachley, P. (1971). In *The Use of Lithium in Psychiatry,* ed. B. Maletzky
and P.H. Blachley (CRC Press, Cleveland, Ohio) p. 47
Mason, J.W. (1975). In *Emotions – Their Parameters and Measurement,* ed. L. Levi (Raven
Press, New York) p. 143
Mason, J.W., Hartley, L.H. and Kotchen, T.A. (1973). *J. Clin. Endocrinol. Metab., 14,* 1567
Messiha, F.S., Agallianos, J. and Clower, C. (1970). *Nature 225,* 868
Nicholson, J.L. and Altman, J. (1972). *Brain Res., 44,* 13
Patel, A.J., Lewis, P.D., Balazs, R., Bailey, P. and Lai, M. (1979). *Brain Res., 172,* 57
Perrild, H., Madsen, S.N. and Hansen, J.E.M. (1978). *Brit. Med. J., 2,* 1108
Poitou, P. and Bohuon, C. (1975). *J. Neurochem., 25,* 535
Post, R.M., Kotin, J., Goodwin, F.K. and Gordon, E.K. (1973). *Amer. J. Psychiat., 130,* 67
Prange, A.J., Jr. and Wilson, I.C. (1972). *Psychopharmacology 26,* 82
Prange, A.J., Jr., Wilson, I.C., Rabon, A.M. and Lipton, M.A. (1968). *Excerpta Medica*
International Congress Series 180, 532
Prange, A.J., Jr., Wilson, I.C., Rabon, A.M. and Lipton, M.A. (1969). *Am. J. Psychiat., 126,*
457
Prange, A.J., Jr., Meek, J.L. and Lipton, M.A. (1970). *Life Sci., 9,* 901
Prange, A.J., Jr., Nemeroff, C.B., Lipton, M.A., Breese, C.R. and Wilson, I.C. (1978). In
Handbook of Pharmacology, ed. L.L. Iversen, S.D. Iversen and S.H. Snyder (Plenum
Press, New York) p. 1
Rastogi, R.B. and Singhal, R.L. (1974a). *Brain Res., 81,* 253
Rastogi, R.B. and Singhal, R.L. (1974b). *J. Pharmacol. Exp. Therap., 191,* 72
Rastogi, R.B. and Singhal, R.L. (1976a). *J. Pharmacol. Exp. Therap., 198,* 609
Rastogi, R.B. and Singhal, R.L. (1976b). *Life Sci., 18,* 851
Rastogi, R.B. and Singhal, R.L. (1977a). *Can. J. Physiol. Pharmacol., 55,* 490
Rastogi, R.B. and Singhal, R.L. (1977b). *J. Pharmacol. Exp. Therap., 201,* 92
Rastogi, R.B. and Singhal, R.L. (1978). *Arch. Pharmacol., 304,* 9
Rastogi, R.B. and Singhal, R.L. (1979). *Psychopharmacology 62,* 287
Rastogi, R.B., Singhal, R.L. and Hrdina, P.D. (1975). *Neuropharmacology 4,* 747
Rastogi, R.B., Lapierre, Y.D. and Singhal, R.L. (1976a). *J. Neurochem., 26,* 443
Rastogi, R.B., Lapierre, Y.D. and Singhal, R.L. (1976b). *J. Psychiat. Res., 13,* 65
Rastogi, R.B., Agarwal, R.A., Lapierre, Y.D. and Singhal, R.L. (1977a). *Eur. J. Pharmacol.,*
43, 91
Rastogi, R.B., Hrdina, P.D., Dubas, T. and Singhal, R.L. (1977b). *Brain Res., 123,* 188
Rastogi, R.B., Lapierre, Y.D. and Singhal, R.L. (1979). *J. Psychiat. Res., 15,* 7
Rastogi, R.B., Lapierre, Y.D. and Singhal, R.L. (1980a). *Proceedings of the Second Annual*
Meeting of the Canadian College of Neuropsychopharmacology, Hamilton, April 26-7,
1979, p.26
Rastogi, R.B., Singhal, R.L. and Lapierre, Y.D. (1980b). *Psychoparmacology, 72,* 85
Rinieris, P., Christodonlou, G.N., Souvatzoglou, A., Koutras, D.A. and Stefani, C. (1980).
Neuropsychobiol., 6, 29
Schalock, R.L., Brown, W.J. and Smith, R.L. (1979). *Developm. Psychobiol., 12,* 187
Schildkraut, J.J. Logue, M.A. and Doge, G.A. (1969). *Psychopharmacologia 14,* 135
Schou, M., Amdisen, A., Eskajaer-Jensen, S. and Olsen, T. (1968). *Brit. Med. J., 3,* 710
Segal, D.S., Callaghan, M. and Mandell, A.J. (1975). *Nature 254,* 58
Shopsin, B., Blum, M. and Gershon, S. (1969). *Comp. Psychiat., 10,* 215
Singhal, R.L. and Rastogi, R.B. (1978). *Adv. Pharmacol. Chemother., 15,* 203
Smith, D.F. (1976). *Pharmacol. Res. Comm., 8,* 575

Smith, D.F. (1976). *Pharmacol. Res. Comm., 8,* 575

Sobrian, S.K., Pappas, B.A., Edson, N., Rastogi, R. and Singhal, R.L. (1976). *Res. Comm. Psychol. Psychiat. Behav., 3,* 419

Stein, L., Wise, C.D. and Belluzzi, J.D. (1975). In *Mechanism of Action of Benzodiazepines,* ed. E. Costa and P. Greengard (Raven Press, New York) p. 29

Strombom, U., Svensson, T.H., Jackson, D.M. and Engstrom, G. (1977). *J. Neural Trans., 41,* 73

Svensson, T.H. (1976). *Proceedings of Tenth Congress of College of International Neuropsychopharmacology,* p. 104

Tata, J.R., Ernster, L., Lindberg, O., Arrhenims, E., Pedersen, S. and Hedman, R. (1963). *Biochem. J., 86,* 408

Toth, S. and Csaba, B. (1966). *Experientia 22,* 755

Transbol, I., Christiansen, C. and Baastrup, P.C. (1978). *Acta Endocr., 87,* 759

Weichsel, M.E. (1974). *Brain Res., 78,* 455

Wheatley, D. (1972). *Arch. Gen. Psychiat., 26,* 229

Whybrow, P.C. and Ferrell, R. (1974). In *The Thyroid Axis, Drugs and Behaviour,* ed. A.J. Prange, Jr. (Raven Press, New York) p. 5

Wise, C.D., Berger, B.D. and Stein, L. (1972). *Science 171,* 180

11 BEHAVIOURAL STUDIES WITH LITHIUM IN RATS: IMPLICATIONS FOR ANIMAL MODELS OF MANIA AND DEPRESSION

Philip E. Harrison-Read

TABLE OF CONTENTS

I	INTRODUCTION	225
II	SHORT REVIEW OF SOME EFFECT OF LITHIUM ON BEHAVIOUR IN RODENTS AND THE NEUROCHEMICAL MECHANISMS INVOLVED	227
	A. The effects of lithium on exploratory activity and sensory analysis	227
	B. Catecholaminergic mechanisms involved in the action of lithium on drug-induced motor hyperactivity	228
	C. Lithium-induced alterations in serotonergic mechanisms	230
III	RECENT EXPERIMENTS	233
	A. General observations during lithium pretreatment	233
	B. Exploratory behaviour in the Y maze after short-term and long-term lithium pretreatment	239
	(a) Y-maze testing	239
	(b) Locomotion and rearing in the Y maze	239
	(c) Investigation of the pots in the Y maze	239
	(d) Alternation of Maze-arm choices	241
	(e) Response to stimulus change	243
	(f) The involvement of serotonergic mechanisms in the altered response to stimulus change after lithium pretreatment	244
	C. The effect of lithium on exploratory choice behaviour in the T maze: the optimum-arousal concept	248
	D. Behavioural evidence for serotonergic supersensitivity after long-term lithium pretreatment	252
	E. Is activity in dopaminergic pathways increased by long-term pretreatment with lithium?	257
IV	SUMMARY AND CONCLUSIONS: OUTLINE FOR A NEW MODEL OF THE THERAPEUTIC ACTIONS OF LITHIUM	258
	REFERENCES	260

I. INTRODUCTION

The use of lithium salts in psychiatry has revolutionized the management of manic-depressive disorders (Fieve, 1977). Lithium is effective in normalizing rather than suppressing behavioural and emotional disturbances in three out of four cases of mania, and it may be helpful in some cases of depression as well (Schou, 1968; Shopsin et al., 1979). Most interesting of all is the proven ability of maintenance lithium therapy to reduce the severity and frequency of both manic and depressive episodes (Schou, 1968; Fieve, 1977). These therapeutic properties offer an exciting challenge for research, because knowledge of the way in which lithium works is likely to throw light on the nature and causes of manic-depressive disorders.

The lithium ion shares many of the properties of sodium, potassium, calcium and magnesium (Schou, 1957; Bunney and Murphy, 1976; Hendler, 1978), and so has numerous and diverse effects on biological systems. In deciding which of these are likely to be relevant to lithium's therapeutic actions, most investigators have used the catecholamine and indoleamine hypotheses of affective disorders as their main frame of reference. The proven therapeutic effectiveness of lithium has in turn stimulated much research into disturbances of electrolyte metabolism, which may be at the root of altered monoamine functions in affective disorders (see reviews by Shaw, 1973; Hendler, 1978).

By identifying effects in animals which are unique to lithium, it may be possible to devise new theories about the causes and pathogenesis of manic-depressive disorders. The opportunity to investigate these factors directly in patients is severely limited by ethical and practical difficulties, so precise and readily testable hypotheses based on animal experiments are essential. If it is held that manic-depressive disorders result from specific organic lesions in the brain, or perhaps in other systems of the body, there would seem to be a stronger case for studying the effects of lithium on biochemical and physiological functions in animals, rather than on behaviour. At the more basic biological 'level' represented by the former, it may be easier to predict actions of lithium on abnormal systems from observations on the normal. Also the validity of extrapolating from animals to man may be less questionable because superficial resemblances in the behaviour of different species may belie quite different ethological, physiological and biochemical mechanisms. However, firm evidence for a biochemical lesion in manic-depressive disorders is lacking, and these conditions are still mainly defined in terms of disturbed behaviour and mood. Lithium may exert its therapeutic effects, not by counteracting or compensating for a biochemical lesion, but by modifying or preempting maladaptive responses which have arisen out of interactions between a number of unfavorable circumstances in the internal and external environments. Viewed in this way, lithium's actions, although presumably based on neurochemical mechanisms, may only be fully understood at the level of the behaving organism.

Since there is little evidence for naturally occurring affective disorders in animals, the search for animal analogs of the therapeutic actions of lithium is likely to be difficult. When administered to normal animals or people, the reported effects of lithium on behaviour or mental functions are unimpressive (Johnson, 1975; Schou, 1968, 1976). However the view that the unique therapeutic actions of lithium have no counterparts in normal animal or human behaviour seems unduly pessimistic. Such effects probably do occur, but they are likely to be subtle, and not necessarily similar to the actions which constitute the therapeutic actions of lithium. This should not be surprising in view of the importance of 'behavioural baselines' for determining both qualitative and quantitative effects of drugs in animals (e.g., Dews and DeWeese, 1977).

In an attempt to demonstrate more striking behavioural effects of lithium, it has been tested against a number of drug-induced hyperactivity states which bear a superficial resemblance to mania (Cox *et al.*, 1971). Although the validity of these analogies may be doubtful, any selective effects of lithium which emerge from this kind of study may form the basis of a working hypothesis or model of lithium's therapeutic actions. These hypotheses may guide subsequent animal experiments, and ultimately lead to predictions which can be tested in man.

It behoves investigators who are interested in 'subtle' behavioural actions of lithium to use doses which achieve subtoxic lithium plasma concentrations (a 24-hour average of less than about 1.5 mM; Schou, 1958). There is less clear guidance for fixing the lower limit to lithium plasma levels in animal studies, although it seems common sense not to go below the 0.5-0.8 mM range generally agreed to be the minimum for therapeutic efficacy in man.

The duration of lithium administration is also a crucial variable in animal studies, as it is in the clinical situation. Administration of lithium for five to seven days is necessary before a therapeutic response is apparent in mania (Shopsin *et al.*, 1979). Part of this delay may correspond to the time taken for intracellular concentrations of lithium in the brain to approach a steady state. When rats are given a single dose of lithium, the concentration in brain tissue rises slowly, reaching a peak after about eight hours (Ho *et al.*, 1970a; Mukherjee *et al.*, 1977). Whilst brain levels of lithium are still rising, central effects are not likely to be fully developed, and they are probably easily obscured by the effects of lithium on peripheral systems. Maintenance lithium therapy may induce changes in the brain which take longer to develop than the antimanic effect seen after the first week or so. Long-term effects of lithium may arise as a reaction to the initial perturbations of function produced by lithium administration, and they may be crucial for the prevention of mania and depression during maintenance treatment with lithium.

Obviously, it is important to compare the effects on animal behaviour of short-term and long-term lithium administration, but until recently this approach has been neglected. Early studies of our own on rat behaviour (Cox *et al.*, 1971; Harrison-Read and Steinberg, 1971) demonstrated quantitative differences between acute, short-term (seven days) and long-term (14 days) lithium admini- stration. Our recent behavioural experiments have been concerned with the possibility that the duration of lithium pretreatment affects the action of lithium qualitatively as well. Before describing these experiments, some relevant be- havioural studies reported in the literature will be outlined. This review will be quite selective; for more comprehensive surveys of the literature, the reader is referred to the reviews by Johnson (1975) and Smith (1977a).

II. SHORT REVIEW OF SOME EFFECTS OF LITHIUM ON BEHAVIOUR IN RODENTS AND THE NEUROCHEMICAL MECHANISMS INVOLVED

A. The Effect of Lithium on Exploratory Activity and Sensory Analysis

Starting with Cade's original observation of lethargy and unresponsiveness in guinea pigs given large doses of lithium (Cade, 1978), there have been numerous reports of lithium reducing motor activity in animals. Exploratory behaviour of rats and mice in novel environments (e.g., rearing onto the hind legs) often seems to be affected selectively by lithium (Johnson, 1975; Smith, 1977a). However, the argument that short-term treatment with lithium affects activity by directly impairing animals' ability to analyze and process sensory information (Johnson, 1975, 1978) is not very convincing. For example, habituation of exploratory activity, which requires processing and storage of information about a novel environment (Halliday, 1968), was not altered by lithium. However, on with- drawal of lithium, rats previously habituated to the test chamber increased their activity to the level shown by control rats which had not been tested before (Johnson, 1972a). Information processing may have been reduced by lithium, but this could have resulted indirectly from impaired ability to initiate the motor acts necessary for investigation of environmental stimuli (Johnson and Wormington, 1972). Alternatively, experience acquired under the influence of lithium may not have 'transferred' to the undrugged state.

An increase in activity on exposure to a change in an accustomed environment probably reflects animals' ability to process information about their surroundings, and to respond to and register 'mismatches' between present and past experience (Halliday, 1968). An attempt to investigate the effects of lithium on rats' response to stimulus change was thwarted by the failure of the control group to respond to the stimulus used (Smith, 1976a). Other evidence suggests that, far from reducing the control which environmental stimuli exert over behaviour, lithium pretreatment may intensify the increase in arousal they produce, and so causes defensive suppression of exploratory activity (Barnett, 1975). Thus, lithium chloride (LiCl, 2 mmol/kg) administered over four days of testing reduced rats' activity in a brightly lit open field, but did not do so when the field was illuminated by dim red light (Gray *et al.*, 1976).

In one study, the reduction in motor activity which occurred after single doses of LiCl was not maximal until eight hours after intraperitoneal (i.p.) injections, and correlated with lithium levels in the brain (Mukherjee *et al.*, 1977). However, in other experiments, the reduction in activity following lithium administration reached a maximum within 20 minutes of i.p. injection (Johnson and Wormington, 1972; Johnson, 1972a, b) when brain levels of lithium were likely to be low. This suggests that lithium may also reduce motor activity by a peripheral action, e.g., through the effects of general malaise (Smith, 1977a, b). Although the ability to initiate movements may be impaired by lithium (Schou, 1968), muscle weakness or increased susceptibility to fatigue do not seem to be involved, because rearing height (Johnson and Wormington, 1972) and endurance in a swimming test (Smith and Smith, 1973) were unaffected by acute treatment with lithium.

These actions of lithium on animal behaviour seem to have little relevance to its therapeutic effects, particularly as reductions in exploratory activity become less marked as pretreatment is prolonged (Segal *et al.*, 1975). Although the flight of ideas, pressure of speech and 'fallacious insights' which occur in mania have been attributed to 'overprocessing' of stimulus information (Johnson, 1975, 1978), this concept is too ill-defined to be applicable to studies of exploratory activity in animals. In any case, as mentioned above, there is no sound evidence for a reduction in 'sensory analysis' by lithium. Quite obviously, the need is for more detailed study of the effects of short-term and long-term lithium pretreatment on the central processes which determine animals' responses to novel environments.

B. Catecholaminergic Mechanisms Involved in the Action of Lithium on Drug-induced Motor Hyperactivity

Amphetamine, either alone or in combination with chlordiazepoxide, increases motor activity in rats and mice, probably because it increases the availability at synapses of noradrenaline (NA) or dopamine (DA) or both by facilitating neuronal release and reducing reuptake (Snyder *et al.*, 1974; Segal, 1975). It is easy to draw analogies between the behavioural effects of amphetamine in animals and the increased motor activity and distractibility which are prominent features of mania (Murphy, 1977). Although the increases in motor activity produced by amphetamine and amphetamine-chlordiazepoxide mixtures are more marked when animals are tested in novel rather than familiar environments (Rushton *et al.*, 1968; Davies *et al.*, 1974), a number of observations suggest that animals' 'awareness' of their surroundings, and their ability to process the information they contain, may be impaired by these drug treatments. For example, amphetamine may cause repetition of certain types of motor activity at the expense of behaviour which represents a selective approach to novel stimuli (Kumar, 1969; Robbins and Iversen, 1973). Similarly, animals treated with a d-amphetamine-chlordiazepoxide mixture appear to investigate the same environmental stimuli repeatedly, and fail to habituate to their surroundings

(Rushton *et al.*, 1968; Cox *et al.*, 1971; Davies *et al.*, 1974).

The resemblance between drug-induced motor-hyperactivity states in rodents and the psychomotor excitement typical of manic patients, together with the possible involvement of catecholaminergic mechanisms in both, has been the main impetus behind attempts to reverse this type of animal behaviour with lithium (Johnson, 1975; Murphy, 1977; Smith, 1977a). A number of studies have shown that lithium salts reduce or abolish some drug-induced hyperactivity states without markedly affecting the activity of otherwise untreated animals. However, lithium probably does not reduce hyperactivity by impairing sensory analysis (Johnson, 1975). On the contrary, lithium has been reported to normalize the quality as well as the quantity of exploratory activity in mice given a d-amphetamine-chlordiazepoxide mixture (Davies *et al.*, 1974).

Short-term pretreatment with lithium may alter motor hyperactivity by affecting presynaptic mechanisms in catecholaminergic pathways. Inhibition of catecholamine synthesis with α-methyl tyrosine had a similar effect to lithium in reducing hyperactivity produced by amphetamine (Weissman *et al.*, 1966), amphetamine-chlordiazepoxide mixtures (Davies *et al.*, 1974) and scopolamine (Sanger and Steinberg, 1974). A selective reduction in NA synthesis using dopamine-β-hydroxylase inhibitors was also effective in reducing hyperactiviy in mice given an amphetamine-chlordiazepoxide mixture (Poitou *et al.*, 1975). Pretreatment with lithium for five days abolished the increased locomotor activity in rats produced by L-DOPA (150 mg/kg + a peripheral decarboxylase inhibitor) (Smith, 1976b). The inhibitory effect of single doses of lithium on the hyperactivity produced by d-amphetamine in mice was antagonized by doses of L-DOPA (5-50 mg/kg) which on their own caused inhibition rather than stimulation of activity (Berggren *et al.*, 1978). Acute lithium treatment did not affect the hyperactivity produced by a mixture of apomorphine and clonidine in mice pretreated with α-methyl tyrosine, indicating a lack of effect at or beyond the receptor level. By contrast, when lithium was administered for up to three weeks, an effect on receptor mechanisms may have been responsible for the reduced stereotyped sniffing behaviour produced in rats by apomorphine and d-amphetamine (Flemenbaum, 1977). In another study, stereotyped jumping behaviour produced by a high dose of d-amphetamine was not reduced but prolonged by pretreatment with lithium for four days (Matussek and Linsmayer, 1968). Differences in both the duration of lithium pretreatment and in the nature and neurochemical bases of the amphetamine-induced behaviour (Snyder *et al.*, 1974; Sloviter *et al.*, 1978b) could account for the contrasting results of these studies.

The metabolic precursors of 5-hydroxytryptamine (5-HT, serotonin), L-tryptophan and 5-hydroxytryptophan, had similar effects to lithium in reducing hyperactivity in mice produced by a d-amphetamine-chlordiazepoxide mixture (Poitou *et al.*, 1975), presumably because they caused activation of inhibitory serotonergic pathways (Mabry and Campbell, 1973). Since there may be a transient increase in 5-HT release as lithium levels rise in the brain (Wielosz and Kleinrok, 1979), the reduction in mixture-induced hyperactivity by lithium may have had a serotonergic as well as a catecholaminergic basis.

Neurochemical experiments indicate that acute and short-term treatment with lithium increases intraneuronal deamination and turnover of NA at the expense of its release and extraneuronal metabolism by catechol-O-methyl-transferase (Davis and Fann, 1971; Shaw, 1975). The effects of lithium on turn-over and release of NA are probably secondary to inhibition of intraneuronal storage mechanisms (Kormiskey and Buckner, 1974; Slotkin *et al.*, 1978), which after a few days leads to depletion of synaptosomal NA (Kuriyama and Speken, 1970). Inhibition by lithium of the vesicle-uptake mechanism appears to potentiate the catecholamine-releasing effects of reserpine-like drugs, and results in neurotransmitters 'spilling over' onto receptors. This may explain the transient hyperactivity seen in rats and mice when tetrabenazine and reserpine are admini-stered after lithium pretreatment (Furukawa *et al.*, 1975; Borison *et al.*, 1978).

After synaptosomal depletion of NA, tyrosine hydroxylase may be subject to less end-product inhibition. The resulting increase in NA synthesis may eventually allow repletion of vesicular stores of NA, which could explain why neuronal release of NA is no longer reduced after long-term lithium pretreatment (Corrodi *et al.*, 1969). It may also explain why hyperactivity produced in rats by a d-amphetamine-chlordiazepoxide mixture was not fully reversed by two weeks' pretreatment with lithium (Cox *et al.*, 1971). Increases in tyrosine-hydroxylase activity in the dopaminergic nigrostriatal pathway were found after pretreatment with lithium for eight days, and were also thought to explain the failure of lithium to antagonize amphetamine-induced hyperactivity in rats (Segal *et al.*, 1975).

These behavioural and neurochemical experiments provide strong evidence for an action of lithium on catecholaminergic transmission, although it is difficult to determine the relative importance of dopaminergic and noradrenergic mechanisms in lithium's effects on behaviour. A reduction in catecholaminergic transmission by lithium is in accordance with its antimanic action, since mania is usually explained in terms of overactivity in catecholaminergic pathways. However, since tolerance to the effects of lithium on catecholaminergic transmission appears to develop as the duration of administration increases beyond a week or so, this interpretation must be accepted with some reservation.

C. Lithium-induced Alterations in Serotonergic Mechanisms

Acute and short-term pretreatment with lithium increases the turnover of 5-HT, partly by impairing storage, and consequently increasing intraneuronal deamina-tion by monoamine oxidase (MAO) (Grahame-Smith and Green, 1974; Collard, 1978). The reduced storage capacity for 5-HT after lithium apparently produces a transient 'spillover' of 5-HT onto receptors, manifested in rats by head-shake responses which occur within the first hour or so after lithium administration (Wielosz and Kleinrok, 1979). Acute treatment with high doses of lithium also increases head shakes in reserpinized mice (Yamada and Furukawa, 1979). Sub-sequently there is a persistent reduction in 5-HT release, as demonstrated *in vitro* and *in vivo* by a number of investigators (Katz *et al.,* 1968; Schubert, 1973; Collard, 1978).

The results of a behavioural study in which LiCl (2 mmol/kg) administered for two weeks lowered rats' jump thresholds to footshock also suggest that lithium reduces the availability of 5-HT in the brain (Harrison-Read and Steinberg, 1971). The effect of lithium on footshock thresholds reached a peak after about ten days of treatment, and then tended to decline slightly as treatment continued (unpublished findings). The effect of lithium was similar to that produced by treatments which deplete brain 5-HT (Harvey *et al.,* 1975), and was abolished by giving 5-hydroxytryptophan (5-HTP). Lithium pretreatment also antagonized head shakes in mice given large doses of 5-HTP, probably by increasing the intraneuronal deamination of newly synthesized 5–HT, and thereby reducing its release and effect on receptors (Kiseleva *et al.,* 1970).

In the presence of an MAO inhibitor, the 5-HT storage defect produced by lithium results in sustained 'leakage' of 5-HT out of nerve endings and onto receptors. This is manifested in shaking responses, and in a behavioural syndrome with tremor or clonic convulsions, lateral head weaving, fore-paw treading, hind-limb splay, hyperpyrexia and salivation as its main features (Grahame-Smith and Green, 1974). A general increase in motor activity also occurs and appears to be mediated through catecholaminergic as well as serotonergic mechanisms, since it is abolished by the synthesis inhibitors α-methyl tyrosine and para-chlorophenyl alanine (Grahame-Smith and Green, 1974; Judd *et al.,* 1975). The hyperactivity produced by the putative 5-HT agonist 5-methoxy-N, N-dimethyltryptamine (5MeODMT) was, if anything, decreased in lithium rats, confirming that the main effect of short-term lithium pretreatment is on presynaptic serotonergic mechanisms (Grahame-Smith and Green, 1974). Similar behaviour occurs without lithium pretreatment if either L-tryptophan of L-DOPA are given to rats after an MAO inhibitor (Grahame-Smith, 1971; Green and Grahame-Smith, 1976). There is some evidence that the behavioural syndrome in these cases is mediated exclusively through serotonergic mechanisms (Jacobs, 1974; Sloviter *et al.,* 1978a), although not all the findings are in agreement with this (Deakin and Green, 1978). Again, there seems to be a catecholaminergic link in the mechanisms mediating the hyperactivity-component of the response to these drugs, since stimulation of 5-HT receptors probably activates a dopaminergic pathway (Green and Grahame-Smith, 1976; Eccleston and Nicolaou, 1978).

Short-term pretreatment with lithium also increases the rate of 5-HT synthesis, apparently by increasing the supply of tryptophan to the rate-limiting enzyme tryptophan hydroxylase (Perez-Cruet *et al.,* 1971; Schubert, 1973; Poitou *et al.,* 1974; Knapp and Mandell, 1975). Despite the maintenance of facilitated tryptophan uptake into nerve endings, 5-HT synthesis reverts to normal after lithium pretreatment for three weeks or more (Knapp and Mandell, 1975). 5-HT turnover is no longer increased, but may be decreased in certain areas of the brain (Ho *et al.,* 1970b), whilst the reduction in 5-HT release persists (Corrodi *et al.,* 1969). The return of 5-HT synthesis to normal is probably due to a decrease in the activity of tryptophan hydroxylase, but it is difficult to accept that this occurs because 5-HT neurons are subjected to feedback inhibition due to increased 5-HT release and receptor activation (Knapp and Mandell, 1975). Firstly, as outlined above, the predominant effect of lithium is to reduce 5-HT release. Knapp and Mandell (1975) did not confirm this, but neither did they find increased 5-HT release. Secondly, the putative 5-HT-receptor blocker thorazine (10 mg/kg) failed to antagonize the reduction in tryptophan hydroxylase, as would be expected if the effect were receptor-mediated (Mandell, 1975). Lastly, raphe-cell firing was not reduced after pretreatment with lithium for two to four days (Sheard and Aghajanian, 1970).

According to Mandell and Knapp (1976), after prolonged lithium pretreatment 'it is as though the uptake of tryptophan were fixed at its maximum rate, the intraneuronal enzyme (tryptophan hydroxylase) were fixed at its minimum activity, and the bidirectional adaptive capacity were "used up," having returned the overall rate of synaptosomal conversion of tryptophan to 5-HT to baseline.' These authors go on to suggest that lithium may regulate affective states by preventing excessive increases or decreases in 5-HT synthesis and release. Although this is an interesting idea, the weight of evidence suggests that a *deficiency* in brain 5-HT is responsible for, or predisposes to, both mania and depression (Murphy *et al.*, 1978). Furthermore, the assumption that changes in 5-HT synthesis result in corresponding changes in 5-HT neurotransmission may not be justified. Attempts to show that lithium pretreatment produces a 'normally functioning system, buffered against further changes that other drugs and perhaps even psychological conditions might otherwise induce' were also not very convincing. Although it does not seem possible to cause further alterations in tryptophan uptake and tryptophan-hydroxylase activity beyond those produced by a very high dose of LiCl (10 mmol/kg), a change in one of these parameters in the direction opposite to the effect of lithium would presumably disrupt the balance, and cause a change in 5-HT synthesis. Cocaine had opposite effects to lithium on both tryptophan uptake (which was decreased) and tryptophan-hydroxylase activity (which was increased). However, contrary to expectation, the resultant effect on 5-HT synthesis of either cocaine or lithium was not zero: cocaine decreased synthesis, whereas lithium (10 mmol/kg for three days) increased it, so it is hardly surprising that these effects cancelled out when cocaine and lithium were administered together. Cocaine was not tested in lithium-pretreated rats in which 5-HT synthesis *had* reverted to normal (e.g., after giving LiCl 5 mmol/kg for 21 days; Knapp and Mandell, 1975), so the 'buffering' effect of lithium against drug-induced changes in 5-HT synthesis was not actually put to the test. The relevance of this model to lithium's therapeutic actions must therefore remain uncertain, particularly as it is based on experiments using very high doses of lithium, which may have affected brain functions indirectly through the effects of stress or general toxicity.

III. RECENT EXPERIMENTS

A. General Observations during Lithium Pretreatment

In the experiments to be discussed here, lithium chloride was administered to hooded male rats (Lister strain) by once-daily injections. Although giving lithium in the food results in more stable plasma levels of lithium, rats reject the food because of its bitter taste and so lose weight. The resulting reduction in sodium intake appears to precipitate lithium toxicity. This can be prevented by giving sodium supplements (Schou, 1976), but we have avoided this method of lithium administration because effects of reduced food intake on behaviour cannot be ruled out.

Every morning, isotonic solutions (0.154 M) of LiCl or NaCl were injected intraperitoneally (i.p.), each rat receiving 13 ml/kg body weight (2 mmol/kg). When a smaller dose of lithium was used, the volume was made up with saline. Plasma concentrations of lithium were in the 'therapeutic' range, when sampled three to four hours after the last injection, e.g., mean (± sd) values in groups of eight or nine rats after eight and 18 days of pretreatment were 0.91 ± 0.22 and 0.88 ± 0.15 mM, respectively.

No obvious signs of malaise or toxicity were seen in our rats, but a number of changes in behaviour were noted during lithium pretreatment. Reactivity to handling was increased over the first week or so of treatment (Figure 11.1) and, after about two weeks, rats began to show treading movements with their forelegs when they were disturbed during treatment sessions. Increased salivation was also seen in many rats towards the end of the pretreatment period (Figure 11.2).

Rat treated with lithium gained more weight than controls, and their food intake was greater as well (Figure 11.3). The extra gain in weight due to lithium was more marked in older, initially heavier rats (Figure 11.4). Plenge *et al.,* (1973) have reported that rats treated with subtoxic doses of lithium gain weight more than controls, provided they are not stressed in any way. We observed that, when rats were subjected to the stress of being tested in unfamiliar apparatus, saline controls lost weight slightly, but rats tested five to seven days after substituting lithium for the daily injection of saline – short-term lithium (SL) group – lost weight markedly. By contrast, rats tested after long-term lithium (LL) pretreatment (15 days or more) continued to gain weight as before (Figure 11.5). In view of their increased food intake and weight gain during pretreatment, it seems that our rats were not intoxicated or unduly stressed, provided they were otherwise left undisturbed.

In accordance with the observations of many other workers (Schou, 1976), lithium produced marked polyuria. Fluid intake also increased markedly over the first week of pretreatment, but thereafter polydipsia began to decline (Figure 11.3). Most of the polydipsia was probably secondary to a nephrogenic diabetes insipidus syndrome (Schou, 1976), but there are a number of possible explanations for the subsequent decrease in polydipsia. For example, water excretion and thirst may have declined because of incipient renal failure (Schou,

Figure 11.1: Squeaking and struggling in rats produced by handling and by penetration with a hypodermic needle prior to i.p. injection. Rats' behaviour was
rated 'blind' on a four-point scale (0-3). Mean scores for three pretreatment
groups are plotted against treatment days. Overall reactivity scores were significantly ($p < 0.05$) increased by the higher dose of lithium (Mann Whitney U test).

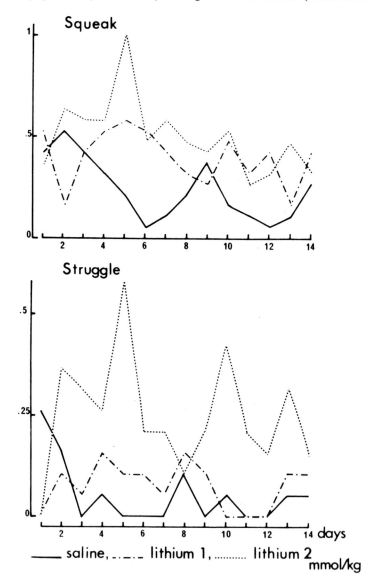

Figure 11.2: Percentage of lithium-pretreated rats (*n* = 40) showing hypersalivation and treading behaviour when observed at the time of the morning injection.

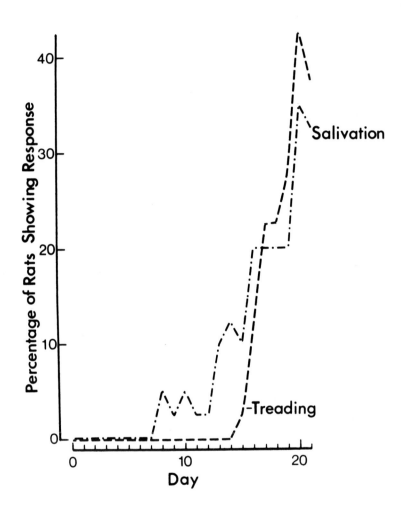

Figure 11.3: Mean changes in food and water intake, and body weight over 16 days of pretreatment with NaCl (dashed curve, filled circles) or LiCl (dotted curve, filled triangles) 2 mmol/kg in each case. Saline and lithium rats were housed separately, three or four to a cage. Room temperature was $21 \pm 2°C$. Artificial lighting was switched on at 6.0 a.m. and off at 6.0 p.m. On the day before pretreatment was begun, rats allocated at random to the saline ($n = 20$) and lithium ($n = 11$) groups weighed (mean $\pm$ SD) 306.4 ± 7.9 g and 307.5 ± 25.1 g, respectively. Daily intake of food (Dixon's 41B pellets) and fluid (tap water) was calculated from the amount consumed by each cage of rats. Mean ($\pm$ SD) food intake per rat on day 0 for the rats in the 'saline cages' ($n = 5$) was 20.3 ± 0.9 g, and for those in the 'lithium cages' ($n = 3$), 21.2 ± 1.1 g. The corresponding values for water intake on day 0 were 25.2 ± 1.8 ml and 25.2 ± 1.6 ml. Dietary sodium intake was about 13 mmol/kg body weight at the start of pretreatment. Lithium markedly increased water intake, which rose to a peak on day 7, and then gradually declined. The trend for food intake and body weight to rise over the treatment period was significantly increased by lithium: $F_{LIN} = 3.20\,(15, 90)$, and $F_{LIN} = 3.68\,(15, 435)$, respectively; $p < 0.001$ in both cases.

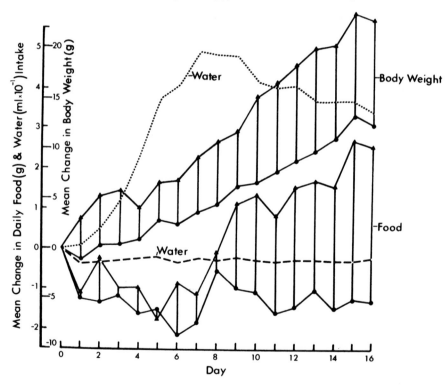

Figure 11.4: The effect of initial body weight on the weight gain produced by lithium. Mean increases in body weight over 15 days of pretreatment with NaCl (filled squares) or LiCl (open squares) (2 mmol/kg i.p. in each case) are plotted against rats' mean weight on day 1. Lithium produced a significant increase in weight gain in two experiments in which mean initial weights were in the region of 300 g. An asterisk denotes $p < 0.05$, and a double asterisk $p < 0.01$; t-tests, two-tailed.

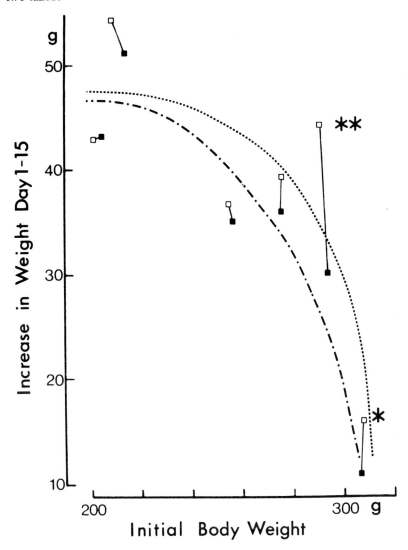

Figure 11.5: Changes in body weight produced by testing rats in novel environments. The graph shows mean (± SEM) body weights on day 19, before the first
test, and the average weight over days 20, 21 and 22, after three successive
exposures to unfamiliar test apparatus. Rats' weights were expressed as differences
from their average weight over days 16 to 18, when they were left undisturbed.
Control (filled circles) and LL (filled triangles) rats received daily injections of
saline and lithium, respectively, throughout. SL (filled squares) rats were injected
with lithium from day 15 onwards. Body weight after behavioural testing was
significantly (a double asterisk denotes $p < 0.01$) reduced in SL rats, and significantly increased (an asterisk denotes $p < 0.05$) in LL rats, when compared with
controls by two-tailed t-tests.

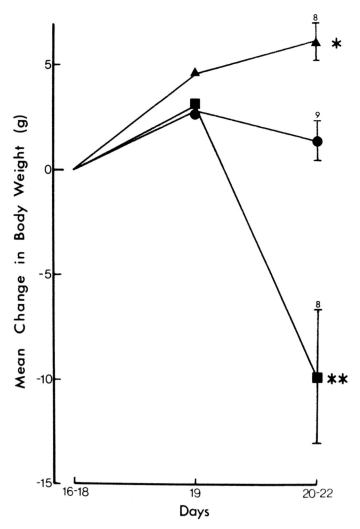

1958). Alternatively, the indirect inhibitory effect of lithium on vasopressin-stimulated adenyl cyclase in renal tubular cells (Schou, 1976) may have been reduced by compensatory changes in the hypothalamus or kidney, or by a fall in tissue levels of lithium (Zilberman *et al.*, 1979) due to increased dietary sodium intake and increased lithium excretion (Schou, 1958, 1976). Since plasma levels of lithium remained more or less stable from the seventh day onwards in our rats, some sort of indirect compensatory change seems the most likely explanation.

B. *Exploratory Behaviour in the Y Maze after Short-term and Long-term Lithium Pretreatment*

(a) Y-maze Testing. Three to five hours after the last morning injection, rats were tested individually in a Y maze (Figure 11.6). In the first study, an empty earthenware pot painted black was positioned at the end of each arm of the maze. A rat was placed in the centre of the maze and allowed to explore freely for two, four or six minutes. Various exploratory behaviours were scored by an observer sitting quietly and unobtrusively above the maze: the number and sequence of entries made into the arms of the maze with all four feet, the number of rears onto the hind-legs and the time spent sniffing the outside of the pots and dipping into them with the snout. At the end of this habituation trial, one of the black pots, selected at random, was exchanged for a white one, and the rat was returned to the maze for a further three minutes. Some rats were not habituated to the maze with all black pots, but were tested for the first time with two black and one white pot.

(b) Locomotion and Rearing in the Y Maze. SL rats pretreated with lithium for five days tended to make fewer entries than control (S) rats (Table 11.1). The locomotor activity (entries) of SL rats in the first minute or so of the trial was not different from that of controls, but their activity fell more steeply over subsequent minutes (Figure 11.7). This was also shown by the steeper linear regression slopes for activity between the first and third minutes of the trial (Table 11.2). Rearing activity was also reduced in SL rats, but this was less obviously the result of increased fall-off in activity over the trial (Figure 11.7; Table 11.2). Pretreatment with lithium for 19 days had inconsistent effects on entries and rears (Table 11.1).

The fall-off in activity over the trial probably indicated that rats explored less as they processed and stored information about their surroundings. Viewed in this way, the increased fall-off in activity shown by the SL rats suggests that they habituated to novelty, and therefore processed information, at a faster rate than controls or LL rats. However, other explanations such as increased fatigue cannot be ruled out.

(c) Investigation of the Pots in the Y Maze. Short-term and long-term lithium pretreatment had different effects on the amount of time rats spent investigating the pots during the habituation trials: sniffing time tended to be decreased in SL

Figure 11.6: Y maze used to measure rats' exploratory behaviour. The maze was placed on the floor of the thermostatically controlled, sound-protected room in which rats were normally housed, directly beneath a 150-W ceiling light (the luminance of the maze floor was about 25 lux). The arms of the maze were 38 cm long, 13 cm wide and 33 cm high, and were painted grey with a matching floor. Background white noise from an air conditioner masked any extraneous sounds.

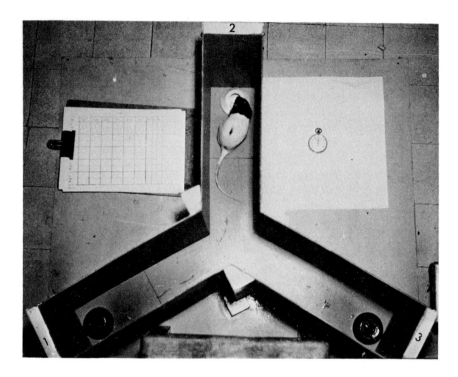

rats, whereas it was increased by LL pretreatment (Table 11.1). About half of the LL rats made distinctive treading movements with their forelegs as they moved around and vigorously sniffed at the novel environment. Many also licked the floor of the maze, and rubbed their snouts along it, giving the impression that they were 'trying harder' to obtain information about their surroundings, particularly as the treading and snout-rubbing behaviour tended to decrease as LL rats became familiar with the maze.

The behaviour of LL rats was reminiscent of that occurring in rats which have been blinded or have had their vibrissae removed (Komisaruk, 1977). One can speculate that LL rats experienced 'stimulus hunger' because of reduced ability to process novel sensory information. This would also imply a deficiency in the ability to habituate. Although there was some indication that this was the case (Figure 11.7), it was not a consistent finding (Table 11.2).

Table 11.1: Exploratory-activity Scores (Means ± SEM) of Groups of Rats Given Saline (S), Short-term Lithium (SL) and Long-term Lithium (LL) Pretreatment*

Trial length (min)	S	SL	LL
		Entries per min	
2	3.0 ± 0.5	3.4 ± 0.3	3.1 ± 0.4
4	3.0 ± 0.2	2.1 ± 0.2***	3.2 ± 0.5
6	3.4 ± 0.2	2.7 ± 0.5	2.9 ± 0.2
		Rears per min	
2	5.2 ± 0.7	6.8 ± 0.4	5.9 ± 0.5
4	5.5 ± 0.5	5.0 ± 0.4	6.7 ± 0.6
6	6.9 ± 0.5	4.1 ± 0.7***	5.1 ± 0.3**
		Time at pots (sec) per min	
2	1.1 ± 0.3	1.2 ± 0.3	1.9 ± 0.3
4	1.6 ± 0.5	1.4 ± 0.3	2.1 ± 0.7
6	4.5 ± 0.4	3.1 ± 0.5	5.9 ± 0.7

* Rats were tested for two, four or six minutes in an unfamiliar Y maze containing three black pots; each group contained eight to ten rats. In the case of the four- and six-minute trials SL rats tended to have lower scores on all measures, but initial activity – reflected by the scores in the two-minute trials – was not reduced. LL rats showed inconsistent alterations in entries and rears, but time spent investigating the pots was increased. The overall effect of lithium was not significant for entries ($F = 1.01$) or rears ($F = 1.21$; DF = 2,73 in both cases). For rears, however, there was a significant interaction between pretreatment and trial length: $F = 5.48$ (4, 73); $p < 0.001$. The overall effect of lithium on time spent investigating the pots – $F = 5.18$ (2, 73), $p < 0.01$, analysis of transformed scores (log sec + 1) – was mainly due to the higher scores of the LL rats – $F = 4.69$ (1, 49). $p < 0.05$. Individual comparisons with S groups were made by two-tailed t-tests.
** $p < 0.05$.
*** $p < 0.01$.

(d) Alternation of Maze-arm Choices. The orderliness of the sequence in which maze arms are entered is usually taken to reflect rats' systematic exploration of the environment as they walk around it (Halliday, 1968). A high percentage of arm alternations usually indicates that rats are avoiding more recently visited parts of the maze, presumably in order to investigate the less familiar. Typically, normal rats show fairly high alternation, which tends to increase further as they lose their fear of novelty. The sequence of arm entries then becomes less orderly again, as rats become familiarized with the maze (Lester, 1966). In S rats, alternation increased over a six-minute trial, but there was a tendency for alternation to fall in association with a significant increase in the rate of habituation in SL rats (Table 11.3). The LL group showed a high percentage of alternation in the first two minutes of the trial, but very low alternation in the last two minutes, with normal habituation of locomotor activity. It is possible that the high initial

Figure 11.7: The effect of lithium on rats' exploratory activity in the Y maze. Entries (locomotion) and rears of S (continuous curve), SL (dotted curve) and LL (dashed curve) rats were recorded over successive 30-second intervals during a four-minute trial in an unfamiliar Y maze containing three black pots. There was a significant overall effect of lithium pretreatment on entries — $F = 3.85$ $(2, 26)$, $p < 0.05$ — mainly as a result of reduced activity in the SL group: $F = 11.78$ $(1, 18)$, $p < 0.01$. The overall effect of lithium on rears just failed to reach significance: $F = 3.17$ $(2, 26)$, $p = 0.06$.

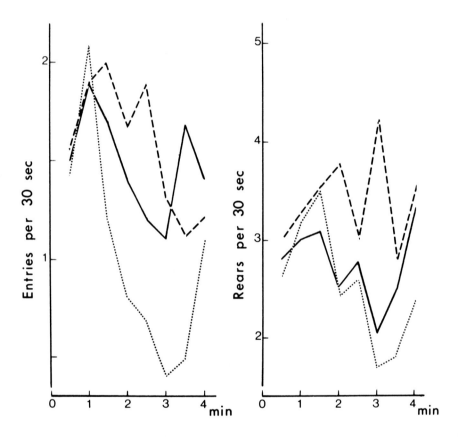

alternation in LL rats reflected perseveration of the direction (left or right) in which rats turned to make a new entry. This may have represented an attempt by LL rats to cope with their reduced ability to process and register novel sensory information, much as one moves systematically around the walls of an unfamiliar darkened room in order to find the light switch. Presumably this strategy failed to increase rats' understanding of the environment, and so their orderly movement around the maze gave way to entering maze arms at random.

Table 11.2: Linear Regression Slopes for the Fall in Activity Scores between the First and Third Minute of the Y-maze Trial*

Trial length (min)	S	SL	LL
		Entries	
4	-0.21 ± 0.13	-0.39 ± 0.05	-0.12 ± 0.07
6	-0.16 ± 0.07	-0.42 ± 0.08**	-0.19 ± 0.07
		Rears	
4	-0.23 ± 0.15	-0.39 ± 0.10	-0.14 ± 0.17
6	-0.09 ± 0.15	-0.01 ± 0.12	-0.29 ± 0.10

* The means (± SEM) are for groups of eight to ten rats. Slopes tended to be increased in SL rats, suggesting increased within-trial habituation. For the results in the second row, F = 4.53 (2, 21), $p < 0.05$; for those in the third row, F = 3.63 (2, 26), $p < 0.05$.
** $p < 0.05$, t-tests (two tailed), S vs SL.

Table 11.3: Mean (± SEM) Spontaneous Alternation Scores during a Six-minute Y-maze Trial*

	Percentage arm alternation		
Two-minute periods	S	SL	LL
First	52.5 ± 3.0	57.7 ± 5.6	66.2 ± 5.4**
Second	58.0 ± 5.5	54.3 ± 2.8	47.6 ± 4.1
Third	63.5 ± 4.2	49.0 ± 10.2	41.4 ± 3.5***

* The number of times a rat entered the least recently visited arm of the maze in each two-minute period was expressed as a proportion of the total number possible. Proportions were converted to percentages using the arcsine transformation — percentage = arcsine $\sqrt{}$ (proportion) — in order to reduce the disproportionate effect of high or low scores on the variances. Lithium pretreatment significantly affected the way spontaneous alternation altered over the trial: F = 3.85 (4, 40), p = 0.01. There were seven or eight animals in each group.
** $p < 0.05$.
*** $p < 0.01$, for comparisons with the S group during the same period of the trial (t-tests, two-tailed).

(e) Response to Stimulus Change. In order to gauge how much information rats obtained as they explored the Y maze, their response to a white pot introduced after various periods of habituation to the maze containing three black pots was assessed. Without previous habituation, on average, all rats spent as much time at the white pot as at either of the two black pots. (The mean ± SEM percentage times at the white pot for S, SL and LL groups were 32.6 ± 4.1, 36.8 ± 3.9 and 33.3 ± 2.8, respectively, with 17 or 18 rats per group). The total time spent

exploring the pots was about equal in S and SL rats, but LL rats spent on average 5.0 ± 1.4 seconds longer at the pots than controls ($p < 0.01$). S and LL rats only showed a preference for the new pot if they had previously been habituated to the maze for six minutes, whereas in SL rats a marked preference for the white pot occurred after habituation for only two minutes (Figure 11.8). Presumably SL rats were more aware of, and more responsive to, the greater novelty of the white pot. This was supported by the result of another experiment, in which SL rats showed a marked preference for a novel pot, regardless of whether it was black or white (Figure 11.10).

SL rats also showed a preference for the white pot after an habituation trial in an *empty* Y maze (Figure 11.9). Presumably they were responding to the extra novelty value of the white pot which resulted from its incongruity in the presence of two black pots. This incongruity only appeared to direct rats' attention after the even greater incongruity of other maze stimuli relative to rats' previous experience had been reduced by familiarization with the maze. These results suggest that SL rats show increased attention and approach to novel stimuli when presented against a background of familiarity. It may be that increased speed and efficiency of habituation to background stimuli allows SL rats to be more aware of stimulus change or incongruity when it occurs.

Novelty appears to have a reward value which is related, up to a point, to the amount of arousal it produces (Berlyne, 1967). Increases in animals' level of arousal due to other factors — e.g., internal 'drive' states such as hunger — also seem to increase the reward value of novelty and stimulus change (Barnett, 1975). It is possible that SL rats find novelty more rewarding because their level of arousal is already higher than that of controls. However, when rats were tested in the Y maze after 'stressful' experience in a brightly lit open field, the tendency for exploration of the incongruous white pot to be increased was seen in S as well as SL rats (Figure 11.9), suggesting that both groups were equally aroused by the open-field trial. Rather than showing a greater increase in arousal to all stimuli, it seems likely that only novel stimuli have extra 'arousal potential' in SL rats, probably because more efficient information processing and habituation increases the contrast between familiar and unfamiliar, and makes novelty more rewarding. The findings with LL rats were in complete contrast. These rats showed no preference for the incongruous white pot under any of the three conditions (Figure 11.9). This supports the idea that there is a failure of information processing in LL rats.

(f) The Involvement of Serotonergic Mechanisms in the Altered Response to Stimulus Change after Lithium Pretreatment. The contrasting effects of SL and LL pretreatment in these experiments suggests a parallel with the findings in neurochemical experiments of qualitative differences between the effects of acute and chronic lithium administration (see Section IIC). Since SL pretreatment seems to reduce 5-HT availability at synapses, an attempt was made to reverse the effect of SL on the response to stimulus change by giving the 5-HT

Figure 11.8: The response to stimulus change in rats pretreated with saline (filled circles) or with lithium for five days (filled squares) or 19 days (filled triangles). Rats were habituated to the maze containing three black pots for various periods, and then retested for three minutes in the Y maze after replacing one of the pots with a white one. Time spent exploring the white pot was calculated as a percentage – arcsine $\sqrt{}$ (proportion) – of the total time at all three pots, and expressed as the difference from the mean score of the rats in each pretreatment group which had not been exposed to the maze before (zero habituation). The points represent mean ($\pm$ SEM) percentages, with the number of rats per group shown in each case. The overall effect of lithium pretreatment – $F = 6.94$ (2, 66), $p < 0.01$ – was mainly due to the increased time spent at the novel pot by the five-day lithium group. A double asterisk denotes $p < 0.01$ with respect to the zero-habituation group; a dagger denotes $p < 0.05$ and a double dagger $p < 0.01$ with respect to the corresponding saline group; two-tailed t-tests.

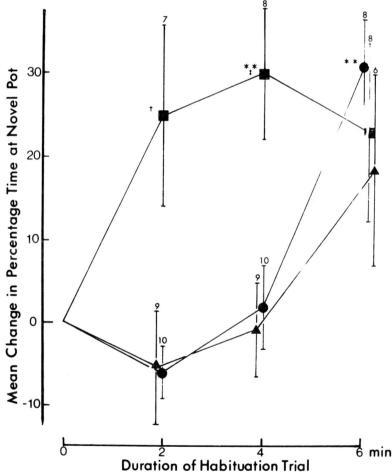

Figure 11.9: The effect of previous experience on rats' response to an incongruous object. Time spent investigating the white pot in a Y maze which also contained two black pots was expressed as a percentage of the time at all three pots (means ± SEM; the number of rats per group is shown above the error bars). Rats pretreated with lithium for five days (SL group) investigated the white pot more than controls (S group) if they had been previously exposed to the empty maze for three minutes (t-test, two-tailed; a double asterisk denotes $p < 0.01$). For both S and SL groups, previous experience in the brightly illuminated open field (luminance = 190 lux) increased preference for the white pot relative to naive controls: $F = 5.48$ $(1, 29)$, $p < 0.05$. Investigation of the white pot by S rats with experience of the open field was significantly (a double-dagger denotes $p < 0.01$) greater than that shown by S rats previously exposed to the empty maze. Neither of the two groups of 'experienced' rats pretreated with lithium for 19 days (LL groups) showed increased preference for the white pot, nor did any of the naive rats.

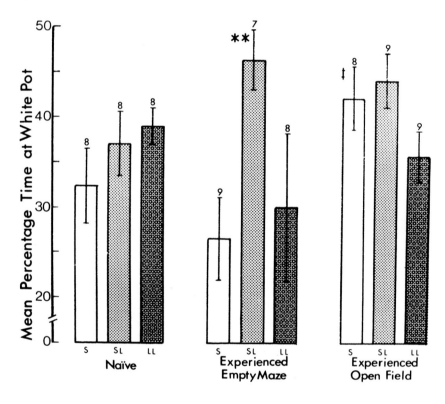

Figure 11.10: The effect of lithium and L-tryptophan on rats' response to stimulus change. Time spent investigating the novel pot (black or white) was expressed as a percentage – arcsine $\sqrt{}$ (proportion) – of the time spent at all three pots in the maze. All rats had previously been habituated for three minutes to the maze containing three pots of the same colour. One hour before testing, rats were injected i.p. with L-tryptophan (100 mg/kg) or with the vehicle solution alone (0.5% polysorbate-80 in saline). Rats had been pretreated for 15 days with saline (S) or with lithium (LL), or with lithium for the last five days only (SL). Mean (± SEM percentages are shown, with the number of rats in each group given above the error bars. Increased investigation of the novel pots by SL rats (an asterisk denotes $p < 0.05$, while a double asterisk denotes $p < 0.01$) was abolished by L-tryptophan (a dagger denotes $p < 0.05$). Overall, there was a significant interaction between the effects of lithium and tryptophan: $F = 6.03 \, (2, 89), p < 0.01$.

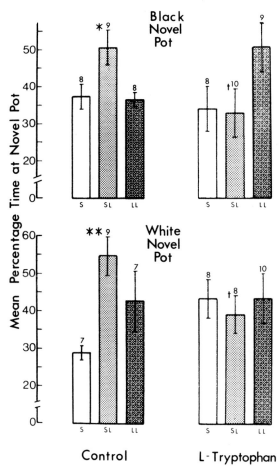

precursor L-tryptophan. A dose of 100 mg/kg completely reversed the effect of SL, but left S rats relatively unaffected (Figure 11.10). In contrast, the putative 5-HT antagonist methysergide mimicked the effect of SL in increasing rats' response to stimulus change (Table 11.4). These findings give strong support to the hypothesis that serotonergic transmission is reduced in SL rats.

Table 11.4: Effect of Lithium and Methysergide on Rats' Preference for the Novel Pot in the Y Maze*

	Percentage time at novel pot		
S		SL	
Control	Methysergide	Control	Methysergide
32.4 ± 4.1 (7)	49.7 ± 4.0 (16)***	49.9 ± 4.5 (9)**	49.6 ± 4.5 (13)**

* The time spent investigating the white pot which had replaced a familiar black one was expressed as a proportion of the time spent at all three pots. Percentages were obtained by arcsine transformation. One hour before testing, saline (S) and short-term lithium (SL) rats were pretreated either with methysergide (1 mg/kg i.p.) or with saline as a control. Mean (± SEM) percentages are shown, with the number of rats in each group given in brackets. Both lithium and methysergide increased rats' preference for novelty, but their effects were not additive.
** $p < 0.05$.
*** $p < 0.01$, t-tests (two-tailed) for comparisons with the saline group.

C. The Effect of Lithium on Exploratory Choice Behaviour in the T Maze: The Optimum-arousal Concept

In order to check that the increased preference for the changed pot shown by SL rats was a response to its increased novelty value, another experiment was carried out in which rats were required to choose between stimuli which differed only with respect to their relative familiarity. This was achieved by first allowing a rat to inspect the two side-arms of a T maze through transparent partitions. One arm was black and the other white, with the white arm on the left for half the subjects, and on the right for the other half. After about three minutes, the rat was removed, and one of the arms was switched so that both were now the same colour, either black or white. The position of the changed arm was decided at random, but it was on the right and left with equal frequency. The partitions were raised and the rat was allowed to make an entry into one of the arms. Any preference for the changed or unchanged arm shown by a group of rats as a whole could only occur if rats were drawing on their previous experience of the maze when choosing between the arms.

Control rats showed a significant tendency to choose the changed arms, regardless of its position or colour (Table 11.5). The overall lack of preference

Table 11.5: Percentage of Rats Entering the Changed Arm on a Free-choice Trial in a T Maze with Both Arms the Same Colour*

Arm colour on choice trial	Percentage choosing changed arm		
	S	SL	LL
Black	73.3 (15)	78.6 (14)***	71.4 (14)
White	66.7 (15)	28.6 (14)***	61.5 (13)
Total	70.0 (30)**	53.6 (28)	66.7 (27)

* The number of rats in each group is shown in brackets. Saline control (S) rats showed an overall preference for change. Rats given long-term lithium pretreatment (LL) were similar but the results were not significant. Short-term lithium (SL) rats also showed a preference for change when choosing between two black arms, but avoided change when the arms were both white.
** Significant ($p < 0.05$) preference for changed arm: $\chi^2 = 4.0$ (1).
*** Significant ($p < 0.05$) difference in choice behaviour due to arm colour: $\chi^2 = 5.2$ (1).

for either alternative shown by the SL group concealed a preference for the changed arm when both arms were black, and avoidance of change when the arms were both white. The choice behaviour of LL rats was similar to that of controls, but their preference for the changed arm was not statistically significant.

In a second experiment, the saline control group failed to show any preference for the changed arm (Table 11.6). All the rats in this experiment had been handled and injected for only six days, and were probably less stressed and aroused than the rats in the previous study. According to Berlyne (1967), the reward value of stimuli, as measured, for example, by the strength of animals' preference for them, is related by an inverted U-shaped function to the prevailing level of arousal. The latter represents the sum of the preexisting level of arousal and the arousal evoked by the stimuli themselves. Major determinants of this arousal potential of stimuli are physical intensity and so-called collative properties such as contrast, surprise and novelty. Animals appear to seek out or avoid novelty in order to maintain an optimum level of arousal. Since the preexisting level of arousal was low in the present experiment, the arousal potential of the novel arm probably represented insufficient reward value to induce a reliable preference in S rats.

However, under these circumstances, SL rats showed an overall preference for the changed arm, although this was less marked when the arms were white. The effect of methysergide was similar, but smaller than that of SL. When methysergide was given to SL rats, choice of the changed arm fell to the level of chance when both arms were white. The 'mismatch' between present and previous experience represented by the changed arm probably increased the arousal level of SL rats towards optimum, and so had a rewarding effect. When the level of arousal of SL

Table 11.6: The Effect of Lithium and Methysergide on the Percentage of Rats Entering the Changed T-maze Arm*

Arm colour on choice trial	Percentage choosing changed arm			
	S		SL	
	Control	Methysergide	Control	Methysergide
Black	50.0 (12)	66.7 (15)	83.3 (12)	83.3 (12)
White	42.0 (12)	66.7 (15)	69.2 (13)	54.0 (13)
Total	45.8 (24)	66.7 (30)	76.0 (25)**	68.0 (25)

* Control (S) rats showed no preference for change. Short-term lithium (SL) rats showed increased preference for change, especially when the choice was between two black arms. Methysergide (1 mg/kg) caused a nonsignificant increase in S rats' preference for the changed arms, but tended to decrease SL rats' choice of the changed arm when the alternatives were both white. Numbers of rats in each group are shown in brackets.
** Significant preference for changed arm: $\chi 2 = 5.8$ (1); $p < 0.02$.

rats was already raised by the brightness of the white arms, the increased arousal potential of the changed arm caused the optimum level to be exceeded, resulting in reduced preference for it (Figure 11.11).

The arousal potential, or novelty value, of stimuli needed to raise the level of arousal towards the optimum is apparently less in SL rats. In line with this conclusion, we reported previously (Harrison-Read, 1978) that novel stimuli (flashing lights) of low intensity evoked an exploratory response in SL rats but not in controls, whereas the reverse was the case when bright flashes were used. The increased response to flashing lights in SL rats was abolished by pretreatment with L-tryptophan, whereas that in control rats was unaffected. LL rats showed no response to either kind of novel stimulus, supporting the idea that novelty fails to engage attention or increase arousal in these rats. The increased sniffing, snout rubbing and treading behaviour of LL rats may represent a vain attempt to raise their level of arousal towards the optimum.

Of course these results are open to other interpretations. For example, SL pretreatment could increase the reward value of novel stimuli directly, without necessarily increasing their arousal potential, and so lower the point on the arousal dimension at which the optimum occurs. Obviously, further information about the effects of lithium on arousal is needed (e.g., from neurophysiological experiments) but, whatever the precise mechanism involved, short-term lithium pretreatment appears to increase rats' ability to analyze information about novel environments. This is in marked contrast to earlier interpretations of lithium's actions on exploratory behaviour (Johnson, 1975). However, it is consistent with the view that lithium normalizes hyperactivity states in animals which may model many of the features of mania (Murphy, 1977), and which may be characterized by inability to attend selectively and systematically to environmental stimuli.

Figure 11.11: Hypothetical inverted U-shaped curve relating the reward value to the arousal potential of novel stimuli. The final level of arousal evoked by a novel stimulus (e.g., the arm of the T maze) depends on its novelty value and its physical intensity (i.e., brightness). Short-term lithium (SL) pretreatment effectively increases the novelty value, and therefore the arousal potential of a stimulus change. Methysergide (Me) has a similar but smaller effect. In all groups, choice between two white arms (open squares) evokes more arousal than choice between two black arms (filled squares). The vertical axis gives the percentage of rats choosing the changed arm of the T maze, expressed as a difference from the 50% chance level (see Table 11.6).

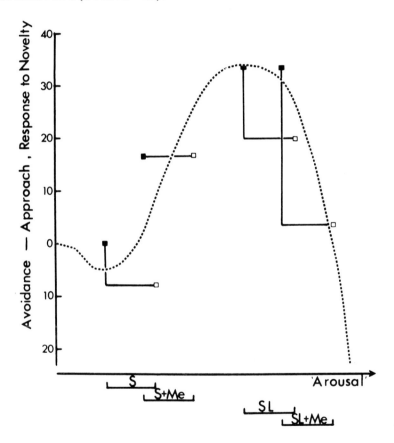

D. Behavioural Evidence for Serotonergic Supersensitivity after Long-term Lithium Pretreatment

In order to explain the contrasting effects of SL and LL on rats' responses to environmental change, we have argued that reduced 5–HT release may lead to supersensitivity in serotonergic pathways after prolonged lithium pretreatment (Harrison-Read, 1978). This was investigated further by pretreating rats with LiCl (2 mmol/kg) for five or 21 days, and then giving them fenfluramine (10 mg/kg), DL-5-HTP (200 mg/kg; preceded by carbidopa, a peripheral decarboxylase inhibitor, 25 mg/kg), or 5MeODMT (1.75 mg/kg), (Harrison-Read, 1979). Fenfluramine releases 5-HT from nerve endings; 5-HTP increases synthesis and release of 5-HT; and 5MeODMT is a putative 5-HT agonist. The intensity of the 'serotonin syndrome' produced by fenfluramine was significantly increased in both groups of lithium pretreated rats: many showed clonic fits, and about 20% died within 24 hours (Figure 11.12). Fenfluramine may release 'unbound' 5-HT, the amount of which is increased by lithium as a result of a vesicular storage defect. By contrast, the syndrome produced by 5-HTP was only significantly increased in LL rats. A similar trend was seen with 5MeODMT, but because some LL rats showed virtually no response, the difference between S and LL groups was only statistically significant when a test for 'extreme reactions' was used (Moses test; Siegal, 1956). The findings with 5-HTP and 5MeODMT support the hypothesis that a delayed increase in the sensitivity of serotonergic systems occurs during lithium pretreatment, probably because of changes at or beyond the receptor level. However it is difficult to account for the LL rats which showed subnormal responses to 5MeODMT. Concentrations of lithium in the plasma of LL rats were about 1.2 mM in both 5MeODMT 'responders' and 'nonresponders,' so the difference between these rats was unlikely to reflect a direct action of lithium. By contrast, in the LL group given 5-HTP, there was a positive correlation between plasma concentrations of lithium (mean ± SEM = 1.1 ± 0.1 mM; $n = 6$) and the serotonin syndrome scores ($r = 0.92$; $p < 0.01$), which raises the possibility that lithium may have direct as well as indirect effects on serotonergic sensitivity.

Examination of individual scores of S and LL rats given various doses of 5MeODMT shows that extreme responses (high and low) to 5MeODMT in lithium pretreated rats became more obvious as the dose of 5MeODMT was increased (Figure 11.13). A possible explanation is that supersensitivity in 5-HT pathways may increase susceptibiltiy to desensitization at high agonist concentrations. As shown in Figure 11.14, partial blockade of 5-HT receptors using l-propranolol (Deakin and Green, 1978) reduced the scores of S rats given 5MeODMT in doses of 2 and 4 mg/kg. There was also a downward shift in rating scores, and an absence of 'nonresponders' to 5-MeODMT in the LL group given propranolol. Figure 11.15 shows the scores of 5MeODMT 'responders' obtained at four-minute intervals after injection of 5MeODMT. The behavioural scores showed an overall increase in the LL groups, but there was no indication

Figure 11.12: Effect of lithium on the response to drugs which affect behaviour through serotonergic mechanisms. Groups of nine or ten rats pretreated with saline (S), short-term lithium (five days; SL) or long-term lithium (21 days; LL) were rated 'blind' for tremor or clonic fits, fore-paw treading, hind-limb splay, head-weaving, Straub tail, hypersalivation and/or ejaculation. Ratings for each category were summed, giving a maximum score of twelve for each rat. Rats were rated eight minutes after 5MeODMT, and 40 and 120 minutes after fenfluramine and 5-HTP, the greater score for each category being used to make up the total in the latter two cases. Mean rating scores of SL and LL rats are expressed as differences from mean ratings of the S groups. There were significant differences between scores after fenfluramine (filled triangles) and 5—HTP (filled squares) (Kruskall-Wallis one-way ANOVAR; $p < 0.01$ and $p < 0.05$, respectively). LL rats had higher scores than controls after fenfluramine and 5-HTP, whereas the scores of SL rats were only higher after fenfluramine (Mann-Whitney U test, two-tailed; an asterisk denotes $p < 0.05$, while a double asterisk denotes $p < 0.01$). The effect of lithium pretreatment on scores afer 5MeODMT (filled circles) was not significant overall, but there was a significant number of extreme responses (high and low) in the LL group compared with controls (Moses test; a dagger denotes $\rho < 0.05$).

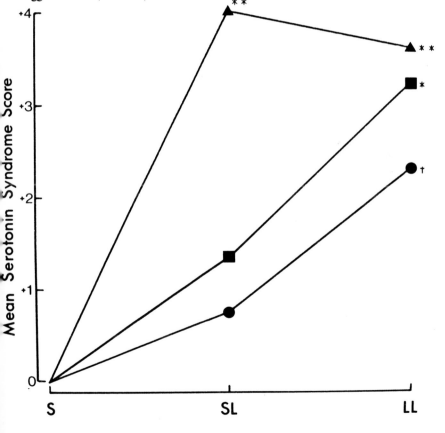

Figure 11.13: Individual scores for the behavioural syndrome produced by the 5-HT agonist 5MeODMT. The scores of rats pretreated with lithium for 21 days or more (filled circles) were distributed differently from those of saline pretreated rats (open circles) (Kolmogorov-Smirnov two-sample test, two-tailed; $p < 0.05$). The scores of lithium rats showed extreme reactions relative to those of saline rats at doses of 1.25 mg/kg ($p < 0.05$), 1.75 mg/kg ($p < 0.05$) and 2.00 mg/kg ($p < 0.01$) (Moses test). Lithium pretreatment significantly increased the response to a dose of 1.75 mg/kg 5MeODMT ($p < 0.05$, Mann-Whitney U test).

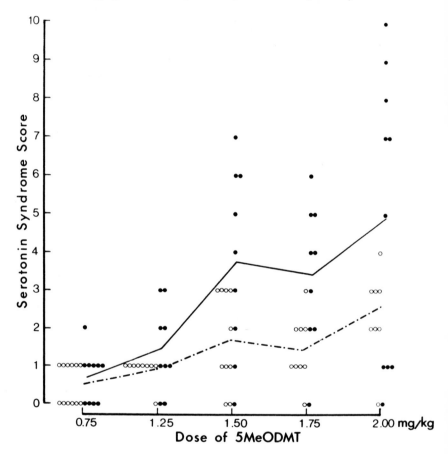

Figure 11.14: Individual scores for the behavioural syndrome produced by two doses of 5MeODMT, with (open squares) or without (filled squares) additional treatment with l-propranolol (5 mg/kg). Rats were tested with both doses of 5MeODMT in a counterbalanced order between days 21 and 24 of saline or lithium pretreatment. Behaviour was scored at four-minute intervals between four and 28 minutes after i.p. injection of 5MeODMT, and the scores totalled for each rat. Propranolol significantly reduced the total scores of saline rats ($p <$ 0.02 for 2 mg/kg 5MeODMT; $p < 0.01$ for 4 mg/kg; Mann-Whitney U tests, two-tailed), but the effect of propranolol was not significant in lithium rats. The scores of lithium rats given 4 mg/kg 5MeODMT were significantly greater than those of saline rats, with ($p < 0.001$) or without propranolol ($p < 0.05$). Lithium rats given 2 mg/kg 5MeODMT without propranolol had a significant ($p < 0.05$) number of extreme scores compared with the corresponding saline group (Moses test).

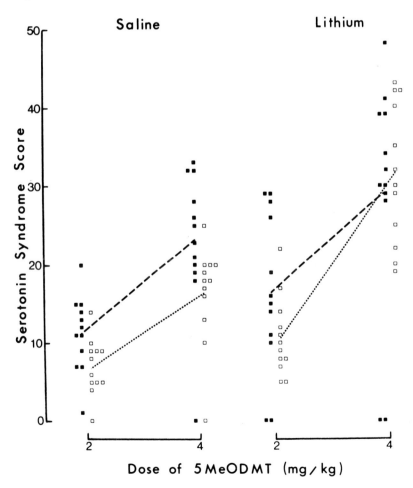

Figure 11.15: Changes in behavioural scores over time after two injections of 5MeODMT. Rats which showed no response to the first injection were excluded. Dotted lines represent the data for rats given l-propranolol 30 minutes before testing. As well as increasing the total scores after the first doses of 5MeODMT (2 or 4 mg/kg; see Figure 11.14), lithium pretreatment decreased the size of the response to the second dose (2 mg/kg) relative to that of the first (increased desensitization index; for 2 mg/kg 5MeODMT, $p < 0.05$; for 4 mg/kg, $p < 0.01$; Mann-Whitney U tests, two-tailed). Pretreatment with propranolol decreased desensitization in the saline rats after the 4 mg/kg dose of 5MeODMT ($p < 0.01$), and abolished the increased desensitization after 4 mg/kg 5MeODMT in lithium rats.

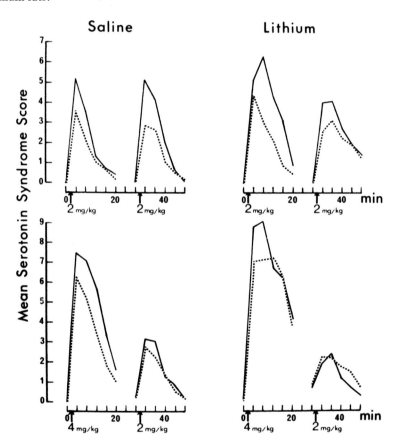

of an increased fall-off in scores, as might be expected if desensitization were occurring. When another injection of 5MeODMT (2 mg/kg) was given 28 minutes after the first, the response in S rats was similar to that produced by the first 2 mg/kg dose, but smaller when preceded by a dose of 4 mg/kg. This suggests that some degree of desensitization occurred in S rats given the higher dose of

5MeODMT. Desensitization in LL rats given 4 mg/kg was more marked than in controls ($p < 0.01$), and desensitization was also apparent in LL rats when the first dose of 5MeODMT was 2 mg/kg ($p < 0.05$). As well as reducing scores overall, propranolol reduced desensitization in the S rats given an initial dose of 4 mg/kg 5MeODMT ($p < 0.01$). Desensitization was also reduced by propranolol in LL rats, but it was still significantly ($p < 0.02$) greater than that in controls with 4 mg/kg 5MeODMT. Further evidence for greater desensitization in LL rats was obtained in another experiment, in which a test dose of 5MeODMT was preceded by increasing doses of 5MeODMT (0.125, 0.5, 1.0 mg/kg), given at twelve-minute intervals. After this desensitization treatment, the behavioural response of LL rats to the 2-mg/kg test dose was, if anything, smaller than that of the S group.

These results suggest that a change analagous to 'denervation supersensitivity' develops in serotonergic pathways during prolonged pretreatment with lithium. A number of pre- and post synaptic mechanisms could account for this, but perhaps the most plausible explanation for both the increased serotonergic sensitivity and the increased susceptibility to desensitization is that the propeties of 5-HT receptors are altered by long-term administration of lithium. It is probable that this effect is mainly indirect, because supersensitivity to 5MeODMT persists for up to one week after withdrawing lithium (unpublished results). Obviously it will be necessary to study receptor properties more directly than can be achieved by behavioural experiments if this hypothesis is to be confirmed and extended.

E. Is Activity in Dopaminergic Pathways Increased by Long-term Pretreatment with Lithium?

Some of the behavioural changes observed in LL rats resemble those caused by stimulation of dopaminergic pathways (e.g., excessive sniffing and 'nosing', inattention to stimulus change), and so it is possible that long-term pretreatment with lithium increases dopaminergic activity, perhaps indirectly through effects on serotonergic transmission. This seems plausible, because serotonergic and dopaminergic pathways are known to interact (Eccleston and Nicolaou, 1978), and there is evidence for increased release and turnover of dopamine in the striatum (Hesketh *et al.*, 1978) and in the tuberoinfundibular system (Corrodi *et al.*, 1969) after pretreatment with lithium for three weeks.

This possibility was investigated by seeing if the motor hypoactivity produced in rats by small doses of apomorphine could be prevented by long-term pretreatment with lithium. Low doses of apomorphine are believed to activate presynaptic dopamine receptors selectively, and so affect behaviour by reducing dopaminergic transmission (Ströbom, 1976). As shown in Table 11.7, the reduction in rats' locomotor activity produced by 0.125 mg/kg apomorphine was completely absent in LL rats. Locomotor hyperactivity and stereotyped sniffing due to stimulation of postsynaptic dopamine receptors by higher doses of apomorphine were unaffected by lithium pretreatment. Unless lithium produced a selective

Table 11.7: Changes in Locomotor Activity Produced by Increasing Doses of Apomorphine*

Control			Apomorphine		
Test	Locomotion		Dose of apomorphine	Locomotion	
	S	LL	mg/kg	S	LL
1	-6.4 ± 1.8	-6.1 ± 3.5	0.125	-16.8 ± 1.4 ↓	+0.4 ± 2.8 ↔
2	-12.8 ± 2.7	-10.5 ± 3.8	0.25	-2.4 ± 2.7 ↑	+0.8 ± 3.5 ↑
3	-14.8 ± 2.3	-10.4 ± 3.2	0.5	+6.0 ± 5.6 ↑	+7.2 ± 5.1 ↑
4	-16.1 ± 2.4	-11.5 ± 3.6	1.0	+13.1 ± 9.4 ↑	+8.6 ± 5.0 ↑
5	-14.5 ± 1.6	-11.5 ± 3.7	2.0	+10.9 ± 9.4 ↑	+12.2 ± 3.9 ↑
N =	8	8		8	8

* The activity of each rat (floor sections crossed) was recorded for two minutes in an open field, and all scores obtained subsequently were expressed as a difference from this pre-injection test score (values shown are means ± SEM). Rats were retested at twelve-minute intervals after receiving increasing doses of apomorphine dissolved in saline (2 ml/kg s.c.). Controls received only saline. The open field was a circular arena 58 cm in diameter, with a wall 33 cm high. Saline (S) rats showed marked hypoactivity with the lowest dose of apomorphine, but locomotor activity was increased with higher doses. Long-term lithium (LL) pretreatment prevented the hypoactivity produced by the lowest dose of apomorphine – $F = 11.94$ (1, 28), $p < 0.01$ – without appreciably affecting the hyperactivity occurring at higher doses: $F = 0.20$ (1, 28), NS. The p values for the results obtained after apomorphine injection are as follows: for the S group in tests 2 and 5, and the LL group in test 2, $p < 0.05$; for the S group and the LL group in tests 2 and 4, $p < 0.01$; for the S group in test 1 and the LL group in test 5, $p < 0.001$; t-tests, two-tailed, apomorphine vs control. For the LL group in test 1, $p < 0.001$; t-test, two-tailed, S vs LL (apomorphine groups).

blockade of presynaptic dopamine receptors, resistance to the inhibitory affect of apomorphine probably resulted from an increase in the level of activity in presynaptic dopaminergic neurons. An incidental finding was that the lower doses of apomorphine reduced the snout-rubbing and treading movements which were typical of otherwise untreated LL rats, suggesting that these behaviours were a reflection of increased dopaminergic activity. Unlike the sniffing, nosing and licking which occurred in all rats at higher doses of apomorphine, these behaviours in LL rats were less obviously stereotyped, and seemed to be fully integrated into rats' response to their surroundings.

IV. SUMMARY AND CONCLUSIONS: OUTLINE FOR A NEW MODEL OF THE THERAPEUTIC ACTIONS OF LITHIUM

Short-term pretreatment with lithium in doses giving plasma concentrations within the therapeutic range appears to increase rats' awareness and response to novel stimuli; and to 'normalize' the impairment of sensory analysis produced by amphetamine and related drug treatments. It seems likely that reduced transmission in serotonergic and catecholaminergic pathways is responsible for these effects of lithium. After prolonged administration of lithium, direct effects on noradrenergic systems are less apparent, but activity in dopaminergic pathways may be increased, perhaps indirectly through the influence of supersensitive serotonergic systems. The resulting effect on behaviour suggests reduced ability to recognize and process the information content of novel stimuli. Neurotransmission in 5-HT pathways may be stabilized as well as enhanced by long-term lithium pretreatment. On the one hand, the effect of excessive increases in 5-HT may be counteracted by the increased susceptibility to desensitization and, on the other, further decreases in 5-HT release are unlikely, because release is already fixed at a low level.

Evidence for a serotonergic involvement in the affective disorders is fragmentary, but there is some suggestion that a deficiency in 5-HT neurotransmission is a necessary, if not the sole, etiological factor in many cases of manic-depressive disorders (Murphy *et al.,* 1978). It is conceivable that by elevating and stabilizing serotonergic activity with lithium, attacks of both mania and depression can be prevented. Alterations in serotonergic activity do not show a close correlation with mood state, so it seems probable that lithium-induced changes in 5-HT mechanisms alter mood indirectly, possibly through actions on catecholaminergic (dopaminergic?) pathways. Bunney *et al.* (1977) have argued that mood swings in manic-depression are the reflection of cyclical changes in catecholamine release and receptor sensitivity. Reduced catecholaminergic transmission is held to be responsible for depressive symptoms, which remit as compensatory increases in catecholamine-receptor sensitivity occur. Should catecholamine release suddenly increase, marked amplification of neurotransmission will occur because of receptor supersensitivity, and mood will swing into a manic phase. Bunney *et al.* (1977) suggest that lithium prevents these mood swings by exerting a direct stabilizing effect on catecholamine receptors. However dopamine receptors were unaffected following withdrawal of lithium after long-term pretreatment, although the increase in receptor numbers resulting from prolonged haloperidol administration was prevented (Pert *et al.,* 1978). Rather than affecting postsynaptic dopamine receptors, our studies suggest that long-term administration of lithium may increase, and possibly stabilize, the level of activity in nerve cells which release dopamine.

In contrast to the effects of long-term administration, short-term pretreatment with lithium appears to reduce catecholaminergic transmission. Reduced access of neurotransmitter to supersensitive receptors probably accounts for the initial antimanic effect which becomes apparent as lithium levels in the brain reach a steady state. Following an improvement in manic symptoms, catecholaminergic activity may then gradually increase on maintenance lithium therapy, and so desensitize supersensitive catecholamine receptors. The mood-stabilizing effect of lithium may be explained on the basis of protection against decreases or further increases in catecholaminergic and serotonergic transmission. If nothing else, this concept of lithium's action in manic-depression brings together elements of both the catecholamine and indoleamine hypotheses of affective disorders, and takes cognizance of the fact that the duration of pretreatment is a crucial factor in determining the effects of lithium in the clinical situation.

ACKNOWLEDGEMENTS

I should like to thank Hannah Steinberg for help and encouragement, A. Davies, D.H. Jenkinson and R.A. Webster for advice and ICI Ltd, Sandoz Products Ltd and Servier Laboratories Ltd for gifts of drugs. This work was aided by grants from the Central Research Fund of the University of London and the Wellcome Trust.

REFERENCES

Barnett, S.A. (1975). *The Rat. A Study in behaviour* (University of Chicago Press, Chicago)
Berggren, U., Tallstedt, L., Ahlenius, S. and Engel, J. (1978). *Psychopharmacology 59,* 41
Berlyne, D.E. (1967). In *Nebraska Symposium on Motivation,* ed. D. Levine (University of Nebraska Press, Lincoln, Nebraska)
Borison, R.L., Sabelli, H.C., Maple, P.J., Havdala, H.S. and Diamond, B.I. (1978). *Psychopharmacology 59,* 259
Bunney, W.E., Jr. and Murphy, D.L. (1976). *Pharmakopsych., 9,* 142
Bunney, W.E., Jr., Post, R.A., Andersen, A.E. and Kopanda, R.T. (1977). *Comm. Psychopharmacol., 1,* 393
Cade, J.F.J. (1978). In *Lithium in Medical Practice,* ed. F.N. Johnson and S. Johnson (MTP Press, Lancaster, England) p. 5
Collard, K.J. (1978). *Br. J. Pharmacol., 62,* 137
Corrodi, H., Fuxe, K. and Schou, M. (1969). *Life Sci., 8,* 643
Cox, C., Harrison-Read, P.E., Steinberg, H. and Tomkiewicz, M. (1971). *Nature 232,* 336
Davies, C., Sanger, D.J., Steinberg, H., Tomkiewicz, M. and U'Prichard, D.C. (1974). *Psychopharmacology 36,* 263
Davis, J.M. and Fann, W.E. (1971). *Ann. Rev. Pharmacol., 11,* 285
Deakin, J.F.W. and Green, A.R., (1978). *Bri. J. Pharmacol., 64,* 201
Dews, P.B. and DeWeese, J. (1977). In *Handbook of Psychopharmacology,* Vol. 7, ed. L.L. Iversen, S.D. Iversen and S.H. Snyder (Plenum Press, New York) p. 107
Eccleston, D. and Nicolaou, N. (1978). *Br. J. Pharmacol., 64,* 341
Fieve, R.R. (1977). In *Handbook of Studies in Depression,* ed. G.D. Burrows (Excerpta Medica, Amsterdam) p. 217
Flemenbaum, A. (1977). *Biol. Psychiat., 12,* 563
Furukawa, T., Ushizima, I. and Nobufumi, O. (1975). *Psychopharmacology 42,* 243
Grahame-Smith, D.G. (1971). *J. Neurochem., 18,* 1053
Grahame-Smith, D.G. and Green, A.R. (1974). *Br. J. Pharmacol., 52,* 19
Gray, P., Solomon, J., Dunphy, M., Carr, F. and Hession, M. (1976). *Psychopharmacology 48,* 277
Green, A.R. and Grahame-Smith, D.G. (1976). *Nature 260,* 487
Halliday, M.S. (1968). In *The Analysis of Behavioural Change,* ed. L. Weiskrantz (Harper and Row, New York) p. 107
Harrison-Read, P.E. (1978). In *Lithium in Medical Practice,* ed. F.N. Johnson and S. Johnson (MTP Press, Lancaster, England) p. 289
Harrison-Read, P.E. (1979). *Br. J. Pharmacol., 66,* 144
Harrison-Read, P.E. and Steinberg, H. (1971). *Nature New Bio., 232,* 120.
Harvey, J.A., Schlosberg, A.J. and Yunger, L.M. (1975). *Fed. Proc., 34,* 1796

Hendler, N.H. (1978). In *Handbook of Psychopharmacology,* Vol. 14, ed. L.L. Iversen, S.D. Iversen and S.H. Snyder (Plenum Press, New York) p. 233

Hesketh, J.E., Nicolaou, N.M., Arbuthnott, G.W. and Wright, A.K. (1978). *Psychopharmacology 56,* 163

Ho, A.K.S., Gershon, S. and Pinckney, L. (1970a). *Arch. Int. Pharacodyn., 186,* 54

Ho, A.K.S., Loh, H.H., Craves, F., Hitzeman, R.J. and Gershon, S. (1970b). *Eur. J. Pharmacol., 10,* 72

Jacobs, B.L. (1974). *Psychopharmacology 39,* 81

Johnson, F.N. (1972a). *Experientia 28,* 533

Johnson, F.N. (1972b). *Nature 238,* 333

Johnson, F.N. (1975). In *Lithium Research and Therapy,* ed. F.N. Johnson (Academic Press, London) p. 315

Johnson, F.N. (1978). In *Lithium in Medical Practice,* ed. F.N. Johnson and S. Johnson (MTP Press, Lancaster, England) p. 305

Johnson, F.N. and Wormington, S. (1972). *Nature New Biol., 235,* 159

Judd, A., Parker, J. and Jenner, F.A., (1975). *Psychopharmacology 42,* 73

Katz, R.I., Case, T.N. and Kopin, I.J. (1968). *Science 162,* 466

Kiseleva, I.P., Lapin, I.P., Oxenkrug, G.F. and Samsonova, M.L. (1970). *CINP Seventh Congress Abstracts* (Prague) p. 239

Knapp, S. and Mandell, A.J. (1975). *J. Pharmacol. Exp. Ther., 193,* 812

Komisaruk, B.R. (1977). *Prog. Psychobiol. Physiol. Psychol., 7,* 55

Komiskey, H. and Buckner, C.K. (1974). *Neuropharmacology 13,* 159

Kumar, R. (1969). *Psychopharmacology 16,* 54

Kuriyama, K. and Speken, R. (1970). *Life Sci., 9,* 1213

Lester, D. (1966). *Nature 220,* 232

Mabry, P.D. and Campbell, B.A. (1973). *Brain Res., 49,* 381

Mandell, A.J. (1975). In *Neurobiological Mechanisms of Adaptation and Behaviour,* ed. A.J. Mandell (Raven Press, New York) p. 1

Mandell, A.J. and Knapp, S. (1976). *Pharmakopsychiatry 9,* 116

Matussek, N. and Linsmayer, M. (1968). *Life Sci., 7,* 371

Mukherjee, B.P., Bailey, P.T. and Pradhan, S.N. (1977). *Neuropharmacology 16,* 241

Murphy, D.L. (1977). In *Animal Models in Psychiatry and Neurology,* ed. I. Hanin and E. Usdin (Pergamon Press, New York) p. 211

Murphy, D.L., Campbell, I.C. and Costa, J.L. (1978). *Prog. Neuro-psycho-pharmacol., 2,* 1

Perez-Cruet, J., Tagliamonte, A., Tagliamonte, P. and Gessa, G.L. (1971). *J. Pharmacol. Exp. Ther., 178,* 325

Pert, A., Rosenblatt, J.E., Sivit, C., Pert, C.B. and Bunney, W.E., Jr. (1978). *Science 201,* 171

Plenge, P.K., Mellerup, E.T. and Rafaelson, O.J. (1973). *Int. Pharmacopsychiatry 8,* 234

Poitou, P., Guerinot, F. and Bohuon, C. (1974). *Psychopharmacology 38,* 75

Poitou, P., Boulu, R. and Bohuon, C. (1975). *Experientia 31,* 99

Robbins, T., and Iversen, S.D. (1973). *Psychopharmacology 28,* 155

Rushton, R., Steinberg, H. and Tomkiewicz, M. (1968). *Nature 220,* 885

Sanger, D.J. and Steinberg, H. (1974). *Eur. J. Pharmacol., 28,* 344

Schou, M. (1957). *Pharmacol. Rev., 9,* 17

Schou, M. (1958). *Acta pharmacol. Toxicol., 15,* 70

Schou, M. (1968). *J. Psychiat. Res., 6,* 67

Schou, M. (1976). *Ann. Rev. Pharmacol. Toxicol., 16,* 231

Schubert, J. (1973). *Psychopharmacology 32,* 301

Segal, D.S. (1975). *Science 190,* 475

Segal, D.S., Callaghan, M. and Mandell, A.J. (1975). *Nature 254,* 58

Shaw, D.M. (1973). *Bioch. Soc. Trans., 1,* 23

Shaw, D.M. (1975). In *Lithium Research and Therapy,* ed. F.N. Johnson (Academic Press, London) p. 411

Sheard, M.H. and Aghajanian, G.K. (1970). *Life Sci., 9,* 285

Shopsin, B., Georgotas, A. and Kane, S. (1979). In *Manic Illness,* ed. B. Shopsin (Raven Press, New York) p. 177

Siegal, S. (1956). *Nonparametric Statistics for the Behavioural Sciences* (McGraw-Hill, New York)

Slotkin, T.A., Seidler, F.J., Whitmore, W.L., Salvaggio, M. and Lau, C. (1978). *Mol. Pharmacol.*, *14*, 868

Sloviter, R.S., Drust, E.G. and Conner, J.D. (1978a). *J. Pharmacol. Exp. Ther.*, *206*, 339

Sloviter, R.S., Drust, E.G. and Connor, J.D. (1978b). *J. Pharmacol. Exp. Ther.*, *206*, 348

Smith, D.F. (1976a). *Experientia 32*, 1320

Smith, D.F. (1976b). *Pharmacol. Res. Comm.*, *8*, 575

Smith, D.F. (1977a). *Lithium and Animal Behaviour* (Churchill Livingstone, London)

Smith, D.F. (1977b). *Psychopharmacology 53*, 103

Smith, D.F. and Smith, H.B. (1973). *Psychopharmacology 30*, 83

Snyder, S.H., Banerjee, S.P., Yamamura, H.I. and Greenberg, D. (1974). *Science 184*, 1243

Ströbom, U. (1976). *Naunyn-Schmiedeberg's Arch. Pharmacol.*, *292*, 167

Weissman, A.B., Koe, K. and Tenen, S.S. (1966). *J. Pharmacol. Exp. Ther.*, *151*, 339

Wielosz, M. and Kleinrok, Z. (1979). *J. Pharm. Pharmacol.*, *31*, 410

Yamada, K. and Furukawa, T. (1979). *Psychopharmacology 61*, 255

Zilberman, Y., Kaptulnik, J., Guerstein, G. and Lichtenberg, D. (1979). *Pharmacological Res. Comm.*, *11*, 467

12 EXPERIMENTAL MODELS OF MENTAL ILLNESS: SEPARATION-INDUCED DEPRESSION IN PRIMATES

Pavel D. Hrdina and Margaret E. Henry

TABLE OF CONTENTS

I	INTRODUCTION	265
II	ANIMAL MODELS OF MENTAL ILLNESS: DEPRESSION	265
III	PRIMATE MODELS OF DEPRESSION	267
	A. Pharmacologically Induced Depression	267
	B. Separation-induced Depression	267
	(a) Mother-infant Separation	267
	(b) Peer-peer Separation	268
	C. Pharmacological Modification of Separation-induced Depression	269
IV	HORMONAL RESPONSES IN ANIMAL MODELS OF DEPRESSION	270
	A. Adrenal Responses to Separation	270
	B. Other Hormonal Responses during Altered Behavioural States	271
V	SUMMARY AND CONCLUSIONS	273
	REFERENCES	273

I. INTRODUCTION

Despite the impressive amount of research on the subject during the past decade or so, our understanding of the etiology of the affective disorders and of biological mechanisms underlying these conditions remains unclear. One of the limitations in investigating the biological and social variables of affective disorders has been the lack of a suitable animal model which would mimic the symptoms present in man and offer an opportunity of making a direct study of the associated biochemical changes in the brain, which, for ethical and practical reasons, cannot be studied in man. This chapter will attempt to evaluate various animal models of depression, and present data from our own investigation utilizing the separation paradigm in nonhuman primates.

II. ANIMAL MODELS OF MENTAL ILLNESS: DEPRESSION

Investigation of the biological basis of this mental disorder, in conjunction with social variables, in an adequate animal model would provide valuable information which may be relevant to some currently held concepts. A model system, according to McKinney and Bunney (1969), should meet the following requirements: (1) the symptoms of 'depression' should be reasonably analogous to those seen in human depression; (2) there should be observable behavioural changes which can be objectively evaluated; (3) independent observers should agree on objective criteria for drawing conclusions about the subjective state; (4) treatments effective in reversing depression in humans should reverse the changes seen in animals; (5) the model system should be reproducible by other investigators. The experimental production of such a model in animals would also allow an investigation of brain biogenic substances and other biological variables while the animal is 'depressed', and of the eventual reversal of changes in these parameters with appropriate treatment.

Several major avenues have been followed to induce altered behavioural states in animals which appear to be objectively similar to depression in humans. These approaches include pharmacologically induced depressive states, learned helplessness, separation paradigms and social isolation rearing.

One of the most prevalent experimental models of depression was the reserpine model in rats. As McKinney (1976) pointed out, the rat model has severe limitations in terms of its behavioural analogies to human depression and its nonsocial nature. The learned-helplessness model was proposed originally by Seligman (Seligman and Maier, 1967; Seligman et al., 1968; Seligman and Groves, 1970;

Seligman, 1972). This term, as used by Seligman *et al.* (1968), describes the maladaptive, passive acceptance of avoidable electric shocks. Initially, the animals are subjected to a series of random unavoidable and inescapable shocks. Later, the contingencies change, and escape is possible by jumping a barrier. The subjects fail to learn that barrier-jumping produces shock termination. This phenomenon, primarily demonstrated in dogs, has also been shown in rats, cats, fish, mice and men. In Seligman's view, the interference with adaptive responding produced by inescapable shock parallels the process underlying clinical reactive depression, in which the individual loses control over the reinforcers in the environment. Negative expectations regarding the effectiveness of one's control over the environment lead to passivity and reduce response initiation (seen clinically as retardation of psychomotor activity and thought processes). It is interesting to note that the norepinephrine depletion produced by uncontrollable shock parallels in some aspects the catecholamine hypothesis of depression (Schildkraut and Kety, 1967), in which lack of norepinephrine is postulated to be the biological basis of human depression.

It is now fairly well accepted that human children may be profoundly disturbed by a period of separation from the mother figure (Bowlby, 1952; Ainsworth, 1962). However, the dynamics of the disturbance and the marked individual differences in the severity of the effects are imperfectly understood (Ainsworth, 1962; Spencer-Booth and Hinde, 1971). This is partially due to the retrospective nature of human studies and to the ethical constraints of relevant experimentation. The possibility therefore arises that description of attachment bonds in subhuman species would have heuristic value. A great number of studies have in fact demonstrated that relatively brief separation from the mother can produce at least transient effects in rats (Hofer, 1975a, b), dogs (Senay, 1966) and monkeys (Jensen and Tolman, 1962; Seay *et al.*, 1962; Seay and Harlow, 1965; Kaufman and Rosenblum, 1967a, b, 1969; Spencer-Booth and Hinde, 1967; Rosenblum and Kaufman, 1968; Spencer-Booth and Hinde, 1971; Hrdina and von Kulmiz, 1978).

The technique of social isolation was one of the first methods used to induce psychopathology in animals (Harlow *et al.*, 1955; McKinney, 1974). By subjecting rhesus monkeys to total social isolation for at least the first six months after birth, the formation of social bonds was prevented. These monkeys displayed persistent and pronounced social, sexual and maternal inadequacies: remaining withdrawn in a huddled, self-clasping posture and engaging in stereotypy and self-aggression. The social-isolation syndrome produced by this model has been compared to various clinical syndromes such as autism, schizophrenia and depression. McKinney (1974) urges caution in attributing clinical labels to this syndrome however, as this model for human psychopathology fails to meet several of the criteria outlined previously.

III. PRIMATE MODELS OF DEPRESSION

Nonhuman primates appear to provide the most feasible model of human depression, because of their phylogenetic relation to man, their tendency to form close social bonds, the existence of a fund of knowledge about their social behaviour and the availability of suitable laboratory techniques for well-controlled longitudinal studies.

A. *Pharmacologically Induced Depression*

A possible animal model of experimental depression which could provide parallels with human depressive illness is pharmacologically-induced 'depression' in primates. According to the 'biogenic-amine hypothesis' of affective disorders, a selective depletion of brain catecholamines should produce behavioural changes resembling those in human depression. Indeed, when α-methylparatyrosine (α-MpT), a selective depletor of norepinephrine and dopamine, was administered to adult stump-tail monkeys, the animals displayed marked changes in behaviour five to six weeks after initiation of α-MpT treatment (Redmond *et al.*, 1971): decreased total social interaction and initiative, lack of concern with the environment and retarded motor activity. Similar effects with α-MpT were observed by McKinney's group (Akiskal and McKinney, 1973). Moreover, reserpine, which has been extensively used in rodents for studies on antidepressant drugs, has been shown to produce, in monkeys, symptoms (decrease in locomotion and visual exploration, and an increase in huddling behaviour) resembling the 'despair' phase seen in both human and primate infants during separation (McKinney *et al.*, 1971). One obvious disadvantage which is common for all pharmacological models is that drugs used as tools often have multiple effects which necessarily introduce additional variables. Therefore, it would appear to be preferable to have experimental models in which the genuine factors leading to the precipitation of psychopathological changes in humans could be simulated. Of such models of depression, the model of separation from an object of attachment is the oldest and best documented.

B. *Separation-induced Depression*

(a) Mother-infant Separation. Since the early 1960s, several separation studies of different designs have been reported. Many of these have used higher-order monkeys and the maternal-separation paradigm. The original focus of attention was the nature of the infant's bond with its mother, and the effect of renewing this bond after separating the pair. Harlow and his coworkers (Seay *et al.*, 1962) were the first to investigate the maternal bond in a primate laboratory by carrying out controlled behavioural studies of mother-infant separation. Previous to this, the phenomenon of separation had been well described in another context by Spitz (Spitz, 1945; Spitz and Wolf, 1946), who was investigating the morbid effects of institutionalization on human infants, and described the syndrome of anaclitic depression. Robertson and Bowlby (1952) postulated that the reaction

occurred in distinct stages. The initial 'protest' stage was characterized by high agitation, the subsequent 'despair' stage by weeping and reduced activity and the final 'detachment' phase by withdrawal from any activity in the environment, and by rejection of any attempts by the mother to reestablish the relationship.

The maternal-separation experiments with monkey subjects, which have been replicated a number of times, report strikingly similar responses (Seay and Harlow, 1965; Hinde *et al.,* 1966; Kaufman and Rosenblum, 1967a, b). Typically, the infants reacted to the separation from their mother with an agitated fear response (protest phase) followed by a suppression of usual behaviours (despair phase) and a gradual recovery. In the midsixties the consensus in the literature was that human anaclitic depression was approximated well by maternal separation in monkeys. In fact, the agreement was so great that the emphasis changed from investigating the mother-infant bond and its disruption (sometimes referred to as attachment and separation from the object of attachment) *per se* to having the paradigm serve as an animal model for this situation, the ontogeny of depression being left somewhat glossed over. A very cogent critique of models of depression in general is given by Akiskal and McKinney (1973) and of separation in particular by Chappell and Meier (1975). The latter authors criticize previous studies for their lack of uniformity in procedure and the resultant difficulty in directly comparing their results.

In one series of separation experiments in our primate laboratory (Hrdina and von Kulmiz, 1978), the four macaque infants showed, in response to maternal separation, behavioural changes similar to those reported in previous separation studies (Seay and Harlow, 1964; Hinde *et al.,* 1966; Hinde and Spencer-Booth, 1971). The design of our experiment called for testing the effects of removing only the mother from the environment, leaving all previous conditions the same (e.g., unlimited peer contact). For this reason, the course of the reaction to separation in our study was not as severe as indicated by some earlier authors (Kaufman and Rosenblum, 1967a, b). Three of the four infants occasionally reached a state of agitation and even panic, and showed very little play activity during separation, but generally were not in a state of extreme despair or 'depression.' We have therefore concluded that unlimited contact with peers during separation provided a mechanism of compensation for the absence of mothers during their time of distress, allowing them, to some extent, to cope better with the separation condition, and thus they demonstrated less despair. The fourth infant, however, who underwent social isolation as well as maternal separation, did indeed show the classic signs of despair.

(b) Peer-peer Separation. The mother-infant-separation paradigm does have some inherent limitations: first, the nature of the mother-infant affectional bond during the first year of the infant's life is rapidly changing (Hansen, 1966). This makes repetitive separations difficult to evaluate, and requires a large number of subjects of similar age to generate meaningful data if controlled pharmacological studies are to be employed. Second, while the mother-infant-separation paradigm

may be relevant to anaclitic depression in human infants and children, its significance as an experimental model of adult depression may be questioned. In fact, Spitz has been criticized for his term 'anaclitic' depression by some authors who feel that infants do not have a sufficiently developed 'psyche' to experience the depth of feelings attributed to human depression (Rie, 1966).

Repetitive peer separation in young or adolescent rhesus monkeys has recently been proposed as an alternative to the mother-infant-separation paradigm for experimental studies on the effects of separation from an attachment object (Suomi *et al.*, 1970; Bowden and McKinney, 1972; McKinney *et al.*, 1972) or as an experimental model of depression (Lewis *et al.*, 1976; Suomi *et al.*, 1978). Suomi *et al.* (1970) found that repetitive, short-term (four days) peer separation in young (under one year of age) macaques produced behavioural symptoms (an initial increase in locomotion, vocalization and environmental exploration, followed by a phase of high levels of passive and self-directed behaviours) similar to those exhibited by infants separated from their mothers. In addition, the subjects did not adapt to the separation procedure and showed similar reactions to repeated separations. Lewis *et al.* (1976), in fact, found intensification of reactions as the number of separations increased. Other studies of adolescent rhesus monkeys who underwent peer separation found only a uniphasic reaction: behaviours characteristic of the protest stage were apparent throughout the entire separation period, with no evidence of a despair stage (McKinney, 1976). This model thus seems promising, particularly for evaluation of the effects of drugs, in that it offers the advantage of repeated treatments and crossover designs.

C. Pharmacological Modification of Separation-induced Depression

The effects of drugs on isolation-induced, disturbed behaviours in primates have been reported. Trifluoperazine, dextroamphetamine sulphate and oxazepam were found to be ineffective in modifying the behavioural deficits in socially isolated chimpanzees (Menzel *et al.*, 1963; Turner *et al.*, 1969), whereas administration of chlorpromazine to rhesus monkeys subjected to partial social isolation resulted in a significant decrease in self-disturbance behaviours including self-bite, huddling and stereotypy (McKinney *et al.*, 1973). In the latter study, chlorpromazine failed to increase the frequency of active social encounters in most of the sujects. Noble *et al.* (1976) report that socially isolated rhesus monkeys responded to diazepam treatment with the reappearance of some social behaviours in addition to a significant reduction in self-disturbance behaviours.

In our laboratory, seven infant macaques separated from their mothers showed reproducible behavioural changes characterized by reduction of active social contact and play behaviours, and a concomitant rise in behaviours indicative of social disengagement and passivity. Treatment of these infants with a specific antidepressant drug, desmethylimipramine (DMI), prevented or even reversed some of the behavioural changes produced by separation (Hrdina *et al.*, 1979). It is of interest to note that the effects of DMI were detected in categories of behaviour most affected by the separation procedure ('active contact,' 'play initiation' and 'social exploration'). During the separations, the DMI-treated infants showed higher frequency than control animals of behaviours indicating active social participation. On the other hand, incidence of distress ('vocalization') and self-oriented behaviours ('body play') was suppressed in subjects receiving the antidepressant drug. It indeed appears as if DMI acted as a behaviour 'normalizer' to lessen the separation-induced behavioural responses. A summary of the results is shown in Figure 12.1. A separate experiment in another group of three infant macaques has shown that i.m. administration of 5 mg/kg of DMI per day for five days resulted in drug plasma levels ranging from 149 ± 32 to 47 ± 10 ng/ml at four and 22 hours after the last injection, respectively. Thus, the dose schedule of DMI used in our study was likely to produce plasma levels which are within the range of those usually attained in depressed humans during effective therapy with tricyclic antidepressants (Yates *et al.*, 1963).

Demonstration that another clinically effective antidepressant drug (imipramine) was able to differentially modify behavioural responses of young macaques subjected to repetitive peer separations (Suomi *et al.*, 1978) provided further support for the usefulness of the separation model in primates to pharmacological studies on depression.

IV. HORMONAL RESPONSES IN ANIMAL MODELS OF DEPRESSION

A. Adrenal Responses to Separation

Among 'biological markers' of psychoses, the determination of patterns of hormonal disturbances present in patients with affective disorders could provide useful information not only for the diagnosis but also for making inferences about the function of the central nervous system in these states. A well-documented neuroendocrine disturbance in depressed patients is elevated plasma cortisol and the abnormal suppression response of the hypothalamic-pituitary-adrenal axis to dexamethasone. It has been shown (Carroll *et al.*, 1976) that an escape from dexamethasone suppression occurred in patients with endogenous depression but rarely in patients with depressive neurosis (Carroll, 1978). Increases in excretion of urinary 17-hydroxycorticosteroids (Rose *et al.*, 1969) and plasma cortisol were also reported to occur in infant rhesus monkeys (Hill *et al.*, 1973; Meyer *et al.*, 1975; Smotherman *et al.*, 1979) or squirrel monkeys (Coe *et al.*, 1978; Mendoza *et al.*, 1978) in response to separation from their mothers or other attachment object. An important finding emerging from these studies was that the effect of separation on the pituitary-adrenal system was not due to disturbances involved in the separation procedure, but was most probably a function of the separation state (Mendoza *et al.*, 1978).

In most of these studies, the adrenal response to short-term separation was investigated. Only Meyer *et al.* (1975) have measured plasma cortisol in infants over eight weeks of separation and reported that, following separation, mother-reared infants had considerably higher plasma cortisol levels than those observed before separation or those noted in the group of surrogate-peer-reared infants.

Figure 12.1: Effect of chronic desipramine (DMI) treatment (5 mg/kg/day i.m.) on behavioural responses of infant macaques to maternal separation. Columns represent mean percentage change in frequency or duration of several categories of behaviour, taking values of preceding baseline period as 100%.Note that treatment with the antidepressant has markedly diminished or reversed most of the behavioural alterations induced by separation. In particular, the increase in distress (vocal) and self-directed (body play) behaviours, as well as suppression of social behaviours (social exploration, active contact) were prevented or antagonized.

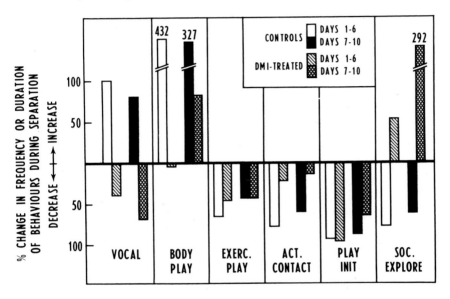

It appeared that mother-reared infant monkeys responded in a stronger manner to the loss of their attachment object than surrogate-peer-reared infants. However, the data on cortisol levels might have been confounded by the order in which animals were tested, and the results do not warrant a definite conclusion as to the effect of separation on adrenal function.

B. Other Hormonal Responses during Altered Behavioural States

Another hormone to be extensively studied in affective disorders is growth hormone (GH), whose regulation has been demonstrated to be abnormal in some depressed patients. In the human adult, GH release physiologically occurs in response to a falling blood sugar, certain amino acids, exercise and stress. From the perspective of the biogenic-amine hypothesis, it would seem plausible to expect that GH responses might be inhibited in depressed patients. In this regard, GH response to insulin hypoglycemia has been studied by several workers (Sachar *et al.*, 1971, 1973; Carroll, 1972). These reports indicate diminished or absent GH responses to insulin-induced hypoglycemia in unipolar depressed patients, while in bipolar and neurotic depressions the response was normal or even enhanced. 5-Hydroxytryptophan (5-HTP), the amino-acid precursor of serotonin, has been reported to cause a rise in plasma GH levels in normal subjects; depressed patients, however, showed an inadequate response to 5-HTP (Takahashi *et al.*, 1974).

In a series of experiments on squirrel monkeys (Brown *et al.*, 1971) and rhesus monkeys (Meyer and Knobil, 1967), it was determined that GH hypersecretion is stimulated by a variety of psychological stimuli such as capture and chair restraint. Information on GH responsiveness to separation from attachment objects in monkeys is lacking, and further studies are required to shed light on the regulation of GH release in this model of depression.

Plasma prolactin (PRL) has also been studied in depressed patients. Increases in prolactin secretion normally occur at the end of pregnancy in conjunction with lactation. Prolactin release also typically occurs in response to stress, insulin hypoglycemia and L-DOPA. In one investigation of prolactin levels in normal and depressed patients (both unipolar and bipolar), baseline prolactin concentration was found to be elevated in the depressed patients (Sachar *et al.*, 1973). The authors suggest this may be due to increased emotional stress in the depressed patients.

Nonprimate studies on prolactin release clearly indicate that acute and prolonged stress can modify the secretion of PRL. However, rats exhibit a different response pattern during short-term as opposed to long-term stress. Acute stressors have been repeatedly shown to elevate PRL levels (Krulich *et al.*, 1974; Turpen *et al.*, 1976), whereas there is more recent evidence that chronic stress causes a diminution of PRL levels (Taché *et al.*, 1978). It is apparent that further investigation of PRL response to acute and chronic stress is required, preferably in species more closely related to man.

There have been conflicting reports that thyrotropin-releasing hormone (TRH) has an antidepressant effect. Two reports have indicated a transient alleviation of depressive symptoms in unipolar depressed patients (Kastin *et al.*, 1972; Prange *et al.*, 1972). Other investigators have reported negative results (Coppen *et al.*, 1974; Ehrensing *et al.*, 1974). Furlong *et al.* (1976) reported equivocal results and suggest that previous conflicting reports about TRH's antidepressant effects may stem from the study of heterogeneous groups of depressed patients. They suggest that there may be a specific depressive subgroup that is responsive to TRH. Despite the lack of unanimity regarding the antidepressant effect of TRH, there has been repeated evidence that TSH and PRL responses are severely diminished in response to administration of TRH in a significant number of depressed patients compared to normal controls (Coppen *et al.*, 1974; Ehrensing *et al.*, 1974). This altered TSH response has been correlated with the severity of depression (Ehrensing *et al.*, 1974; Furlong *et al.*, 1976). Du Ruisseau *et al.* (1978) reported that TSH levels in rats are very sensitive indicators of stress, and that stress-induced increases are pronounced and are positively correlated to the intensity and duration of the stressing agent (except exposure to cold, which is idiosyncratic and reduces the level of TSH).

Although the decreased libido characteristic of depressive illness had led to speculation that sex hormones may be affected in depression, there is no conclusive evidence yet that they are in fact secreted abnormally. Studies have reported either no change in plasma LH concentrations from illness to recovery in postmenopausal depressed women (Sachar *et al.*, 1972) or reduced plasma LH levels in depressed postmenopausal women compared to normal postmenopausal women (Altman *et al.*, 1975). The response of LH levels of stress in rats indicate a biphasic pattern; that is, levels of LH increase with acute stress, but levels decrease below control values after chronic stress (Du Ruisseau *et al.*, 1978). Plasma testosterone levels measured in a group of depressed men showed no significant alteration before and after recovery (Sachar *et al.*, 1973), even in the case of subjects who manifested considerable emotional distress and increased cortisol levels during their illness. This was contrary to numerous reports on both monkeys and young men under stress who show increased corticosteroid output and significantly reduced plasma and urinary testosterone (Rose, 1969; Rose *et al.*, 1971, 1975; Kreuz *et al.*, 1972). Clearly, more evidence is necessary to confirm abnormal sex-hormone secretion in human affective disorders and to establish an adequate animal model.

V. SUMMARY AND CONCLUSIONS

Separation is an event frequently associated with depressive states. Recent results from several laboratories, including ours, indicate that the separation-induced 'depression' in primates is a suitable model for controlled studies in which the factors precipitating aberrant emotional behaviour can be manipulated at the experimenter's will. Such manipulations, for ethical reasons, cannot be done in humans. Separation studies in nonhuman primates thus offer a utilisable animal model for investigating certain aspects of depression. Further studies of pharmacological modification and physiological and biochemical correlates of separation-induced behavioural disorder could help to elucidate its underlying neurobiological mechanisms.

ACKNOWLEDGEMENT

The work reported in this chapter has been supported by Ontario Mental Health Foundation grants 714-77/79 and 767-79/81.

REFERENCES

Ainsworth, M.D. (1962). In *Deprivation of Maternal Care: A Reassessment of Its Effects*, Public Health Papers, No. 14, (WHO, Geneva)

Akiskal, H.S. and McKinney, W.T. (1973). *Science 182,* 20

Altman, N., Sachar, E.J., Gruen, P.H., Halpern, F.S. and Eto, S. (1975). *Psychosom. Med., 37,* 274

Bowden, D.M. and McKinney, W.T. (1972). *Develop. Psychobiol., 5,* 353

Bowlby, J. (1952). *Maternal Case and Mental Health,* Monograph No. 2 (WHO, Geneva)

Brown, G.M., Schalch, D.S. and Reichlin, S. (1971). *Endocrinology 89,* 694

Carroll, B.J. (1972). In *Depressive Illness: Some Research Studies,* ed. B. Davies, B.J. Carroll and R.M. Mowbray (Charles C. Thomas, Springfield, Illinois) p. 149

Carroll, B.J. (1978). In *Depressive Disorders,* ed. S. Garattini (Schattaner Verlag, Stuttgart) p. 231

Carroll, B.J., Cuttis, G.C. and Mendels, J. (1976). *Arch. Gen. Psychiat., 33,* 1039

Chappell, P.E. and Meier, G.W. (1975). *Biol. Psychiat., 10,* 643

Coe, C.L., Mendoza, S.P., Smotherman, W.P. and Levine, S. (1978). *Behav. Biol., 22,* 256

Coppen, A., Montgomery, S. and Peet, M. (1974). *Lancet ii,* 433

Du Ruisseau, P., Tache, Y., Brazeau, P. and Collu, R. (1978). *Neuroendocrinology 27,* 257

Ehrensing, R.H., Kastin, A.J., Schalch, D.S., Friesen, H.G., Vargas, J.R. and Schally, A.V. (1974). *Am. J. Psychiat., 131,* 714

Furlong, F.W., Brown, G.M. and Beeching, M.F. (1976). *Am. J. Psychiat., 133,* 1187

Hansen, E.W. (1966). *Behavior 27,* 107

Harlow, H.F., Dodsworth, R.O. and Harlow, M.K. (1955). *Proc. Acad. Sci., 54,* 90

Hill, S.D., McCormack, S.A. and Mason, W.A. (1973). *Develop. Psychobiol., 6,* 421

Hinde, R.A. and Spencer-Booth, Y. (1970). *J. Child. Psychol. Psychiat., ii,* 159

Hinde, R.A. and Spencer-Booth, Y. (1971). *Science 173,* 111

Hinde, R.A., Spencer-Booth, Y. and Bruce, M. (1966). *Nature 210,* 1021

Hofer, M.A. (1975a). *Psychosom. Med., 37,* 245

Hofer, M.A. (1975b). *Biol. Psychiat., 10,* 149

Hrdina, P. and von Kulmiz, P. (1978). In *Depressive Disorders,* ed. S. Garattini (Schattauer Verlag, Stuttgart) p. 211

Hrdina, P., von Kulmiz, P. and Stretch, R. (1979). *Psychopharmacology 64,* 89

Jensen, G.D. and Tolman, C.W. (1962). *J. Comp. Physiol. Psychiat., 55,* 131

Kastin, A.J., Ehrensing, R.H., Schalch, D.S. and Anderson, M.S. (1972). *Lancet ii,* 740

Kaufman, I.C. and Rosenblum, L.A. (1967a). *Science 155,* 1030

Kaufman, I.C. and Rosenblum, L.A. (1967b). *Psychosom. Med., 29,* 648

Kaufman, I.C. and Rosenblum, L.A. (1969). *Ann. NY Acad. Sci., 159,* 681

Kreuz, L.E., Rose, R.M. and Jennings, J.R. (1972). *Arch. Gen. Psychiat., 26,* 479

Krulich, L., Hefco, E., Illner, P. and Read, C.B. (1974). *Neuroendocrinology 16,* 293

Lewis, J.K., McKinney, W.T., Young, L.D. and Kraemer, G.W. (1976). *Arch. Gen. Psychiat., 33,* 699

McKinney, W.T. (1974). *Arch. Gen. Psychiat., 31,* 422

McKinney, W.T. (1976). In *Depression,* ed. D.M. Gallant (Spectrum Publications, New York) p. 1

McKinney, W.T. and Bunney, W.E. (1969). *Arch. Gen. Psychiat., 21,* 240

McKinney, W.T., Suomi, S.J. and Harlow, H.F. (1971). *Am. J. Psychiat., 127,* 1313

McKinney, W.T., Suomi, S.J. and Harlow, H.F. (1972). *Arch. Gen. Psychiat., 27,* 200

McKinney, W.T., Young, L.D., Suomi, S.J. and Davis, J.M. (1973). *Arch. Gen. Psychiat., 29,* 490

Mendoza, S.P., Smotherman, W.P., Miner, M.T., Kaplan, J. and Levine, S. (1978). *Develop. Psychobiol., 11,* 169

Menzel, E.W. (1963). *J. Comp. Physiol. Psychol., 56,* 329

Meyer, V. and Knobil, E. (1967). *Endocrinology 80,* 163

Meyer, J.S., Novak, M.A., Bowman, R.E. and Harlow, H.F. (1975). *Develop. Psychobiol., 8,* 425

Noble, A.B., McKinney, W.T., Mohr, C. and Moran, E. (1976). *Am. J. Psychiat., 133,* 1165

Porsolt, R.D., Le Pichon, M. and Jalfre, M. (1977). *Nature 266,* 730

Porsolt, R.D., Bertin, A., Blauet, N., Deniel, M. and Jalfre, M. (1979). *Eur. J. Pharmacol., 57,* 201

Prange, A.J., Wilson, I.C., Lara, P.P., Alltop, L.B. and Breese, G.R. (1972). *Lancet ii,* 999

Redmond, E., Maas, J.W., Kling, A. and Dekirmenjian, H. (1971). *Psychosom. Med., 33,* 97

Rie, H.E. (1966). *J. Am. Acad. Child Psychiat., 5,* 653

Robertson, J. and Bowlby, J. (1952). *Courrier 2,* 131
Rose, R.M. (1969). *Psychosom. Med., 31,* 405
Rose, R.M., Mason, J.W. and Brady, J.V. (1969). In *Proceedings of the 2nd International Congress of Primatology,* Vol. 1, ed. C.R. Carpenter (Karger, Basel) p. 211
Rose, R.M., Holaday, J.W. and Bernstein, I.S. (1971). *Nature 231,* 366
Rose, R.M., Bernstein, I.S. and Gordon, T.P. (1975). *Psychosom. Med., 37,* 50
Rosenblum, L.M. and Kaufman, I.C. (1968). *Am. J. Orthopsychiat., 35,* 418
Sachar, E.J., Finkelstein, J. and Hellman, L. (1971). *Arch. Gen. Psychiat., 25,* 263
Sachar, E.J., Schalch, D.S., Reichlin, S. and Platman, S.S. (1972). In *Recent Advances in the Psychobiology of Depressive Illnesses,* ed. T.W. Williams, M.M. Katz and J.A. Shield Publication No. 70-9053 (US Department of Health, Education and Welfare, Washington, DC) p. 229
Sachar, E.J., Frantz, A.G., Altman, N. and Sassin, J. (1973). *Am. J. Psychiat., 130,* 1362
Schecter, M.D. and Chance, W.T. (1979). *Eur. J. Pharmacol., 60,* 139
Schildkraut, J.J. and Kety, S.S. (1967). *Science 156,* 21
Seay, B. and Harlow, H.F. (1965). *J. Nerv. Ment. Dis., 140,* 434
Seay, B., Hansen, E. and Harlow, H.F. (1962). *J. Child Psychol. Psychiat., 3,* 123
Seligman, M.E.P. (1972). *Am. Rev. Med., 23,* 407
Seligman, M.E.P. and Maier, S.F. (1967). *J. Exp. Psychol., 74,* 1
Seligman, M.E.P. and Groves, D. (1970). *Psychonom. Sci., 19,* 191
Seligman, M.E.P., Maier, S.F. and Geer, J.H. (1968). *J. Abnorm. Soc. Psychol., 73,* 256
Senay, E.C. (1966). *J. Psychiat. Res., 4,* 65
Smotherman, W.P., Hunt, L.E., McGinnis, L.A. and Levine, S. (1979). *Develop. Psychobiol., 12,* 211
Spencer-Booth, Y. and Hinde, R.A. (1967). *J. Child Psychol. Psychiat., 7,* 179
Spencer-Booth, Y. and Hinde, R.A. (1971). *Animal Behav., 19,* 174
Spitz, R.A. (1945). *Psychoanalyt. Study Child 1,* 53
Spitz, R.A. and Wolf, K.M. (1946). *Psychoanalyt. Study Child 2,* 313
Suomi, S.J., Harlow, H.F. and Domek, C.J. (1970). *J. Abnorm. Psychol., 76,* 161
Suomi, S.J., Seaman, S.F., Lewis, J.K., Kelizio, R.D. and McKinney, W.T. (1978). *Arch. Gen. Psychiat., 35,* 321
Taché, Y., Du Ruisseau, P., Ducharme, J.R. and Collu, R. (1978). *Neuroendocrinology 26,* 208
Takahashi, R., Nagayama, H., Kido, A. and Morita, T. (1974). *Biol. Psychiat., 9,* 191
Turner, C.H., Davenport, R.K. and Rogers, C.M. (1969). *Am. J. Psychiat., 125,* 85
Turpen, C., Johnson, D.C. and Dunn, J.D. (1976). *Neuroendocrinology 20,* 339
Yates, C., Todrick, A. and Tait, A. (1963). *J. Pharm. Pharmacol., 15,* 432

13 THE ROLE OF NEUROENDOCRINE FUNCTION IN ANOREXIA NERVOSA

Paul E. Garfinkel, Gregory M. Brown and Padraig L. Darby

TABLE OF CONTENTS

I	INTRODUCTION	279
II	MENSTRUAL FUNCTION AND FERTILITY	280
III	HYPOTHALAMIC FUNCTION	281
IV	GONADOTROPINS	281
V	GROWTH HORMONE	284
VI	PITUITARY-THYROID FUNCTION	285
VII	PITUITARY-ADRENAL REGULATION	286
VIII	PROLACTIN	287
IX	MELATONIN	288
X	TESTOSTERONE	289
XI	CARBOHYDRATE METABOLISM	289
XII	THERMOREGULATION AND WATER CONSERVATION	289
XIII	MONOAMINE METABOLISM	290
XIV	SUMMARY AND CONCLUSIONS	290
	REFERENCES	291

I. INTRODUCTION

Anorexia nervosa is characterized by vigorous pursuit of a thin body size and an exaggerated fear of weight gain despite emaciation. While once considered a rare disorder, recent studies have confirmed clinical impressions that anorexia nervosa has become a relatively common problem of adolescence. The prevalence in adolescent girls has been estimated to be about 0.5-1% (Crisp, 1977). While rare in males, it does occur.

The anorexic's pursuit of thinness begins with dieting not unlike that of her peers. Often the girl is slightly overweight and may begin to diet in response to comments about her size. A particular part of her body (e.g., stomach or thighs) may feel too large and the initial intent is to 'lose a few pounds.' At first she will cut out sweets, desserts and snacks, selectively avoiding carbohydrates in general. When her weight goal has been attained, she still feels somewhat overweight and decides to increase her avoidance of foods until she displays a stubborn refusal to eat anything but small quantities of food. While the term 'anorexia' implies a loss of appetite, often the individual's appetite is maintained until late in the starvation process. Most patients report a normal hunger awareness but also express a fear of giving in to the impulse to eat (Garfinkel, 1974). By contrast, satiety perception is extremely distorted. Patients report severe bloating, nausea and distention after even small amounts of foods.

A large subgroup of anorexia nervosa patients alternate between not eating and gorging. Often they are predisposed to obesity, through a history of familial and personal obesity (Garfinkel *et al.,* 1979). These patients fluctuate between an intense sense of self-control and loss of control. Usually, they restrict food intake when they are busy; when alone, however, they will overeat to the point of exhaustion. Foods then eaten are usually those 'forbidden' by the diet: rich cakes, desserts and ice-cream. The quantities of food eaten during these periods are immense; for example, dozens of doughnuts may be consumed. Patients will often then vomit or misuse large quantities of laxatives with the intention of preventing absorption of the ingested food. The amount of money needed to buy foods can become considerable, and some patients begin to steal to support this behaviour.

In spite of the progressive and severe weight loss, many anorexic women are not aware of their emaciated state (Bruch, 1973). They may deny many aspects of the disorder, especially the fact that their changed bodies are not healthy or beautiful. This body-image disturbance is of delusional proportions. Denial of illness also extends to a lack of awareness of internal bodily perceptions, including fatigue, and patients often enjoy excess energy until rather late in the illness.

Although energetic, many anorexics restrict their activities to exercise and schoolwork, while becoming progressively more socially withdrawn.

The physical signs of anorexia nervosa include the emaciated appearance and lanugo hair. Often there is some hair loss from the scalp and carotene pigmentation may be evident on the soles and palms. Bradycardia and hypotension can be marked, and bruises from subsequent falls may be evident. In contrast to primary hypopituitarism, the secondary sexual characteristics are normal.

II. MENSTRUAL FUNCTION AND FERTILITY

Amenorrhea is a characteristic feature of anorexia nervosa. While this is usually a secondary amenorrhea, depending on the age of onset of the illness, some girls may present with primary amenorrhea; for example, Hurd *et al.* (1977) found 11% to be prepubertal in onset. For many patients, loss of menses occurs shortly after the onset of the weight loss (Fries, 1977; Hurd *et al.*, 1977). However, in a significant proportion (estimates vary between 7 and 24%), amenorrhea actually appears to precede the weight loss (Fries, 1977). Obviously these findings are limited by the imprecise retrospective reports of the onset of the dieting behaviour. Nevertheless, for some patients at least, amenorrhea appears to be the first manifestation of the illness. Whether this is due to the emotional upheaval that precedes the dieting behaviour or an independent hypothalamic abnormality is not clear.

For most anorexic patients, amenorrhea is related to the degree of weight loss. This is also true for the female population as a whole. For example, in a random sample of a Swedish population, Pettersson *et al.* (1973) found that 6% of females under 25 had secondary amenorrhea for more than three months. In a follow-up investigation, the amenorrheic women in this sample were found to be significantly underweight when compared to controls. Richardson and Pieters (1977) also found that adolescent girls who had primary amenorrhea weighed less and had significantly less body fat than their menstruating peers. Weight alone cannot be the critical factor, however, as evidenced by the report of Feicht *et al.* (1978) on female athletes; while 19 out of 54 were amenorrheic, those in the amenorrheic sample did not weigh less than the others. Thus amenorrhea may relate more to the percentage of body fat, extent of exercise or the degree of stress that the individual is under than to the actual weight. Therefore, while weight is one important variable in amenorrhea, it is not the only one.

Patients with anorexia nervosa usually, but not always, resume menstruation after their weight is restored. This is true even for patients who have been amenorrheic for many years. Overall, Theander (1970) and Dally (1969) estimated that 76% and 59% respectively of surviving patients resumed normal menses. Garfinkel *et al.* (1977) found that about 50% of their patients were menstruating regularly two years after treatment. Return of menses is quite common but not invariable when patients weigh more than 90% of average for their age and height.

These findings are in accord with the concept proposed by Frisch and McArthur (1974) that the onset and maintenance of regular menstrual function are both dependent upon maintenance of a minimum weight for height. This minimum weight for height is proposed as representing a critical fat storage, and implies that a particular body composition of fat/lean or fat/body weight may be an important determinant for reproductive ability in the human female. Frisch and McArthur (1974) estimated that, for the onset of menarche, fat must constitute an estimated 17% of body weight. Similarly, these authors consider that a body composition of about 23% fat is necessary for the maintenance of regular ovulation. However, not all studies on anorexia nervosa are in accord with this hypothesis; for example, Katz *et al.* (1978) found that the return of menses did not show a simple relationship to weight or fatness. Similarly, Richardson and Pieters (1977) describe adolescent girls with amenorrhea whose body fat was greater than 23%. These data emphasize that the concept of a critical fat/lean ratio is not the only factor determining menses; Frisch and McArthur (1974) clearly state 'other factors such as emotional stress affect the maintenance or onset of menstrual cycles. Therefore, menstrual cycles may cease without weight loss and may not resume even though the minimum required weight is attained.' According to this theory, then, the minimum weight is necessary but not sufficient to support regular menstruation.

Little is known about the fertility of anorexic patients. However, there is no evidence to suggest they are not normally fertile after weight restoration and resumption of menses. Dally (1969) reported that about 50% of his married patients had children at the time of follow-up. Similarly Theander (1970), in the longest follow-up to date, observed that 47 of his anorexic patients had married; of these, 34 had borne at least one child.

III. HYPOTHALAMIC FUNCTION

Patients with anorexia nervosa have been repeatedly found to display abnormal hypothalamic function. Possible mechanisms underlying these disturbances include: (1) that hypothalamic changes reflect starvation or caloric deprivation; (2) that emotional disturbance may be directly responsible for hypothalamic alterations; (3) that both the emotional disturbance and the altered hormonal state reflect a basic primary hypothalamic disorder. In the following review we note that the various hormones are affected preferentially by weight, caloric intake and emotional distress, but that these factors account for what is currently known about the origin of hypothalamic disturbances.

IV. GONADOTROPINS

Studies in markedly underweight patients with anorexia nervosa demonstrate low plasma gonadotropins (Brown *et al.*, 1977; Hurd *et al.*, 1977). A linear relationship of resting plasma luteinizing hormone (LH) with body weight has been shown, with LH levels normalizing as weight is regained (Brown *et al.*, 1977; Sherman and Halmi, 1977; Wakeling *et al.*, 1977; Jeuniewic *et al.*, 1978). In contrast to weight, there is no relationship between resting plasma LH levels and caloric intake or duration of amenorrhea (Brown *et al.*, 1977). In addition, baseline FSH values are reduced and return to normal with weight gain (Sherman and Halmi, 1977).

Studies of 24-hour secretory patterns of serum gonadotropins demonstrate that anorexic patients with weight loss have 24-hour patterns resembling those found in normal prepubertal or pubertal girls. These patterns are found in both restricting and bulimic groups of patients (Katz *et al.*, 1978). They are characterized either by low LH levels throughout the 24 hours, as is normally found in prepubertal girls, or by decreased LH secretory activity during waking with higher mean LH concentration during sleep, as is found in early to midpubertal girls (Katz *et al.*, 1976; Brown *et al.*, 1979). The possibility that these developmentally inappropriate, hypothalamically mediated circadian patterns of LH secretion might be found in other types of amenorrhea has been minimized by studies on women with other causes of secondary amenorrhea (Katz *et al.*, 1976).

This LH pattern may not be purely a manifestation of weight loss. Katz *et al.* (1978) reported that the immature circadian pattern persisted in women who had partially or totally achieved average weight but who were otherwise symptomatic. They found that the degree of immaturity of the pattern did not correlate reliably with the duration of illness, percentage body fat or extent of weight loss. It is important to note, however, that those women who were restudied when both weight-recovered and in clinical remission showed an adult LH pattern. Possibly, those weight-recovered but symptomatic anorexics may persist in dietary patterns so abnormal as to maintain deficiencies in critical nutrients, which may alter neurotransmitter function, and the LH pattern is sensitive to these. However, it is also plausible that the LH pattern is sensitive to the particular psychological distress of the anorexic rather than dietary factors. A third intriguing possibility remains: that individuals who are at risk for primary anorexia nervosa do not develop normal adult circadian LH patterns as they pass through puberty and that this may represent one constitutional predisposition to the disorder.

Investigators have shown that the LH response to LH-releasing hormone (LHRH) is greatly reduced in patients with anorexia nervosa. This response is also correlated with body weight, so that those patients with the greatest loss in body weight have the smallest LH rise in response to LHRH (Mecklenberg *et al.*, 1974; Brown *et al.*, 1977; Sherman and Halmi, 1977; Warren, 1977). Normal release of LH after LHRH can occur in patients with immature circadian patterns (Katz *et al.*, 1977). Some authors have emphasized that there is a linear relationship between the LH response to LHRH and body weight (Brown *et al.*, 1977; Jeuniewic *et al.*, 1978). Other investigators have quantified the LH and FSH responses to LHRH using planimetry and have obtained evidence that, although the return of FSH responsiveness is linear and correlates directly with increase in body weight, the regression analysis of LH response produces a linear correlation on a logarithmic scale. This relationship, therefore, is exponential, such that the LH responsiveness to LHRH increases dramatically at approximately 15% below ideal weight for height (Warren, 1977). Whichever interpretation fits the data better, it is now clear that the FSH responses to LHRH are normalized prior to the return of LH responses.

The repeated administration of LHRH will cause a return of the LH response toward normal (Sherman and Halmi, 1977). The pretreatment LHRH response in anorexia nervosa with minimal LH responses and normal FSH responses is comparable to that of a normal prepubertal child (Nillius and Wide, 1977). Repeated treatment with LHRH in patients with anorexia nervosa can restore normal ovarian steroid secretion and a normal ovarian cycle with follicular growth and maturation, ovulation and corpus-luteum formation (Nillius and Wide, 1977). During repeated therapy there is a progressive increase in the LH response to LHRH concomitant with a progressive decrease in the FSH response, and the prepubertal-like response pattern reverts to normal (Nillius and Wide, 1977). The results from treatment with prolonged LHRH strongly support the concept of a deficiency of endogenous LHRH secretion in anorexia nervosa. In addition to alterations in the total response of gonadotropins to LHRH, some investigators have emphasized a delay in the time course of response of plasma LH and FSH to LHRH (Vigersky and Loriaux, 1977). Beumont *et al.* (1978) have studied the response to a four-hour infusion of LHRH rather than a single bolus, as done in previous studies. In patients below 70% of average weight, maximal responses occurred only in the first hour, while in those with a higher weight response was biphasic and maximal levels were achieved in the fourth hour (Beumont *et al.*, 1978). This delay appears to be related to weight loss, since delay is also found in patients with weight loss and secondary amenorrhea who do not have anorexia nervosa (Vigersky *et al.*, 1977). As for the LHRH results, data obtained from responses to clomiphene citrate suggest that the dysfunction is at the level of the hypothalamus. The LH response to clomiphene is impaired in anorexia nervosa only in those patients who are underweight (Brown *et al.*, 1977; Jeuniewic *et al.*, 1978). After clomiphene, the occurrence of menstruation has been unpredictable (Jeuniewic *et al.*, 1978).

The regulation of normal menstruation depends on an intact hypothalamic-pituitary gonadal axis. The cyclic pattern of pituitary gonadotropin release is determined in large part by the complex negative- and positive-feedback effects of the gonadal steroids, particularly estrogen, on the hypothalamic-pituitary unit. Knobil *et al.* (1972) have shown that the negative and positive effects of estrogen on gonadotropin release can be demonstrated experimentally by the administration of estrogen during the follicular phase of the menstrual cycle. This may be used as a tool for assessing the gonadotropin-releasing function of the complete hypothalamic-pituitary unit. Wakeling *et al.* (1977) found that in contrast to normal women, patients with anorexia nervosa prior to weight gain failed to show feedback effects on LH after estrogen administration. In patients following weight restoration, there was an increase in LH resting levels to normal, and negative-feedback effects of estrogen were then apparent. The positive-feedback effects of estrogen, however, remained impaired in the majority of patients. Although the ability to respond to the positive-feedback effect was delayed and more impaired than the negative-feedback effect, it was clearly weight-related.

Taken together, the above findings suggest that, in recovery from anorexia nervosa, there is a return of hypothalamic-pituitary-gonadal activity in a definite sequence which recapitulates puberty (Donavon and Van Der Werfften Bosch, 1965). With marked weight loss there is amenorrhea, associated with low circulating levels of gonadotropins and estrogens. At this time, LH responses to LHRH and clomiphene are absent, and FSH responses are either greatly reduced or absent. With weight gain, the basal levels of LH and FSH rise, and FSH responses to LHRH increase, in some cases being exaggerated. With continued weight gain, responses normalize so that LH responses to LHRH increase, while FSH responses decrease. At about this time the hypothalamus begins to respond normally to negative-feedback effects of estrogen; subsequently the capacity of the hypothalamus to respond to positive-feedback effects of estrogen returns. Clomiphene responses are also normalized. In some patients however, return of menstruation is further delayed, and may not occur even when normal weight is regained.

V. GROWTH HORMONE

A number of investigators have reported that resting growth-hormone (GH) levels are elevated in some patients with anorexia nervosa (Garfinkel *et al.*, 1975; Brown *et al.*, 1977; Casper *et al.*, 1977; Sherman and Halmi, 1977; Vigersky and Loriaux, 1977). With recovery, plasma GH levels are normalized (Garfinkel *et al.*, 1975; Brown *et al.*, 1977). This fall in GH levels occurs prior to weight gain, and is unrelated to body weight; rather, it is more closely tied to the patient's caloric intake (Garfinkel *et al.*, 1975). Changes in plasma LH and in plasma GH are therefore related to two different factors; plasma LH changes are related to loss of body weight and are reversed by an increase in body weight, while GH levels are altered by caloric intake.

Following a glucose load, patients with elevated resting GH levels have been reported to show a drop in GH (Garfinkel *et al.*, 1975; Sherman and Halmi, 1977); however, a paradoxical rise following oral glucose has also been recorded (Casper *et al.*, 1977). The reason for this discrepancy is not clear, but may be related to the type of glucose solution used (Garfinkel *et al.*, 1975; Casper *et al.*, 1977). Harrower *et al.* (1977) described that in anorexia nervosa, as in acromegaly, bromocryptine reduces the raised GH concentrations that occur during the oral glucose-tolerance test. A reduced GH response to apomorphine has also been found in anorexia nervosa (Casper *et al.*, 1977; Sherman and Halmi, 1977), with normalization of the response following weight restoration (Casper *et al.*, 1977).

However, the flattened response to L-DOPA seen in anorexia nervosa does not appear to normalize following weight gain (Sherman and Halmi, 1977). The GH response to insulin is also described as attenuated in anorexia nervosa (Devlin, 1975). On the other hand, normal GH responses to arginine have been reported (Neri *et al.*, 1972). In normal subjects, GH secretion is unaltered by the administration of thyrotropin-releasing hormone (TRH), while in anorexia nervosa a prompt release of GH occurs (Maeda *et al.*, 1976). This response, however, is not specific to anorexia nervosa. It has also been reported in depression (Maeda *et al.*, 1975). A normal rise in GH has been reported during sleep in anorexia nervosa (Vigersky and Loriaux, 1977; Brown *et al.*, 1979).

The data on GH in anorexia nervosa suggest that the raised basal levels are related directly to starvation. However, GH responses to provocative tests have been variable; reasons for this are not clear but may reflect patients being studied at various phases of the disorder or the clinical heterogeneity of the disease.

VI. PITUITARY-THYROID FUNCTION

Thyroxine (T_4) levels in anorexia nervosa are usually considered to be within the normal range (Brown *et al.*, 1977) but have also been described as being lower than in controls (Wakeling *et al.*, 1979). Wakeling *et al.* (1979) did not find a rise in their patients' T_4 levels after four to six weeks of weight gain. A correlation of resting T_4 and cortisol levels has been noted (Brown *et al.*, 1977). Thyroxine-binding protein, as assessed by T_3-resin uptake, is also normal (Brown *et al.*, 1977).

Serum triiodothyronine (T_3) levels are reduced in anorexia nervosa (Burman *et al.*, 1977; Hurd *et al.*, 1977; Moshang and Utiger, 1977). This decrease in T_3 is associated with an increase in the inactive reverse form of T_3 (Burman *et al.*, 1977). Absolute levels of T_3 are linearly correlated with body weight expressed as a percentage of ideal and both reduced T_3 and elevated reverse T_3 are normalized by weight gain (Wakeling *et al.*, 1979). Reversible changes in the levels of thyroid hormones occur with acute starvation and caloric deprivation (Vagenakis, 1977). A teleological explanation has been proposed for the low T_3 and increased reverse T_3 in states of inanition; the target cells for thyroid hormones have regulatory mechanisms based upon metabolic needs (Moshang and Utiger, 1977). If conversion of T_4 to T_3 at the cellular levels is regulated by need, then deprivation of calories would alter cellular mechanisms involved in regulating the deiodination of T_4. This would reduce T_3 production and increase reverse T_3, which is a much less calorigenic hormone. If these mechanisms also exist within the pituitary, the pituitary thyrotroph might not interpret low levels of T_3 as being insufficient for needs and TSH levels would not increase. In accord with this, Gardner *et al.* (1979) have recently reported that fasting is associated with a lower set point of the pituitary thyrotrophs.

Resting levels of thyrotropin (TSH) appear to be within normal limits (Brown *et al.,* 1977; Moshang and Utiger, 1977; Vigersky and Loriaux, 1977; Wakeling *et al.,* 1979). Wakeling *et al.* (1979) have recently shown that TSH levels are not related to body weight. Normal TSH levels demonstrate that the T_4 and T_3 changes are not due to a primary failure of the thyroid gland.

Thyrotropin responses to TRH are of normal magnitude in anorexia nervosa (Brown *et al.,* 1977), again demonstrating that the pituitary is intact. As in normal subjects, the response correlates with the resting TSH level (Brown *et al.,* 1977). The maximal response to TRH has also been correlated with body weight (Wakeling *et al.,* 1979). A delay in the TSH response to TRH with normal magnitude has been noted by several investigators (Vigersky and Loriaux, 1977; Wakeling *et al.,* 1979) and occurs with simple weight loss in the absence of anorexia nervosa (Vigersky *et al.,* 1977). The increase in serum T_3 which normally follows TRH administration is not diminished (Burman *et al.,* 1977).

Low values of T_3 and T_4, together with normal TSH concentrations, have been shown to occur in euthyroid patients with a variety of chronic illnesses (Carter *et al.,* 1974). In anorexia nervosa, the changes in the hypothalamic-pituitary-thyroid axis are likely to occur as an adaptation to chronic illness and starvation, and are unlikely to be causally associated with the amenorrhea. Since these thyroid changes represent an adaptation to starvation, administration of thyroid hormone to correct these changes has no place in the management of anorexia nervosa.

VII. PITUITARY-ADRENAL REGULATION

Morning plasma cortisol levels have generally been found to be elevated in anorexia nervosa (Brown *et al.,* 1977), while normal levels have also been reported (Vigersky and Loriaux, 1977). A correlation of resting plasma morning cortisol and plasma thyroxine has been reported. This may be related to alterations in binding globulins (Brown *et al.,* 1977).

The diurnal variation of plasma cortisol has been described as flattened (Garfinkel *et al.,* 1975; Vigersky and Loriaux, 1977). Boyar *et al.* (1977) found that the circadian rhythm for cortisol was preserved, but at a much higher level than that of normal controls. The absolute cortisol production rate (CPR) has been reported to be normal (Boyar *et al.,* 1977) but was found to be increased if cortisol production is calculated relative to body size (Walsh *et al.,* 1978). There is indirect evidence that CPR is normally proportional to body size (Walsh *et al.,* 1978) and, therefore, in comparing the CRPs of emaciated patients with anorexia nervosa to those of normal-weight controls, it is appropriate to express the CPR relative to body size. These results show an increased rate of cortisol secretion in anorexia nervosa. While the 24-hour excretion of urinary free cortisol is also increased (Boyar *et al.,* 1977; Walsh *et al.,* 1978), the metabolic clearance rate of cortisol is significantly decreased, and there is also a marked prolongation of the cortisol halflife (Boyar *et al.,* 1977). These results are consistent with a moderate impairment of cortisol metabolism.

Studies relying on urinary steroid excretion have generally reported 17-ketosteroids to be low (Warren and Vande Wiele, 1973; Garfinkel *et al.*, 1975), but 17-hydroxy- and 17-ketosteroids have also been reported as normal (Bethge *et al.*, 1970). While responsiveness to dexamethasone may be normal (Warren and Vande Wiele, 1973), at least in some patients there is an incomplete failure of cortisol suppression (Walsh *et al.*, 1978). Similarly, ACTH stimulation tests generally result in normal to hyperreactive adrenal responses (Warren and Vande Wiele, 1973). Responses to metapyrone have been reported on a small number of patients. These have been generally normal (Warren and Vande Wiele, 1973) but Vanluchene *et al.* (1979) found that the metapyrone response, while quantitatively normal, was delayed.

Several investigators (Boyar *et al.*, 1977; Vanluchene *et al.*, 1979) have found that emaciated anorexic patients have an increased ratio of cortisol metabolites, tetrahydrocortisol to tetrahydrocortisone. The same abnormalities are found in patients with hypothyroidism (Boyar *et al.*, 1977). As previously noted, starving anorexic patients show low T_3, as do individuals with starvation from other causes. That these alterations in cortisol metabolism are probably related to the low levels of T_3 has recently been demonstrated by Boyar *et al.* (1977), who administered T_3 to anorexia nervosa patients and found that the cortisol abnormalities were reversed.

Many of the changes in adrenal function seen in anorexia nervosa have also been described for starvation from other causes. For example, decreased urinary excretion of adrenal corticosteroids, diminished rates of cortisol metabolism, elevated levels of plasma cortisol and incomplete suppression by dexamethasone (Smith *et al.*, 1975) have all been described. However, in malnutrition, unlike anorexia nervosa, urinary free cortisol is about normal and the cortisol production rate has been found to be reduced (Smith *et al.*, 1975). It may be, then, that the elevation of cortisol production rate relative to body size and the increased urinary free cortisol in anorexia nervosa indicate an increased activity of the hypothalamic-pituitary-adrenal system that is out of proportion to the emaciation. A variety of poorly understood factors, including the patient's emotional distress, increased physical activity and sleep disturbance, may account for these alterations in the hypothalamic-pituitary-adrenal axis.

VIII. PROLACTIN

Since hyperprolactinemia is a common cause of secondary amenorrhea, it is reasonable to determine whether abnormal prolactin secretion plays a role in the amenorrhea of anorexia nervosa. A series of studies have reported normal resting prolactin levels (Mecklenberg *et al.*, 1974; Vigerksy and Loriaux, 1977; Wakeling *et al.*, 1979). In addition, Wakeling *et al.* (1979) clearly demonstrated that there is no relationship between basal prolactin and body weight in anorexia nervosa. Similarly, there is no relationship between prolactin levels and those of the gonadotropins or estradiol (Wakeling *et al.,* 1979). Hafner *et al.* (1976) have shown that anorexic patients display a normal prolactin response to chlorpromazine. Moreover, TRH-stimulation tests have resulted in normal increments in prolactin, with a delayed response (Vigersky and Loriaux, 1977) or totally normal responses, not influenced by weight (Wakeling *et al.* 1979).

Kalucy *et al.* (1976) found anorexic subjects to have low nocturnal prolactin levels. This absence of a sleep-induced rise in prolactin in some anorexic subjects has been confirmed by Brown *et al.* (1979). It is not known whether this represents a pubertal pattern as does the LH rhythm. Since TRH can normally stimulate prolactin in these patients, this suggests that the decreased prolactin secretion during sleep is of suprahypophyseal origin. That this may be due to dietary intake is likely in view of the recent report (Hill and Wynder, 1976) that changing four healthy nonobese nurses from a western diet to a vegetarian diet resulted in a reduction in nocturnal release of prolactin.

The available data on prolactin in anorexia nervosa strongly suggest that basal prolactin secretion is essentially normal, and prolactin responds to provocative tests normally. It is unlikely therefore that prolactin plays a role in the amenorrhea of these individuals. The failure of patients to resume menstruation and cyclical output of gonadotropins after weight restoration is clearly not associated with high circulating prolactin. The one consistent prolactin abnormality, low nocturnal levels, is probably the result of inadequate dietary intake.

IX. MELATONIN

Melatonin may be of interest in anorexia nervosa for two reasons: (1) it has a clearly defined diurnal rhythm which may be altered as are LH and prolactin; and (2) it is known to influence the reproductive hormones. In the adult human female, a relationship between melatonin and the gonadotropins has been reported. Morning melatonin levels fluctuate through the menstrual cycle, being highest early in the follicular phase and late in the progestational phase (Wetterberg *et al.,* 1976). A marked 24-hour rhythm in blood melatonin exists in adult humans, with levels higher at night-time than in the day (Vaughan *et al.,* 1976). There seems to be a close link between melatonin rhythms and a relatively stable circadian clock. Recently, Brown *et al.* (1979) assessed the circadian rhythm of melatonin secretion in five emaciated subjects with primary anorexia nervosa and in one with both anorexia nervosa and Turner's syndrome. All subjects showed a nocturnal rise in melatonin. By contrast, three of these subjects showed flattened nocturnal prolactin levels and all five primary anorexics displayed an immature circadian LH pattern. Recent studies on a control group of women under investigation for menstrual irregularities have revealed that the nocturnal melatonin rise is considerably lower in all subjects (Brown and Woolever, unpublished observation), suggesting that the nocturnal melatonin rise in anorexia nervosa is exaggerated.

X. TESTOSTERONE

Males with anorexia nervosa, especially of the food-restricting type, show complete impotence and absence of sexual activity and interest. Males with the bulimic variety are usually also impotent. Following weight restoration, there is a gradual return of normal sexual activity.

Both reduced urinary testosterone in male patients (Garfinkel *et al.*, 1975) and an elevated serum testosterone with a negative correlation to degree of weight loss in females (Baranowska and Zgliczynski 1979) have been described. Both abnormalities tend to normalize with weight restoration. In anorexia nervosa the ratio of 5-α-reductase activity to 5-β-reductase activity is significantly reduced, such that the urinary metabolites of testosterone exhibit a preponderance of etiocholanolone at the expense of androsterone (Boyar and Bradlow, 1977). A similar alteration is seen in hypothyroidism. That the thyroid abnormality is responsible for the changes in androgen metabolism is evidence from the report by Boyar *et al.* (1977) noting a reversal of these changes in anorexia nervosa by administration of exogenous T_3.

XI. CARBOHYDRATE METABOLISM

Vigersky *et al.* (1976) found 56% of their anorexic patients to have fasting hypoglycemia. In response to a glucose load, anorexics have a significantly greater degree of hypoglycemia than normals. A response similar to the anorexics is seen after five days of carbohydrate deprivation in normal subjects (Hales and Randle, 1963). Russell and Bruce (1964) have shown this impaired glucose tolerance in anorexia nervosa to be readily reversed by refeeding.

Fasting insulin levels are raised in anorexia nervosa, and the insulin response to a glucose load is sustained (Vigersky *et al.*, 1976). This insulin resistance has been shown to correlate with the severity of the weight loss (Vigersky *et al.*, 1976). Similarly, insulin resistance has been reported in carbohydrate deprivation (Hales and Randle, 1963). The insulin resistance in anorexia nervosa has been shown to persist after weight restoration and resumption of menses, in spite of normalization of the impaired glucose-tolerance curve (Crisp, 1968). This has been the basis for suggesting an inherent insulin resistance in anorexia nervosa; however, it is important to note that Hales and Randle (1963) found a sustained insulin response in normal subjects for 14-35 days after carbohydrate deprivation.

XII. THERMOREGULATION AND WATER CONSERVATION

Failure of individuals to respond normally to acute hypo- or hyperthermia has been described as constituting positive evidence for hypothalamic dysfunction (Mecklenberg *et al.*, 1974), and is not related to amount of adipose tissue

(Vigersky *et al.*, 1977). There is a reduced basal temperature in anorexic patients (Wakeling and Russell, 1970). In response to acute hypothermia, Vigersky *et al.* (1976) found that 69% of their anorexic patients did not display a paradoxical rise in temperature and none had observable shivering. Mecklenberg *et al.* (1974) found the paradoxical rise but none of their five patients shivered in response to cold. The severity of this cold intolerance has been significantly correlated with the degree of weight loss (Vigersky *et al.*, 1976).

In response to a heat stimulus, most anorexic subjects did not display a paradoxical fall in core temperature and showed both an excess rise in core temperature in response to heat (Vigersky *et al.*, 1976) and a significant delay in vasodilation (Wakeling and Russell, 1970). These responses to heat have been found to normalize after weight gain (Wakeling and Russell, 1970) and the severity of the heat intolerance is significantly correlated with degree of weight loss (Vigersky *et al.*, 1976). Moreover, Vigersky *et al.* (1977) have demonstrated abnormal thermoregulation in individuals with amenorrhea and simple weight loss without anorexia nervosa.

XIII. MONOAMINE METABOLISM

The neurotransmitter monoamines (MA) are of interest in anorexia nervosa for several reasons: (1) they exert significant control over the release or inhibition of release of various neurohumors; and (2) they exert significant regulatory control over eating behaviors in animals. Several investigators (Barry and Klawans 1976; Redmond *et al.*, 1976) have recently implicated abnormalities in MA function in the pathophysiology of anorexia nervosa. However, evidence for this is largely indirect. Coppen *et al.* (1976) found free plasma tryptophan to be abnormally low in anorexic subjects, independent of nutritional state. Halmi *et al.* (1978) have recently reported urinary MHPG excretion, as an index of norepinephrine metabolism, to be reduced in anorexic patients; they linked this to depressive symptomatology. This finding of lowered MHPG excretion has been confirmed by our group, as have the lowered plasma free tryptophan levels. Moreover, excretion of HVA and 5-HIAA are not abnormal. At present, the mechanisms for these changes in free tryptophan and MHPG are not understood, nor is it known how these abnormalities relate to the changes in neuroendocrine secretion.

XIV. SUMMARY AND CONCLUSIONS

There is considerable evidence to support the presence of disturbances in hypothalamic function in patients with anorexia nervosa. These occur through a variety of mechanisms and are generally reversible. Certain distrubances relate directly to the degree of weight loss. These include alterations in thermoregulation and water conservation, in TSH responses to TRH, in resting gonadotropin levels and in LH responses to provocative tests. Other hypothalamic dysfunctions are secondary to caloric deprivation *per se*. Thus, alterations in resting plasma growth hormone, plasma T_3 and reverse T_3 all appear to be directly related to caloric intake. Some variations in adrenal steroid and testosterone metabolism reflect these T_3 changes. While many of these disturbances in hypothalamic function are secondary either to weight loss or to caloric restriction, some hypothalamic disturbances appear independent of these parameters. Notable in this regard is the immature pattern of circadian LH secretion. Moreover, amenorrhea may precede the onset of weight loss, and restoration of weight does not invariably lead to resumption of menstruation. These observations highlight the need for further investigation to determine factors underlying the immature LH pattern and amenorrhea in anorexia nervosa.

ACKNOWLEDGEMENTS

This work was funded by Medical Research Council Grant MA-6543. Gregory M. Brown is an Ontario Mental Health Foundation Research Associate.

REFERENCES

Baranowska, B. and Zgliczynski, S. (1979). *Acta Endocrinol. (Copenh.)* 90, 328

Barry, V.C. and Klawans, H.L. (1976). *Lancet 2*, 307

Bethge, H., Nagel, A.M., Solbach, H.G., Weigelmann, W. and Zimmerman, H. (1970). *Mat. Med. Nordm., 22*, 204

Beumont, P.J.V., Abraham, S.F., Argall, W.J. and Turtle, J.R. (1978). *Aust. NZ J. Med., 8*, 509

Boyar, R.M. and Bradlow, H.L. (1977). In *Anorexia Nervosa*, ed. R.A. Vigersky (Raven Press, New York) p. 271

Boyar, R.M., Hellman, L.D., Roffwarg, H., Katz, J., Zumoff, B., O'Connor, J., Bradlow, L. and Fukushima, D.K. (1977). *New England J. Med., 296*, 190

Brown, G.M., Garfinkel, P.E., Jeuniewic, N., Moldofsky, H. and Stancer, H.C. (1977). In *Anorexia Nervosa*, ed. R.A. Vigersky (Raven Press, New York) p. 123

Brown, G.M., Kirwan, P., Garfinkel, P. and Moldofsky, H. (1979). Abstract of Paper Presented at Second International Symposium on Clinical Psycho-neuro-endocrinology in Reproduction (Venice)

Bruch, H. (1973). *Eating Disorders* (Basic Books, New York)

Burman, K.D., Vigersky, R.A., Loriaux, D.L., Strum, D., Djuh, Y-Y., Wright, F.D. and Wartofsky, L. (1977). In *Anorexia Nervosa* ed. R.A. Vigersky (Raven Press, New York) p. 255

Carter, J.N., Corcoran, J.M., Eastman, C.J. and Lazarus, L. (1974). *Lancet 2*, 971

Casper, R.C., Davis, J.M. and Pandey, G.N. (1977). In *Anorexia Nervosa*, ed. R.A. Vigersky (Raven Press, New York) p. 137

Coppen, A.J., Gupta, R.K., Eccleston, E.G. and Wood, K.M. (1976). *Lancet i*, 961

Crisp, A.H. (1968). *Gut 9*, 370

Crisp, A.H. (1977). *Inter. J. Obesity 1*, 231

Dally, P. (1969). *Anorexia Nervosa* (Grune and Stratton, New York)

Devlin, J.G. (1975). *Irish Med. J., 68*, 227

Donovan, B.T. and Van Der Werfften Bosch, J.J. (1965). *Physiology of Puberty*, Monographs of the Physiological Society (Edward Arnold, London)

Feicht, C.B., Johnson, T.S., Martin, B.J., Sparkes, K.E. and Wagner, W.W., Jr (1978). *Lancet ii*, 1145

Fries, H. (1977). In *Anorexia Nervosa*, ed. R.A. Vigersky (Raven Press, New York) p. 163

Frisch, R.E. (1977). In *Anorexia Nervosa*, ed. R.A. Vigersky (Raven Press, New York) p. 149

Frisch, R.E. and McArthur, J.W. (1974). *Science 185*, 949

Gardner, D.F., Kaplan, M.M., Stanley, C.A. and Utiger, R.D. (1979). *New England J. Med., 300*, 579

Garfinkel, P.E. (1974). *Psychol. Med., 4*, 309

Garfinkel, P.E., Brown, G.M., Moldofsky, H. and Stancer, H.C. (1975). *Arch. Gen. Psychiat., 32*, 739

Garfinkel, P.E., Moldofsky, H. and Garner, D.M. (1977). *Can. Med. Assoc. J.*, 1041

Garfinkel, P.E., Moldofsky, H. and Garner, D.M. (1979). *Arch. Gen. Psychiat., 37*, 1036

Hafner, R.J., Crisp, A.H. and McNeilly, A.S. (1976). *Postgrad. Med. J., 52*, 76

Hales, C.N. and Randle, P.J. (1963). *Lancet i*, 790

Halmi, K.A., Dekirmenjian, H., Davis, J.M., Casper, R. and Goldberg, S. (1978). *Arch. Gen. Psychiat., 35*, 458

Harrower, A.D.B., Yap, P.L., Nairn, I.M., Walton, H.J., Strong, J.A. and Craig, A. (1977). *Brit. Med. J., 2*, 156

Hill, P. and Wynder, F. (1976). *Lancet ii*, 806

Hurd, H.P., Palumbo, P.J. and Gharid, H. (1977). *Mayo Clin. Proc., 52*, 711

Jeuniewic, N., Brown, G., Garfinkel, P. and Moldofsky, H. (1978). *Psychosom. Med., 40*, 187

Kalucy, R.C., Crisp, A.H., Chard, T., McNeilly, A., Chen, C.N. and Lacey, J.H. (1976). *J. Psychosom. Res., 20*, 595

Katz, J.L., Boyar, R.M., Weiner, H., Gorzynski, G., Roffwarg, H. and Hellman, L. (1976). In *Hormones, Behavior and Psychopathology*, ed. E.J. Sachar (Raven Press, New York) p. 263

Katz, J.L., Boyar, R.M., Roffwarg, H., Hellman, L. and Weiner, H. (1977). *Psychosom. Med., 39*, 241

Katz, J.L., Boyar, R., Roffwarg, H., Hellman, L. and Weiner, H. (1978). *Psychosom. Med., 40*, 549

Knobil, E., Dietschke, D.J., Yanagi, T., Hotchkiss, J. and Weick, R.F. (1972). In *Gonado-trophins*, ed. R.B. Saxena (Wiley, New York) p. 72

Maeda, K., Kato, Y., Ohgo, S., Chihara, K., Yoshimoto, Y., Yamaguchi, N., Uromaru, S. and Imura, H. (1975). *J. Clin. Endocrinol. Metab., 40*, 501

Maeda, K., Kato, Y., Yamaguchi, N., Chihara, K., Ohga, S., Iwasaki, Y., Yoshimoto, Y., Moridera, K., Kuromaru, S. and Imura, H. (1976). *Acta Endocrinol. (Kbh) 81*, 1

Mecklenberg, R.S., Loriaux, D.L., Thompson, R.H., Anderson, A.E. and Lipsett, M.B. (1974). *Medicine 53*, 147

Moshang, T., Jr. and Utiger, R.D. (1977). In *Anorexia Nervosa*, ed. R.A. Vigersky (Raven Press, New York) p. 263

Neri, V., Ambrosi, B., Beck-Peccoz, P., Travaglini, P. and Faglia, G. (1972). *Folia Endocrinol. (Roma) 25*, 143

Nillius, S.J. and Wide, L. (1977). In *Anorexia Nervosa*, ed. R.A. Vigersky (Raven Press, New York) p. 225

Pettersson, F., Fries, H. and Nillius, S.J. (1973). *Am. J. Obstet. Gynecol., 117*, 80

Redmond, D.J. Jr., Swann, A. and Heninger, G.R. (1976). *Lancet ii*, 307

Richardson, B.D. and Pieters, L. (1977). *Am. J. Clin. Nutr., 30*, 2088

Russell, G.F. and Bruce, J.T. (1964). *Clin. Sci., 26*, 157

Sherman, B.M. and Halmi, K.A. (1977). In *Anorexia Nervosa*, ed. R.A. Vigersky (Raven Press, New York) p. 277

Smith, S.R., Bledsoe, T. and Chhetri, M.K. (1975). *J. Clin. Endocrinol. Metab., 40*, 43

Theander, S. (1970). *Acta Psychiatr. Scand. Suppl., 214*, 5

Vagenakis, A.G. (1977). In *Anorexia Nervosa*, ed. R.A. Vigersky (Raven Press, New York) p. 243

Vanluchene, E., Aertsens, W. and Vandekerckhove, D.V. (1979). *Acta Endocrinol., 90*, 133

Vaughan, G.M., Pelham, R.W., Pang, S.F., Loughlin, L.L., Wilson, K.M., Sandock, K.L., Vaughan, M.K., Koslow, S.H. and Reiter, R.J. (1976). *J. Clin. Endocrinol. Metab., 42*, 752

Vigersky, R.A., Loriaux, D.L., Anderson, A.E., Mechlenburg, R.S. and Vaitukaitis, J.L.

(1976). *J. Clin. Endocrinol. Metab., 43,* 893

Vigersky, R.A. and Loriaux, D.L. (1977). In *Anorexia Nervosa,* ed. R.A. Vigersky (Raven Press, New York) p. 109

Vigersky, R.A., Andersen, A.E., Thompson, R.H. and Loriaux, D.L. (1977). *N. Engl. J. Med., 297,* 1141

Wakeling, A. and Russell, G.F.M. (1970). *Psychol. Med., 1,* 30

Wakeling, A., DeSouza, V.A. and Beardwood, C.J. (1977). *Psychol. Med., 7,* 397

Wakeling, A., DeSouza, V.F.A., Gore, M.B.R., Sabur, M., Kingstone, D. and Boss, A.M.B. (1979). *Psychol. Med., 9,* 265

Walsh, B.T., Katz, J.L., Levin, J., Kream, J., Fukushima, D.K., Hellman, L.D., Weiner, H. and Zumoff, B. (1978). *Psychosom. Med., 40,* 499

Warren, M.P. (1977). In *Anorexia Nervosa,* ed. R.A. Vigersky (Raven Press, New York) p. 189

Warren, M.P. and Vande Wiele, L. (1973). *Am. J. Obstet. Gynecol., 117,* 435

Wetterberg, L., Arendt, J., Paunier, L., Sizonenko, P.C., Van Donsellar, W. and Heyden, T. (1976). *J. Clin. Endocrinol. Metab., 42,* 185

PART FOUR

BIOLOGICAL MARKERS OF ALTERED MENTAL FUNCTION

14 ALTERATION IN BRAIN RECEPTORS IN AFFECTIVE DISORDERS

Michael S. Briley

TABLE OF CONTENTS

I INTRODUCTION   299

II BIOCHEMICAL STUDIES IN DEPRESSION   300

III RECEPTOR-BINDING STUDIES   302
    A. Receptor changes following antidepressant treatment   302
    B. Direct interactions of antidepressant drugs with
        brain receptors   303
    C. Studies using ^{3}H-tricyclic antidepressants   305
    D. High-affinity ^{3}H-imipramine binding   305
    E. Changes in ^{3}H-imipramine binding following various
        antidepressant treatments   309
    F. ^{3}H-imipramine binding in human platelets   309

IV. SUMMARY AND CONCLUSIONS   312

REFERENCES   313

I. INTRODUCTION

Affective disorders are defined by their human symptoms. These symptoms, which are based on subtle behavioural and perceptual changes, are very difficult, if not impossible, to transfer to animals. How, for example, is one to recognise a depressed mouse or an anxious rat? In comparison with disorders which have a simple organic manifestation, such as hypertension, the establishment of an animal model for an affective disorder is extremely difficult. For this reason there are as yet no general screening tests for drugs active in affective disorders. The tests currently used are based on a single aspect of the drugs successful in the treatment of the depression. For instance, the inhibition of reserpine ptosis in the rat is widely used as a test for potential antidepressant activity. This test selects simply for drugs with a direct or indirect noradrenergic-agonist activity, which is a property of many, but certainly not all, clinically effective antidepressant drugs.

Studies aimed at understanding the biochemical lesions at the origin of affective disorders are limited by the research tools available. Unfortunately this range of tools is not very extensive. One approach is to study differences in various biochemical parameters between patients and normal volunteers. Such studies are obviously limited to humans with all the experimental restraints that this implies. It is also possible to study the biochemical actions of drugs shown to be effective in the disorder in question. Unfortunately these drugs, which have invariably been discovered by serendipity, often have a wide spectrum of biochemical and pharmacological activities, of which probably only one is directly related to their beneficial clinical effects.

In the case of anxiety and more recently depression, a new approach has been introduced. Using the radioactively labelled forms of potent and effective drugs, such as the anxiolytic, ^{3}H-diazepam, and the antidepressant, ^{3}H-imipramine, binding studies have succeeded in each case in identifying a specific binding site. The role of these binding sites in the pathogenesis of these disorders is still to be elucidated but the simple existence of these specific binding sites provides a new type of experimental approach to the study of affective disorders.

This chapter will review the contribution of receptor-binding studies to our current understanding of the pathology and therapy of affective disorders and attempt to indicate the future potential of this new biochemical tool.

II. BIOCHEMICAL STUDIES IN DEPRESSION

Biochemical studies in affective disorders have been extensively reviewed in recent years (Van Praag, 1977; Wehr and Goodwin, 1977) and will therefore only be briefly discussed here. In general two strategies have been adopted. One approach is to measure metabolite levels, turnover rates and in some cases enzyme activities in the various human fluids such as blood, urine and cerebro-spinal fluid. Comparisons can then be made between patients suffering from various affective disorders and normal control volunteers. Another common approach is to observe the effects of drugs effective in either improving or aggravating the disease state and to attempt to correlate the resulting biochemical and clinical changes. The latter approach is readily extended to animal studies, whereas the former is obviously limited to clinical biochemistry.

The catecholamine hypothesis of depression (Bunney and Davis, 1965; Schildkraut, 1965) and the indolamine hypothesis (Van Praag, 1974) have focused interest on the changes in the metabolism of noradrenaline and serotonin. In spite of the complication of a large peripheral contribution, the urinary levels of the noradrenaline metabolite, 3-methoxy-4-hydroxyphenylglycol (MHPG) have produced some interesting results. Two typical studies are summarized in Table 14.1. Here both studies show a significant reduction in the levels of urinary MHPG in bipolar endogenously depressed patients, although the findings with unipolar endogenously depressed patients are contradictory.

Table 14.1: Levels of Urinary MHPG in Depressed and Control Subjects*

	Deleon-Jones *et al.* (1975)	Goodwin and Beckmann (1975)
Controls	1348 (n = 21)	1350 (n = 15)
Depressed bipolar	911** (n = 5)	1020** (n = 11)
unipolar	1066** (n = 9)	1623 (n = 19)

* MHPG (3-methoxy-4-hydroxyphenylglycol) levels are given in μg per 24 hours.
** Significantly different ($p < 0.05$) compared to the corresponding controls.

Compatible with such decreased levels of MHPG in some groups of depressed patients was the observation by Greenspan *et al.* (1970) that, following recovery from depression, the levels of MHPG were increased towards the control values and in the manic phase were higher than the controls.

Similar results have also been obtained in samples of cerebro-spinal fluid (CSF), where Post *et al.* (1973) found the levels of MHPG to be lower in a group of undifferentiated depressed patients than in the control population. Postmortem studies have, however, failed to confirm these indications of lower brain nor-adrenaline levels. Moses and Robin (1975), for example, found no differences in the endogenous concentration of noradrenaline or the activities of its principle synthetic or degradative enzymes between the postmortem brains of depressed (suicides) and control subjects.

The use of MHPG levels as a reflection of endogenous brain noradrenaline levels has been criticised since external factors such as the monoamine content of the diet and the degree of physical activity have been shown to profoundly affect the level of urinary MHPG (Wehr and Goodwin, 1977).

Serotonin levels have been most successfully monitored by measuring the concentration of its major metabolite, 5-hydroxy-indoleacetic acid (5HIAA), in the CSF. Comparisons of the levels of 5HIAA in the CSF of depressed and control subjects (Van Praag, 1977) have shown both no difference and a decrease. A possible explanation for this was offered by Asberg *et al.* (1976), who demon-strated that the distribution of 5HIAA levels in the CSF of depressed patients was bimodal. This suggests that there may be a biochemical and pathogenetic heterogeneity in depression. The authors noted that those patients with low levels of 5HIAA had a greater tendency to commit suicide, especially by violent means (Asberg *et al.,* 1976).

The idea of 'high serotonin depression' and 'low serotonin depression' is supported by the finding that probenecid-induced 5HIAA accumulation in depressed patients also shows a bimodal distribution. Probenecid is an inhibitor of the transport of 5HIAA out of the central nervous system. The accumulation of 5HIAA in CSF can thus be used as a sensitive index of serotonin degradation in the brain. In general, the probenecid-induced accumulation of 5HIAA was lower in endogenously depressed patients than in control subjects.

Postmortem data tend, in general, to support the findings in the CSF: serotonin and 5HIAA concentrations in the postmortem brain are slightly lower in suicide victims than in accident mortalities. This difference is more marked when the raphe nuclei are considered alone (Lloyd *et al.,* 1974).

Thus measurements of monoamine metabolites in biological fluids and postmortem brain samples suggest that disturbances in serotonin and possibly noradrenaline neurotransmission may be involved in the pathogenesis of de-pression. Similar indications have come from the studies of drugs acting upon these two systems.

Certain drugs which increase noradrenaline levels, such as monoamine oxidase inhibitors and tricyclic antidepressants, have antidepressant actions. Others, such as amphetamine and cocaine, while they have no beneficial effect in depressed patients, tend to precipitate the manic phase in bipolar depressives. Reserpine, on the other hand, which reduces the level of noradrenaline, dopamine and serotonin, can provoke a depressive episode in susceptible subjects. The substance, α-methyl tyramine, which inhibits tyrosine hydroxylase and thus the synthesis of noradrenaline and dopamine, has unclear effects in depression, but is very effective in reducing mania. Finally, subjects showing low urinary levels of MHPG tend to respond better to antidepressants such as desipramine which are more active at inhibiting the uptake of noradrenaline than of serotonin (Mass *et al.*, 1972).

Parachlorophenylalanine (PCPA), which inhibits tryptamine hydroxylase and hence the synthesis of serotonin, does not produce depression. Pretreatment with PCPA, however, abolishes the therapeutic effects of imipramine. Patients in whom a serotonin deficit is suggested by low CSF levels of 5HIAA tend to respond better to clomipramine, a drug which inhibits primarily the uptake of serotonin (Van Praag, 1977). Those patients with high levels of 5HIAA on the other hand respond best to drugs such as nortriptyline (Asberg *et al.*, 1973).

III. RECEPTOR-BINDING STUDIES

Since the early 1970s the study of the reversible binding of radioactively labelled ligands to neurotransmitter receptors has grown from a useful simple technique into a major branch of neuropharmacology (the theory, technique and applications of receptor binding have been extensively reviewed in a recent monograph; Yamamura *et al.*, 1978). The indications from clinical biochemistry of an involvement of noradrenaline and serotonin in depression have focused the attention of receptor-binding studies in depression mainly on the α- and β-adrenoceptors and the serotonin receptors.

Clinical receptor-binding studies comparing the postmortem binding in the brain of suicides and controls have not yet been undertaken for any receptor. Indeed, nearly all of the studies carried out to date have concentrated on the effects of chronic antidepressant treatments in animals. This approach has, no doubt, been stimulated by the general feeling that the actions of tricyclic antidepressant drugs involve changes in receptor sensitivity. When interpreting such results, however, it should be borne in mind that the receptors of normal animals may not necessarily react in the same way as those of depressed humans!

A. Receptor Changes Following Antidepressant Treatment

The most widely observed receptor change following chronic administration of tricyclic antidepressants is the decrease in maximal binding to β-adrenoceptors (Banerjee *et al.*, 1977; Sarai *et al.*, 1977; Wolfe *et al.*, 1978; Bergstrom and Kellar, 1979a; Pandey *et al.*, 1979a; Rosenblatt *et al.*, 1979; Sellinger *et al.*, 1979; Maggi *et al.*, 1980; Raisman *et al.*, 1980). These decreases, which are specific for the β_1 type of adrenoceptor (Minneman *et al.*, 1979), appear to require the presence of intact noradrenaline nerve terminals (Schweitzer *et al.*, 1979) and are thus probably mediated through the increased levels of noradrenaline following the inhibition of uptake by these drugs.

Recently U'Prichard and Enna (1979) have been able to produce similar but reversible decreases in β-adrenoceptor binding after short-term incubation of rat brain-cortex slices with either desipramine or isoproterenol. This phenomenon, if it is related to *in vivo* 'down-regulation' of the β-adrenoceptor, as suggested by the authors, may prove to be a useful tool for the detailed study of antidepressant-induced changes in the β-adrenoceptor.

The alterations in sensitivity of the noradrenaline-sensitive adenyl cyclase (Vetulani and Sulser, 1975) and of the neuronal firing rate of the noradrenergic cells of the locus coeruleus (McMillen *et al.*, 1980) closely parallel changes in β-receptor binding. This would suggest that such changes in receptor density may be of physiological relevance. Furthermore, a recent study of the noradrenaline-sensitive adenyl cyclase in the leucocytes of depressed patients showed a significantly lower sensitivity to exogenously applied noradrenaline or isoproterenol than in the control population (Pandey *et al.*, 1979b).

Various antidepressant treatments other than tricyclic drugs have also been shown to decrease the binding of the β-adrenoceptor. Chronic administration of atypical antidepressants such as iprindol, mianserine and trazadone (Clements-Jewery, 1978; Pandey *et al.*, 1979a; Sellinger *et al.*, 1979) and chronic electro-convulsive shock (Bergstrom and Kellar, 1979b; Pandey *et al.*, 1979c) have all been shown to reduce β-adrenoceptor binding in the brain. In the case of electro-convulsive shock, this reduction is parallelled by changes in the noradrenaline-sensitive adenyl cyclase (Vetulani and Sulser, 1975).

The α-adrenoceptor in the brain does not appear to be altered by chronic ad-ministration of tricyclic antidepressants (Bergstrom and Kellar, 1979a; Rosenblatt *et al.*, 1979; Raisman *et al.*, 1980) in spite of the fact that noradrenaline levels in the synaptic gap are increased following the inhibition of neuronal uptake. The unique sensitivity of the β-adrenoceptor to 'down-regulation' by tricyclic anti-depressants may possibly involve an apparent direct potentiation of the β-adrenoceptor-mediated noradrenaline-sensitive adenyl cyclase, as has recently been demonstrated *in vitro* (Jones, 1978) and *in vivo* (Jones and Roberts, 1978, 1979). This interaction is unaffected by lesions of the median forebrain bundle (Jones and Roberts, 1979), suggesting that it is probably a direct effect on the postsynaptic β-adrenoceptor/adenyl-cyclase complex.

Whether the serotonin receptor in the brain is altered after chronic treatment with tricyclic antidepressants is not yet clear. Although some groups (Segawa *et al.*, 1979; Maggi *et al.*, 1980) have reported a decrease in ^{3}H-5HT binding after chronic administration of tricyclic antidepressant drugs such as imipramine, desipramine and amitriptyline, others (Bergstrom and Kellar, 1979a; Raisman *et al.*, 1980; Savage *et al.*, 1980) failed to find such changes.

B. Direct Interactions of Antidepressant Drugs with Brain Receptors

The advent of receptor-binding techniques has greatly facilitated the task of determining whether a drug interacts directly at a certain receptor site and with what potency. These studies cannot, however, distinguish between an agonist and an antagonist interaction of a drug with the receptor.

Binding-inhibition studies with antidepressant drugs (Table 14.2) have con-firmed the clinically and pharmacologically observed interactions with the

Table 14.2: Interactions of Antidepressant Drugs with Various Receptors*

Receptor	Adrenoceptors			Serotonin	Opiate	Histamine H_1	Cholinoceptor muscarinic
	α_1	α_2	β				
Ligand	^{3}H-WB4101	^{3}H-CLON	^{3}H-DHA	^{3}H-5HT	^{3}H-NALOX	^{3}H-MEPYR	^{3}H-QNB
				IC_{50} (µM)			
Imipramine	0.16	4.93	13.3	1.08	21.0	0.026	0.078
Desipramine	0.44	9.40	14.2	3.07	25.0	0.250	0.170
Amitriptyline	0.05	0.85	7.0	0.24	33.0	0.004	0.010
Nortriptyline	0.13	2.98	5.4	0.38	34.0	0.046	0.057
Doxepine	0.02	2.00	20.5	0.24	33.0	0.001	0.044
Iprindol	6.15	16.00	17.00	6.00	–	0.100	–
Mianserine	0.06	0.01	11.20	0.09	120.0	–	–

* IC_{50} values are the concentration of drug required to inhibit the binding of the radioactive ligand by 50%. The ligands used were ^{3}H-WB4101, ^{3}H-clonidine (^{3}H-CLON), ^{3}H-dihydroalprenolol (^{3}H-DHA), ^{3}H-5HT, ^{3}H-naloxone (^{3}H-NALOX), ^{3}H-mepyramine (^{3}H-MEPYR) and ^{3}H-quinuclidinyl benzilate (^{3}H-QNB).

Source: Data for ^{3}H-WB4101, ^{3}H-CLON, ^{3}H-DHA and ^{3}H-5HT from Tang and Seeman (1979); data for ^{3}H-naloxone from Beigon and Samuel (1980a); data for ^{3}H-mepyramine from Tran *et al.* (1978); and data for ^{3}H-QNB from Snyder and Yamamura (1977).

muscarinic cholinergic and histamine H_1 receptors. In addition, they show that tricyclic antidepressants are active in the submicromolar range on the α_1-adreno-ceptor and the serotonin receptor but that, in general, they interact only weakly with the α_2- and β-adrenoceptors. In spite of their reported analgesic action (Beigon and Samuel, 1980a), the tricyclics appear virtually inactive on the opiate receptor. These receptor interactions are probably responsible for many of the side effects of these drugs seen clinically, but add little which may help to explain their antidepressant activities.

C. Studies Using ^{3}H-tricyclic Antidepressants

Radiolabelled drugs were first used to label the receptor sites at which they were already known to act. Thus, ^{3}H-haloperidol was used as a ligand for the dopamine receptor (Creese *et al.*, 1975) and more recently ^{3}H-mepyramine as a ligand for the histamine H_1 receptor (Tran *et al.*, 1978). This led subsequently to the idea of studying the binding of radiolabelled drugs whose mechanism of action was not clear, in the hope of labeling the unknown site of action. A major success of this approach came in 1977 when Squires and Braestrup (1977) and Möhler and Okada (1977) found a specific binding site for the benzodiazepine, ^{3}H-diazepam. The binding of ^{3}H-benzodiazepines has since been extensively studied, and has been shown to fulfil many of the criteria for its identification as a receptor.

The wide range of interactions of the tricyclic antidepressants with various receptors hampered several early attempts to find a specific binding site for these drugs. It was found that ^{3}H-amitriptyline binding to rat-brain membranes in the 10-100 nM range involved principally muscarinic cholinergic and histamine receptors (Rehavi and Sokolovsky, 1978). At about the same time, a low-affinity binding site for ^{3}H-imipramine was described in rat brain (O'Brien *et al.*, 1978). This binding site, which had an affinity for ^{3}H-imipramine in the 10 μM range, was probably the same low-affinity site as that studied subsequently by Beigon and Samuel (1980b). Using ^{3}H-desipramine, these authors found that the ligand bound to more than one site in rat brain. The major binding site had an affinity for ^{3}H-desipramine of 4 μ M, with an indication of another site of higher affinity.

D. High-affinity ^{3}H-imipramine Binding

Specific binding of ^{3}H-imipramine in the nanomolar range was first described by Raisman *et al.* (1979a, b). This site, which has an affinity for ^{3}H-imipramine of about 4 nM, is saturable with a maximal binding in the rat brain cortex of about 250 fmoles/mg protein (25 pmoles/g tissue) (Table 14.3). Scatchard analysis gives a straight line, suggesting a single homogeneous population of binding sites (Raisman *et al.*, 1980), with no apparent cooperativity, as confirmed by a Hill coefficient close to unity (n_H = 0.97). The binding is readily reversed by the addition of an excess of unlabelled tricyclic antidepressant. The dissociation affinity constant (K_d) derived from the kinetic constants of association and dissociation is very similar to that obtained from equilibrium experiments by Scatchard analyses (K_d in the cortex from kinetic data, 6.8 nM; K_d in the cortex

Table 14.3: Regional Distribution of ^{3}H-imipramine Binding in Rat Brain and Periphery*

	n	B_{max} (fmoles/mg protein)	K_d (nM)
Hypothalamus	6	317 ± 50	4.1 ± 0.8
Cortex	13	249 ± 23	4.0 ± 0.5
Spinal cord	3	199 ± 63	4.0 ± 1.0
Corpus striatum	7	164 ± 11	2.8 ± 0.5
Midbrain	3	156 ± 35	2.4 ± 0.7
Hippocampus	3	144 ± 26	3.3 ± 0.1
Cerebellar cortex	3	60 ± 11	8.0 ± 3.0
Heart	3	Not detectable	
Vas deferens	3	Not detectable	

* The B_{max} and K_d values for each region were calculated by Scatchard analysis; n is the number of Scatchard plots used for the estimation of each region. The values are given as means ± SEM.

Source: Raisman *et al.* (1980) and unpublished data by Raisman, Briley and Langer.

from equilibrium data, 4.0 nM; Raisman *et al.*, 1980).

These high-affinity binding sites for ^{3}H-imipramine are found in varying densities in different regions of the brain (Table 14.3), the richest region being the hypothalamus and the poorest the cerebellum. None of the peripheral organs tested had any detectable specific binding of ^{3}H-imipramine. Certain of these peripheral organs, such as the vas deferens, possess a very dense noradrenergic innervation and consequently a very effective uptake system for noradrenaline. The absence of any detectable ^{3}H-imipramine binding in these tissues therefore implies that there is no relationship between the neuronal uptake mechanism for noradrenaline and the binding site for ^{3}H-imipramine.

From an extensive study of the inhibition of ^{3}H-imipramine binding by a wide range of drugs (of which a selection are presented in Tables 14.4 and 14.5) it would appear that the ^{3}H-imipramine binding site is not directly related to any of the known neurotransmitter receptors. No class of drug inhibits ^{3}H-imipramine binding with high affinity, although certain individual drugs such as phentolamine and chlorpromazine appear relatively potent (Table 14.5). Various blockers of the uptake of serotonin exhibit affinities for the ^{3}H-imipramine binding site in the same range as the tricyclic antidepressants themselves (Tables 14.4 and 14.5) and it cannot be ruled out that the high-affinity binding site for ^{3}H-imipramine and the uptake mechanism for serotonin may be associated in some way.

Indeed, a comparison of the IC_{50} values for the inhibition of ^{3}H-imipramine binding and the inhibition of the neuronal uptake of serotonin and noradrenaline for 17 antidepressant and nonantidepressant drugs showed a significant correlation

Table 14.4: Inhibition of ^{3}H-imipramine Binding in Rat Cortex by Antidepressant Drugs*

	IC_{50} (nM)
Tricyclic antidepressant drugs	
Imipramine	15
Protriptyline	20
Clomipramine	25
Amitriptyline	25
Desipramine	120
Nortriptyline	200
Doxepine	300
Amoxapine	500
Dibenzepin	1400
Nontricyclic antidepressant drugs	
Zimelidine	1200
Trazodone	1850
Nomifensin	3200
Iprindol	5500
Viloxazine	11500
Mianserine	20000

* The IC_{50} values were determined at 2.0 nM ^{3}H-imipramine.

Source: Raisman *et al.* (1980) and unpublished results by Raisman, Briley and Langer.

between the potencies for the inhibition of ^{3}H-imipramine binding and the inhibition of serotonin uptake but not the uptake of noradrenaline (Langer *et al.*, 1980a).

Tricyclic antidepressant drugs inhibit ^{3}H-imipramine binding with IC_{50} values (the concentration required to inhibit 50% of the specific ^{3}H-imipramine binding) in the range 15 nM to 1.4 μM (Table 14.4). A comparison of the IC_{50} values for the ten drugs shown in Table 14.4 with their mean daily clinical doses as antidepressants gives a highly significant positive correlation (correlation coefficient $r = 0.883$, $p < 0.001$; Briley *et al.*, 1980a). This correlation between the clinical potency and the affinity for the tricyclic-antidepressant binding site suggests that this site might be in some way involved in the therapeutic action of these drugs.

Various nontricyclic drugs which have been suggested as having antidepressant properties were found to be much less active at inhibiting ^{3}H-imipramine binding (IC_{50} values in the range 1.2-20 μM) (Table 14.4). For these drugs, there was no correlation between their clinical and binding potencies (correlation coefficient $r = 0.513$, $p > 0.1$), suggesting that if their postulated antidepressant actions are

Table 14.5: Inhibition of ^{3}H-imipramine Binding in Rat Cortex by Various Drugs*

	IC$_{50}$ (μM)		IC$_{50}$ (μM)
Neurotransmitters		*Inhibitors of noradrenaline uptake*	
Serotonin	1	Nisoxetine	0.2
Histamine	20	Cocaine	0.9
Noradrenaline	>100	Metaraminol	>100
Dopamine	>100	Amphetamine	>100
GABA	>100	*Inhibitors of serotonin uptake*	
Adrenoceptor antagonists		Nitalapram	0.035
Phentolamine	0.2	Fluoxetine	0.2
Prazosin	>100	Femoxetine	2
Yohimbine	>100	*Amino acids and peptides*	
Propranolol	2.5	Aspartate	>100
Atenolol	>100	Glutamate	>100
Metoprolol	>100	Lysine	>100
Serotonin-receptor antagonists		Substance P	>100
Methiothepine	3.2	Met-enkaphalin	>100
Methysergide	>100	*Other drugs*	
Cyproheptadine	>100	Diazepam	>100
Dopamine-receptor antagonists		Progesterone	>100
Chlorpromazine	0.2	Pargyline	>100
Spiroperidol	>100	Reserpine	>100
Thiothixene	>100	Melatonin	>100
Histamine-receptor antagonists			
Cimetidine	>100		
Promethazine	>100		

* The IC$_{50}$ values were determined at 2.0 nM ^{3}H-imipramine.

Source: Raisman *et al.* (1980) and unpublished results by Raisman, Briley and Langer.

confirmed they probably act via a mechanism different from that of the tricyclic antidepressants.

An important criterion for distinguishing a specific site of drug action or a receptor site is the demonstration of stereoselectivity. A recent study using the (z) and (e) isomers of four antidepressant drugs and their derivatives has shown that there is a stereoselectivity for the (z) forms of these drugs. The (z) isomers of 10-hydroxyamitriptyline, 10-hydroxynortriptyline, zimelidine and norzimelidine were found to be from seven to 70 times more potent than the (e) isomers (Langer *et al.*, 1980b).

High-affinity binding of ^{3}H-imipramine has also been found in a variety of other species including man. ^{3}H-imipramine binding in samples of human brain obtained either postmortem or during neurosurgery was found to be very similar to that studied in rat brain (Langer *et al.*, 1981).

E. Changes in ^{3}H-imipramine Binding Following Various Antidepressant Treatments

Tricyclic antidepressants usually need to be given for a period of one to three weeks before any therapeutic effects are seen. Any changes in ^{3}H-imipramine binding following the chronic administration of tricyclic antidepressants might thus be of importance to the therapeutic role of this site.

Chronic treatments of rats or cats with desipramine or imipramine, which have been shown to decrease β-adrenoceptor binding (see Section IIIa), also decreased ^{3}H-imipramine binding in various brain regions (Raisman *et al.*, 1980a; Kinnier *et al.*, 1980, Arbilla *et al.*, 1981). In both species the changes were restricted to decreases in the maximal binding, the K_d remaining unchanged. In certain regions such as the hippocampus (Kinnier *et al.*, 1980) and the cortex (Raisman *et al.*, 1980) β-adrenoceptor binding and ^{3}H-imipramine binding decreased in parallel. In other regions, however, such as the cerebellum, ^{3}H-imipramine binding was unchanged, whereas the β-adrenoceptor was decreased, suggesting that there is no interrelation between the 'down-regulation' of these two sites (Kinnier *et al.*, 1980). Chronic treatment of rats with the nontricyclic antidepressant, iprindol (10 mg/kg i.p., twice daily) for ten days failed to change the level of ^{3}H-imipramine binding (Kinnier *et al.*, 1980).

Other antidepressant therapies, such as electroconvulsive shock, which have been shown to decrease β-adrenoceptor binding, also decrease ^{3}H-imipramine binding. Rats who received daily electroconvulsive shock for ten days had 20-30% less ^{3}H-imipramine binding in the cortex than in the control group of animals (Briley, Green and Langer, unpublished results). Sleep deprivation, especially deprivation of paradoxical or rapid-eye-movement (REM) sleep, has been suggested as a possible therapy for depression. Recent animal studies, which have measured the binding of various ligands in rats deprived of paradoxical sleep for up to 72 hours, have shown decreases in both the β-adrenoceptor and ^{3}H-imipramine binding (Mogilnicka *et al.*, 1980). These changes in ^{3}H-imipramine binding induced by antidepressant drugs and other nondrug therapies support the suggestion that the antidepressant binding site may be involved in the pathogenesis of depression.

F. ^{3}H-imipramine Binding in Human Platelets

^{3}H-imipramine binding in the brain of various species represents a useful tool for the further investigation of affective disorders. The usefulness of this tool was considerably expanded by the recent discovery that a similar high-affinity binding site for ^{3}H-imipramine exists in human blood platelets (Briley *et al.*, 1979).

^{3}H-imipramine binding in human platelets has apparently identical properties to that studied in the rat and human brain. A comparison of the IC_{50} values of a series of 25 drugs for the inhibition of ^{3}H-imipramine binding in rat brain and human platelets gives a highly significant correlation (correlation coefficient $r = 0.81, p < 0.001$), with a slope of approximately unity (Langer *et al.*, 1980c). The existence of identical ^{3}H-imipramine binding sites in brain and platelets is compatible with the suggestion (Sneddon, 1973; Stahl, 1977) that platelets may represent a model for the monoamine (especially serotonin) neuron in the central nervous system.

In a study of ^{3}H-imipramine binding in the platelets of 35 healthy volunteers (Langer *et al.*, 1980c), there was no difference in either the maximal binding (B_{max}) or the affinity constant (K_d) between male and female donors. On the other hand, there were changes in ^{3}H-imipramine binding with age. Over the age range studied (17-97 years) there was a significant reduction of the B_{max} values with age. In addition, the variation between individuals was greater in younger subjects. The B_{max} and K_d values of an individual were found to be relatively stable over a period of several weeks. Over a five-week period, the B_{max} and K_d values of a sample of seven subjects varied by an average of less than 16%.

The possibility of measuring ^{3}H-imipramine binding repeatedly in humans has made it possible to test the working hypothesis, established on the basis of animals studies, that ^{3}H-imipramine binding may be involved in the biochemical changes related to the cause of depression. A study was thus undertaken to compare the binding parameters of the platelets of untreated depressed patients with those of an age-matched control population. The criteria for inclusion of depressed patients in the study are given in Table 14.6. The control population consisted of unmedicated healthy volunteers.

Table 14.6: Criteria Used for the Inclusion of Depressed Patients in the Study of ^{3}H-imipramine Binding in Platelets

Age: 20-65 years old	
Diagnosis:	A 'typical depression' of either the reactive or mono- or bipolar endogenous type
Hamilton score:	Greater than 39
Medication:	Receiving no antidepressant treatment for the preceding month and totally drug-free for 24 hours prior to blood sampling
Hospitalization:	Blood samples were taken on admission of the patients into hospital. The patients were therefore essentially ambulatory

Source: Summarized from Briley *et al.* (1980).

In a preliminary analysis based on the data obtained from 16 female patients, Briley *et al.* (1980) found that the mean B_{max} value for the depressed patients was significantly lower than that of the controls, whereas the K_d values were similar. The latest results of this on-going study, which now includes 27 depressed patients (eight males and 19 females) and 38 controls (14 males and 24 females), are shown in Figure 14.1. The difference in the B_{max} values between the platelets

Figure 14.1: ^{3}H-imipramine binding in platelets from untreated depressed patients and normal controls. Each point represents the B_{max} or K_d value of an individual subject calculated by Scatchard analysis. The bars are the mean values, and the hatched areas cover ± SEM.

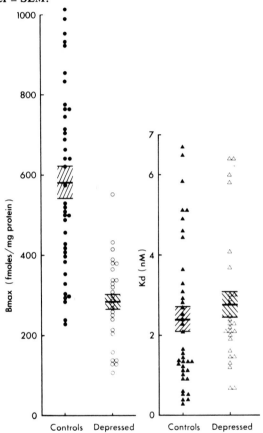

Source: Based on unpublished results of Raisman, Sechter, Briley, Zarifian and Langer.

from depressed and control subjects is highly significant ($p < 0.001$), whereas the mean K_d values are very similar. These results suggest several possibilities and pose several questions, which will now be briefly considered.

Is this difference in ^{3}H-imipramine binding between depressed and control volunteers a reflection of similar differences in the brain? A study is currently under way to study ^{3}H-imipramine binding in the postmortem brain of suicides and depressed patients, in comparison with accident victims who have no history of mental disorders. There are, however, already some indications that the number of ^{3}H-imipramine binding sites in platelets and brain change in parallel. In cats chronically treated with imipramine, the decrease in ^{3}H-imipramine binding seen in the brain was paralleled by a similar decrease in the platelets of the same animals (Arbilla *et al.*, 1981).

Can the level of ^{3}*H-imipramine binding in platelets be used as an index of the severity of the depression?* A comparison of scores on the Hamilton depression-rating scale (HDRS) of depressed patients with their B_{max} values shows no correlation (Briley *et al.*, 1980). Furthermore, when 17 patients were sampled a second time, after nine days of treatment with tricyclic antidepressants, and a third time, when they were considered to be 'cured,' the B_{max} values were found to be very similar to the original 'untreated' values (Raisman, Sechter, Briley, Zarifian and Langer, unpublished results). It remains to be seen whether a fourth, 'cured and unmedicated,' sample will show any differences.

If the level of ^{3}*H-imipramine binding is not an index of the severity of the depression, is it a marker of a biological predisposition or susceptibility to depression?* At this point, one can only speculate as to this possibility. A frequency distribution of the B_{max} values of ^{3}H-imipramine binding in the platelets from the control population shows a possible bimodal distribution (Raisman, Sechter, Briley, Zarifian and Langer, unpublished results). The distribution of the depressed population, however, is clearly symmetrical, corresponding approximately to the lower population of the controls. If these results are confirmed in a larger study, they would indeed be compatible with the hypothesis that low levels of ^{3}H-imipramine binding in platelets might be a biological marker of susceptibility to depression.

Is the ^{3}*H-imipramine binding site a receptor involved in mood regulation?* ^{3}H-imipramine binding obeys many of the criteria established for the binding of a ligand to a receptor site (Burt, 1978). The binding is sensitive to the chronic administration of antidepressant drugs and to certain other antidepressant treatments. In platelets especially, a relationship with depression is strongly suggested. It is obviously still too early to speak of an antidepressant receptor, but the evidence is already strong enough to propose this as a possible working hypothesis.

IV. SUMMARY AND CONCLUSIONS

Advances in certain areas of science are often stimulated by new techniques, not only because they provide new experimental possibilities but also because they tend to attract scientists from other fields who look upon their new subject from a different viewpoint. The affective disorders are an example of this. Analytical advances opened the way for the study of neurotransmitter metabolite levels. More recently, ligand-binding techniques have permitted a more receptor-orientated approach. It is hoped that the discovery of high-affinity ^{3}H-imipramine binding will provide new insights into the molecular pathogenesis and pharmacotherapy of affective disorders.

ACKNOWLEDGEMENTS

The author would like to thank Dr S.Z. Langer for his critical appraisal of the manuscript and Danielle Matherion for typing it.

REFERENCES

Arbilla, S., Briley, M., Cathalu, F., Langer, S.Z., Porain, C. and Raisman, R. (1981) *Brit. J. Pharmacol.* (in press)

Asberg, M., Bertilsson, L., Tuck, D., Cronholm, B. and Sjoqvist, F. (1973). *Clin. Pharmacol. Ther., 14,* 277

Asberg, M., Thoren, P., Traskman, L., Bertilsson, L. and Ringberger, V. (1976). *Science 191,* 478

Banerjee, S.P., King, L.S., Riggi, S.J. and Chanda, S.K. (1977). *Nature 268,* 455

Beigon, A. and Samuel, D. (1980a). *Biochem. Pharmacol., 29,* 460

Beigon, A. and Samuel, D. (1980b). *Biochem. Pharmacol.* (in press)

Bergstrom, D.A. and Kellar, K.J. (1979a). *J. Pharmacol. Exp. Ther., 209,* 256

Bergstrom, D.A. and Kellar, K.J. (1979b). *Nature 278,* 464

Briley, M.S., Raisman, R. and Langer S.Z. (1979). *Eur. J. Pharmacol., 58,* 347

Briley, M., Raisman, R., Sechter, D., Zarifian, E. and Langer, S.Z. (1980a). *Neuropharmacology, 19,* 209

Briley, M.S., Langer, S.Z., Raisman, R., Sechter, D. and Zarifian, E. (1980b). *Science, 209,* 303

Bunney, W.E. and Davis, J.M. (1965). *Arch. Gen. Psychiat., 13,* 483

Burt, D.R. (1978) In *Neurotransmitter Receptor Binding,* ed. H.I. Yamamura, S.J. Enna and M.J. Kuhar (Raven Press, New York) p. 41

Clements-Jewery, S. (1978). *Neuropharmacology 17,* 779

Creese, I., Burt, D.R. and Snyder, S.H. (1975). *Life Sci., 17,* 993

Deleon-Jones, F., Mass, J.M., Dekirmenjian, H. and Sanchez, J. (1975). *Am. J. Psychiat., 132,* 1141

Goodwin, F.K. and Beckmann, H. (1975). *Proc. Amer. Psychiat. Ass., 128,* 96

Greenspan, K., Schildkraut, J.J., Gordon, E.K., Bar, L., Arnoff, M.S. and Durell, J. (1970). *J. Psychiat. Res., 7,* 171

Jones, R.S.G. (1978). *Neuropharmacology 17,* 771

Jones, R.S.G. and Roberts, M.H.T. (1978). *Brit. J. Pharmacol., 65,* 501

Jones, R.S.G. and Roberts, M.H.T. (1979). *J. Pharmacol., 31,* 87

Kinnier, W.J., Chuang, D.M., Farber, L. and Costa, E. (1980). *Fed. Proc., 39,* 1097

Langer, S.Z., Moret, C., Raisman, R., Dubocovich, M.L. and Briley, M. (1980a). *Science, 210,* 1133

Langer, S.Z., Raisman, R. and Briley, M. (1980b). *Eur. J. Pharmacol., 64,* 89

Langer, S.Z., Briley, M.S., Raisman, R., Henry, J.F. and Morselli, P.L. (1980c). *Naunyn-Schmiedeberger's Arch. Pharmacol., 313,* 189

Langer, S.Z., Javoy-Agiel, F., Raisman, R., Briley, M. and Agid, Y. (1981). *J. Neurochem.* (in press)

Lloyd, K.J., Farley, I.J., Deck, J.H.N. and Hornykiewicz, O. (1974). *Adv. Biochem. Psychopharmacol., 11,* 387

McMillen, B.A., Warnack, W., German, D.C. and Shore, P.A. (1980). *Eur. J. Pharmacol., 61,* 239

Maggi, A., U'Prichard, D.C., and Enna, S.J. (1980). *Eur. J. Pharmacol., 61,* 91

Mass, J.M., Fawcett, J.A. and Dekirmenjian, H. (1972). *Arch. Gen. Psychiat., 26,* 252

Minneman, K.P., Dibner, M.D., Wolfe, B.B. and Molinoff, P.B. (1979). *Science 204,* 866

Mogilnicka, E., Arbilla, S., Departere, H. and Langer, S.Z. (1980). *Eur. J. Pharmacol., 65,* 289

Mohler, H. and Okada, T. (1977). *Life Sci., 20,* 2101

Moses, S.G. and Robin, E. (1975). *Psychopharmacol. Commun., 1,* 327

O'Brien, R.A., Spirt, N.M. and Horst, W.D. (1978). *Neuroscience Abstr., 4,* 1366
Pandey, G.N., Heinze, W.J., Brown, B.D. and Davis, J.M. (1979a). *Fed. Proc., 38,* 592
Pandey, G.N., Dysken, M.W., Garver, D.L. and Davis, J.M. (1979b). *Am. J. Psychiat., 136,* 675
Pandey, G.N., Heinze, W.J., Brown, B. and Davis, J.M. (1979c). *Nature 280,* 234
Post, R.M., Gordon, E.K., Goodwin, F.K. and Bunney, W.E. (1973). *Science 179,* 1002
Raisman, R., Briley, M. and Langer, S.Z. (1979a). *Eur. J. Pharmacol., 54,* 307
Raisman, R., Briley, M. and Langer, S.Z. (1979b). *Nature 281,* 148
Raisman, R., Briley, M. and Langer, S.Z. (1980). *Eur. J. Pharmacol., 61,* 373
Rehavi, M. and Sokolovsky, M. (1978). *Brain. Res., 149,* 525
Rosenblatt, J.E., Pert, C.B., Tallman, J.F., Pert, A. and Bunney, W.E. (1979). *Brain Res., 160,* 186
Sarai, K., Frazer, A., Brunswick, D. and Mendels, J. (1977). *Biochem. Pharmacol., 27,* 2179
Savage, D.D., Mendels, J. and Frazer, A. (1980). *J. Pharmacol. Exp. Ther., 212,* 259
Schildkraut, J.J. (1965). *Am. J. Psychiat., 32,* 509
Schweitzer, J.W., Schwartz, R. and Friedhoff, A.J. (1979). *J. Neurochem., 33,* 377
Segawa, T., Mizuta, T. and Nomura, Y. (1979). *Eur. J. Pharmacol., 58,* 75
Sellinger, M.M., Frazer, A. and Mendels, J. (1979). *Neuroscience Abstr., 5,* 1924
Sneddon, J.M. (1973). *Progr. Neurobiol., 1,* 153
Snyder, S.H. and Yamamura, H.I. (1977). *Arch. Gen. Psychiat., 34,* 236
Squires, R.F. and Braestrup, C. (1977). *Nature, 266,* 732
Stahl, S.M. (1977). *Arch. Gen. Psychiat., 34,* 509
Tang, S.W. and Seeman, P. (1979). *Neurscience Abstr., 5,* 353
Tran, V.T., Chang, R.S.L. and Snyder, S.H. (1978). *Proc. Nat. Acad. Sci., 75,* 6290
U'Prichard, D.C. and Enna, S.J. (1979). *Eur. J. Pharmacol., 59,* 297
Van Praag, H.M. (1974). *Pharmacopsychiatry 7,* 281
Van Praag, H.M. (1977). *Biol. Psychiat., 12,* 101
Vetulani, J. and Sulser, F. (1975). *Nature 257,* 495
Wehr, T. and Goodwin, F.K. (1977). In *Handbook of Studies on Depression,* ed. G.D. Burrows (Excerpta Medica, Amsterdam) p. 283
Wolfe, B.B., Harden, T.K., Sporn, J.R. and Molinoff, P.B. (1978). *J. Pharmacol. Exp. Ther., 207,* 446
Yamamura, H.I., Enna, S.J. and Kuhar, M.J. (1978). *Neurotransmitter Receptor Binding* (Raven Press, New York)

15 PSYCHOPHYSIOLOGICAL CORRELATES OF MENTAL DISORDERS

Yvon D. Lapierre and Vernon J. Knott

TABLE OF CONTENTS

I   INTRODUCTION: RATIONALE FOR THE USE OF
    AUTONOMIC PSYCHOPHYSIOLOGICAL CORRELATES   317
II SUMMARY OF THE METHODS EMPLOYED IN
    AUTONOMIC PSYCHOPHYSIOLOGY   318
III PSYCHOPATHOLOGY AND THE AUTONOMIC
    NERVOUS SYSTEM   320
IV ELECTRODERMAL AND CARDIOVASCULAR
    PSYCHOPHYSIOLOGY   323
    A. In Schizophrenia   323
    B. In Anxiety   327
    C. In Depression   330
V   SUMMARY AND CONCLUSIONS: LIMITATIONS OF
    PSYCHOPHYSIOLOGICAL ASSESSMENT OF
    MENTAL DISORDERS   332
REFERENCES   334

I. INTRODUCTION: RATIONALE FOR THE USE OF AUTONOMIC PSYCHOPHYSIOLOGICAL CORRELATES

Observed behaviour is the resultant of complex interactions of physiological and psychological variables. The observed phenomena are generally attributed to the psychological components of emotional and other factors directing motivation. These include perception, cognition, integration and action following the analysis of perceived stimuli. Subserving these psychological processes is the ongoing interaction between the activity of the autonomic (ANS) and central (CNS) nervous systems. The ascending reticular formation feeds into neural structures involved in both emotional and motivational factors of behaviour. The action of these systems are interwoven but can be dissociated partially to determine the correlates of arousal involving autonomic and cortical functions.

The orienting reflex (OR) is a syndromal reaction which can serve as a model linking the central and autonomic nervous systems. As part of this syndrome, CNS activity changes, to produce an increased sensitivity of sense organs and a desynchronization of electroencephalographic activity. This is associated with changes in autonomic functions such as skin conductance, heart rate, peripheral blood flows, respiration and muscular activity. For these to occur, the stimulus must be of a sufficient intensity to be perceived by the sensory organs, and it must have sufficient novelty, complexity and significance to initiate the central mechanisms leading to the development of the syndrome.

As mental processes cannot be objectified directly, the search for peripheral components of this activity has focused upon autonomic and CNS functions. Psychophysiology is a science which has a particular interest in these peripheral or physiological phenomena, which are thought to reflect this inner activity. The procedures generally involved are noninvasive and consist in gathering data reflecting internal processes as they appear on body surfaces. These include electrical phenomena emanating from brain, heart, skin and volumetric and motion changes caused by the lungs and blood flow.

The data obtained from these physiological changes can thus be quantified and submitted more readily to objective scientific methods. They may thus be applied to the assessment and diagnostic differentiation of mental processes, inasmuch as these mental processes or emotional changes modify these functions. These physiological profiles may thus reflect inner emotional activity and lead to a better understanding of perception, cognition and motivation, and thus of behaviour.

II. SUMMARY OF THE METHODS EMPLOYED IN AUTONOMIC PSYCHOPHYSIOLOGY

Broadly speaking, psychophysiological measurement includes methods of recording and processing bioelectric signals from the skin surface of the intact human subject (Sternbach, 1966). As it has been the domain of psychophysiology to examine the relation of these noninvasive measures of physiological status to behaviour and to manipulated or naturally occurring psychological states and events, it is not surprising that the applicability of psychophysiological techniques and methods is particularly appropriate in the psychiatric context. Although the derived measures are entirely indirect indices of implied central states and events, and thus are subject to measurement and interpretative difficulties, significant advances in theory, technique and method in the last 20 years have increased the scientific value of physiological measures, and have provided unique contributions to the study of psychopathology. Like other areas of research, this discipline has increasingly employed more sensitive and sophisticated electronic instrumentation and techniques, among which are biotelemetry, which allows freedom from artificial laboratory environments and permits 24-hour monitoring; the instrumentation tape recorder, which allows bioelectric signals to be stored in an electrical form for later analysis; and the laboratory computer, which permits more sophisticated analysis of the data (Brown, 1967, 1972; Venables and Martin, 1967; Thompson and Patterson, 1974a, b).

Psychophysiology, however, has not abandoned the polygraph as its mainstay and the bioelectric signals which are obtainable from polygraph records typically reflect the activity of the central and peripheral nervous systems in mediating the response of the effector systems and/or CNS activity itself. Although the last decade has observed a shift in interest away from peripheral ANS activity towards a growing emphasis of psychophysiological assessment of brain function as recorded by the electroencephalogram (EEG) and event-related potentials (ERPs), autonomic measures have been the most used in this field, and thus emphasis here will be confined largely to two autonomic measures: electrodermal and cardiovascular activity. Although this focus will omit central measures, as well as a wide variety of other peripheral measures such as respiration, electro-oculography, electromyography, pupillography and salivation, all of which have wide potential interest and applicability, it is hoped that this limited focus will serve as an adequate illustration of two examples of psychophysiological thinking and characteristic methodology.

Two general types of electrodermal activity may be recorded, and details of underlying central and peripheral physiological mechanisms, together with measurement techniques, have been outlined by Venables and Christie (1973). Using an endosomatic method, where the potential generated between the active skin surface of a palmar or plantar area and an abraded reference area is recorded, one derives a measure of skin potential (SP). Using an exosomatic method, where sweat gland activity is recorded as a change in conductivity to an externally applied potential difference across two palmar or plantar sites, one derives a measure of skin conductance (SC) or skin resistance (SR), depending on whether a constant-voltage or constant-current system is used. Both procedures may yield values of tonic resting levels (SPL, SCL, SRL), stimulus-induced phasic responses (SPR, SCR, SRR) and nonspecific responses (NSSPRs, NSSCRs, NSSRRs). Primarily because of difficulties in quantifying amplitude and temporal aspects of the biphasic SPR, the distribution of data is markedly skewed, the majority of electrodermal measures being exosomatic. A general consensus advocates the need for standardized methodology, and Lykken and Venables (1971) have proposed the use of silver/silver-chloride electrodes, the use of physiological levels of NaCl or KCl in electrolyte and the measurement of conductance and a constant-voltage system measuring directly in microohms as necessary steps in standardization. On this basis, results of tonic and, more importantly, phasic aspects of skin conductance, such as amplitude and the various temporal measures including latency, rise time, recruitment and 50% recovery, are made more meaningful, and thus cross-laboratory comparisons in experimental psychopathology are greatly facilitated.

Although the underlying central and peripheral controls of cardiac activity are extremely complex, the basic methodological techniques for obtaining indications of the electrical activity of the heart are relatively simple (Brener, 1967; Schneiderman *et al.*, 1974). As far as the psychophysiologist is concerned, the EKG complex is recorded primarily for the purpose of measuring heart rate (HR). For the most part, exceptions to this rule are limited to the effects of stimuli presented within specific phases of the cardiac cycle and with t-wave amplitude as a measure reflecting potassium concentrations in plasma. Monitoring of heart rate over time yields an abundance of data and, although data reduction may be carried out by hardware or computer, the analysis and interpretation of the data gives rise to numerous problems and controversies (Obrist *et al.*, 1974), which will not be discussed here. Although methods of analysis vary considerably, the majority are directed at assessing either tonic resting levels of heart rate, e.g., mean and variability measures, or stimulus-induced phasic responsivity. The phasic heart-rate response is generally thought to have a triphasic form, and latency and amplitude parameters are quantified with respect to three components, early deceleration, secondary acceleration and late deceleration (Graham and Clifton, 1966; Connor and Lang, 1969), all occurring within approximately twelve cardiac cycles or ten seconds after stimulus onset. Psychophysiologists have employed a variety of other cardiovascular variables such as blood pressure, peripheral blood flow and peripheral vascular resistance. Although autonomic methods of blood-pressure recording for use with the polygraph have met with restricted success, vasomotor activity as observed in peripheral vasoconstriction and dilation may be conveniently measured by photo (Brown, 1967; Weinman, 1967) and pneumatic (Lader, 1967a) plethysmogram techniques. Depending on the time constants of the recording system, one may gather information on relatively slow changes in constriction and dilation as a result of vasomotor tone or on faster blood pulsatile changes due to variations in volume at each beat.

Prior to the last decade, a good majority of psychophysiological research attempted to focus on autonomic correlates of various global states such as stress, anxiety and, most notably, arousal (Johnson, 1974). With growing knowledge of the complexity of peripheral and central factors controlling the various ANS-mediated functions, and awareness that these functions are influenced differently by different parts of the CNS and both can and do respond independently of one another, it becomes increasingly clear that one could no longer rely on such simplistic notions as the unitary concept of arousal (Duffy, 1972). The implication of this knowledge is that there is no such phenomenon as generalized arousal. Neither the sympathetic nor the parasympathetic division of the ANS discharges as a whole, nor does any other system and, since somatic-somatic and somatic-behavioural disassociations can and do occur, ANS, CNS and behavioural arousal are regarded as different forms of arousal (Lang *et al.,* 1972). On this basis, studies of single dependent measures are being avoided, and it has become increasingly obvious that physiological measures respond with both individual and stimulus/task specificity (Engel, 1972). Lacey (1967) has long been a proponent of response specificity with his notion of 'autonomic response stereotype.' This stemmed from data showing that individuals exhibit idiosyncratic 'patterns of response' in the various ANS response systems which are relatively stable across time and situations (Lacey, 1967; Lacey and Lacey, 1970). On this basis, the last decade has observed a shift in the direction and use of psychophysiological research methods, with interests shifting from global states to more specific cognitive, perceptual and behavioural processes such as attention, information processing, learning, memory and physiological responsivity to specific stimulus configurations. This shift has been employed in psychiatry and clinical psychology, and has provided unique contributions to the study of psychopathology.

III. PSYCHOPATHOLOGY AND THE AUTONOMIC NERVOUS SYSTEM

Venables (1971, 1974a, b, 1975a, 1979) has frequently commented on the role of psychophysiology in psychiatry and clinical psychology and has stated that, to date, it is only realistic to consider psychophysiological techniques purely as research tools. It is indeed rare for psychophysiological profiles of a patient to be used in diagnosis or prognosis, and this is in part due to the lack of adequate normative data and base rates of abnormal-responses incidence. However, with the development of measurement techniques and a movement towards standardization of procedures and systems of measurement, there are indications now that psychophysiological techniques will allow the establishment of norms against which functional abnormalities may be recognized and thus will be of use in both a diagnostic and prognostic sense.

The abundance of research on psychophysiological aspects of mental disorders reflects the popularity of these techniques as a means of testing models of psychopathology and exploring underlying mechanisms of behaviour (Alexander, 1972; Fowles, 1975; Lader 1975). In comparison to clinical judgements and, for that matter, other biological measures in the clinic, psychophysiological measures are noninvasive, methodologically simple, make few demands on the patient and have the advantage of providing data that may not be directly observable (Duncan-Johnson, 1979). In a similar vein, Venables (1974a) has commented that the covert nature of psychophysiological measurement is of potential value in psychiatry and clinical psychology and that, in contrast to more overt measures, it enables the investigator to examine aspects of dysfunction possibly before they become apparent to the outside observer, and even perhaps before they become apparent to the subject himself.

Although the most fruitful contribution of autonomic psychophysiology to the elucidation of mental disorders will come from the study of patterns of activity in different ANS response systems rather than from studying these systems separately, single measures have and will continue to provide unique contributions. Gruzelier (1979) and Zahn (1979) have recently discussed the independent multifaceted applications of electrodermal and heart-rate measures in psychopathology, with particular emphasis on schizophrenia. Electrodermal activity has shown potential as an index of hemispheric dysfunction, of the activity of particular brain systems, of symptom severity, prognosis and drug therapy and, most importantly, it holds promise as a basis for subclassification. Similarly, heart rate has shown potential for clinical applications and contributions to theories of psychopathology, in that it has been shown to differentiate schizophrenics from normals, and has supported models of attentional dysfunction in schizophrenia. In addition, heart rate has been indicative of poor neuroleptic response and has predicted the clinical course of a psychotic episode by differentiating fast and slow remitters. It has also been predictive of clinical outcome, even when clinical measures fail, and may have potential in predicting a relapse in remitted schizophrenic patients.

While a great deal of autonomic research has focused on tonic, basal resting levels of electrodermal and cardiac activity, in order to ask whether a particular group of patients is hyper- or hypoaroused relative to normal controls, the most frequent organizing concept that psychophysiologists have employed in considering psychopathology is whether psychiatric patients evidence different levels of phasic responsivity-reactivity than normal controls (Alexander, 1972). A search for defective regulation of physiological responsivity has generated an enormous amount of research, but the last two decades have witnessed a considerable emphasis on the possibility that defects lie in the realm of perceptual processes, and ANS variables are used as a means of monitoring 'openness' and 'input dysfunction' (Alexander, 1972; Venables, 1973). A good bulk of this research has been couched within the framework of the OR and its habituation, since these measures are taken to be indicators of the way that the organism selects, and pays attention to, particular parts of his environment. The important function of the OR appears to be the enhancement of the CNS for receptivity of input and registration of novel, relevant stimuli, while the habituation process is said to serve a 'gating' function by preventing continous, indiscriminate arousal and attentional reactions to the numerous, irrelevant stimuli impinging on the nervous system. As such, both have clear implications for cognitive and perceptual processes in psychopathology (Pavlov, 1927; Sokolov, 1963; Mackworth, 1969). The OR is elicited by novel and low-intensity stimuli and the psychophysiological response pattern is characterized by EEG α-desynchronization, heart-rate deceleration, a SCR, cephalic vasodilation and peripheral vasoconstriction. The OR habituates with repeated presentation of the stimuli, i.e., the degree of response to successive stimuli decreases progressively. Aversive, noxious, high-intensity painful stimuli elicit a defensive response (DR), which is characterized by generalized vasoconstriction, heart-rate acceleration and little or no habituation (Lynn, 1966).

The deceleration and acceleration components of the heart-rate OR and DR closely parallel the work of the Laceys (for example, Lacey and Lacey, 1970) on 'openness to the environment.' Specifically, Lacey has suggested that heart-rate deceleration and increases in skin conductance occur in stimulus situations which require attentiveness or intake of the external environment (i.e., usually in situations not involving decisional mechanisms). In contrast, cardiac acceleration and increases in skin conductance occur when one needs to reject environmental information (i.e., in situations requiring attendance to internal cognitive, decisional processes) or in situations where the external environment is a source of interference and has to be rejected. Commenting on this work, Venables (1971) stated that the 'idea of environmental intake-orientation versus environmental rejection-defence types of response can thus be conceptualized. Patterns of heart rate and skin conductance can thus be used to classify subjects as having environmental intake or rejection types of response.'

Investigation of the OR, DR and intake-rejection patterns of psychophysiological responsivity in psychopathology has been extensively advocated by Venables (1973), and Roessler (1977) has discussed the potential of these and similar procedures in the psychiatric context. Although the great volume of research which has taken this approach has produced contradictory and inconsistent results, there are several trends which suggest that psychiatric patients in general are characterized by a type of responsivity. For example, Dykman *et al.* (1968) published results on an investigation comparing autonomic responsivity of neurotics, organics, schizophrenics and patients with personality disorders to that of normal subjects. In reviewing the study, Venables (1971) stated that there was a strong tendency for the patients as a whole to show a more 'rejecting' pattern of response than normals. This was particularly so among chronic schizophrenics classified as having no affective symptoms. Roessler (1977) attempted a brief summary of research on psychophysiological responsivity in psychiatric patients, and concluded that the greater the degree of psychopathology, the lesser the degree of physiological response and the slower the rate of habituation. On this basis, he suggested that psychopathology was characterized by impaired capacity to discriminate relevant and irrelevant stimuli, and the degree of impairment was greater in persons with more severe pathology.

However, the above are generalities and a detailed examination of the mass of available data yields, for one reason or another, a myriad of contradictory results. In an attempt to depict a clearer picture of psychophysiological correlates of mental disorders, the remaining sections will focus on specific diagnostic groups. For the sake of concision, elaboration of psychophysiological correlates will be limited to three specific disorders — schizophrenia, anxiety and depression.

IV. ELECTRODERMAL AND CARDIOVASCULAR PSYCHOPHYSIOLOGY

A. In Schizophrenia

Early reviews of the psychophysiology of schizophrenia have yielded a decidely mixed, if not chaotic, picture (Epstein and Coleman 1970; Lang and Buss, 1965; Venables 1966). The inconsistency of results is reflected in the divergent conclusions of two reviews on electrodermal activity: Depue and Fowles (1973) concluded that overarousal and hyperresponsivity were characteristic of schizophrenics, while Jordan (1974), in contrast, suggested that underarousal and hyporesponsivity were the distinguishing features. It is not surprising that studies differing widely in their consideration and control of such factors as the influence of medication, stimulus characteristics, recording methods, instructions, patient subdiagnosis and institutionalization should produce conflicting findings regarding the differences between schizophrenics and normals. However, recent studies and reviews have strongly indicated that the inconsistency and lack of unanimity of results may not be due to uncontrolled, unrecognized factors and artifact, but may reflect two or more distinct physiological types of schizophrenia (Venables, 1975b, c, 1977, 1978; Baer and Hines, 1977; Spohn and Patterson, 1979).

Although Bernstein (1964) reported that approximately half of his 'regressed' schizophrenics failed to exhibit SCRs to orienting stimuli, Gruzelier and Venables (1972, 1974) were the first to report on a bimodal distribution of SCRs in a heterogeneous group of schizophrenics. Approximately 50% of the patients exhibited no SCR to simple tone stimuli (nonresponders) and the remaining 50% who did exhibit SCRs (responders) showed large response amplitudes and deficient habituation in that they failed to habituate over 15 trials, i.e., they were hyperresponders. The responding-nonresponding status of the patients was not dependent on medication, length of hospitalization, chronicity or diagnostic subcategory. Although literature reviews by Venables (1977) and Spohn and Patterson (1979) have indicated that a number of additional studies have replicated a proportion of nonresponders approximating 50%, inconsistencies have been reported by Zahn (1976) who found the rate of nonresponders to be 15%, a figure much closer to the rate observed in normal populations (7-10%). On reviewing the methodological and procedural differences which may be at the source of discrepancy, Venables (1977) concluded that a high level of non-responding was a real phenomenon in schizophrenia, and that discrepancies in rates may be due to the nature of the stimuli eliciting the OR. Like the non-responding group, the responding group may also be characterized by a more complicated picture than originally suspected. Contrary to the original study by Gruzelier and Venables (1972), which showed that responding schizophrenics exhibited no habituation, several studies reviewed by Spohn and Patterson (1979) indicate that there are other types of responders, and that not only are there responding schizophrenics who habituate within a normal range but there are also extremely 'fast habituators' who evidence large initial SCRs to the first one or two orienting stimuli and then stop responding (Patterson and Venables, 1978).

There has been a considerable amount of 'neurologizing' accompanying the responding-nonresponding dimension and, although there is a convincing argument that the dimension may reflect limbic forebrain pathology, i.e., nonresponding is a concomitant of amygdala dysfunction and hyperresponding a concomitant of hippocompal dysfunction (Venables, 1973), Spohn and Patterson (1979) caution that a more complicated neurophysiological/neurochemical dysfunction may be involved. Despite the lack of knowledge of central mechanisms, the subclassification of patients into responders, nonresponders, fast habituators, slow habituators and nonhabituators definitely requires further research, since it has been useful in differentiating patients on a variety of clinical, behavioural and physiological measures (Straube, 1979a, b).

As with phasic responsivity, there has been a great deal of inconsistency regarding tonic levels of electrodermal activity in schizophrenia. A possible solution to variability in findings is found in studies by Gruzelier and Venables (1973) and Patterson and Venables (1978), which indicate that SCL is dependent on whether the patients are skin-conductance responders or nonresponders and whether the responding schizophrenic is a fast habituator. In these studies, schizophrenic and normal responders exhibited a higher SCL and a higher incidence of NSSCRs than nonresponders. Schizophrenics who exhibited fast SCR habituation evidenced a SCL which was intermediate in value between responders and nonresponders. Schizophrenics subclassified as paranoid and non-paranoid have also been found to exhibit different tonic levels, in that paranoid schizophrenics exhibit an increase in SCL with increasing behavioural arousal and nonparanoid schizophrenics respond 'paradoxically' by exhibiting decreases in SCL (Gruzelier *et al.*, 1972; Jordan, 1974). Tonic SCL also appears to be a function of laterality (Gruzelier, 1973; Gruzelier and Venables, 1974). Here, electrodermally responding schizophrenics exhibited both higher SCLs and SCRs in the right hand than the left hand; in nonresponding schizophrenics the reverse was shown, SCL being higher on the left hand than on the right hand. No difference was observed in the responsivity of the two hands among normal controls. These findings were suggested to be indicative of lateral left-hemispheric dysfunction in schizophrenia. Laterality differences in SCL and SCR may also be a function of medication since it has been reported (Gruzelier and Hammond, 1977) that chlorpromazine normalized the imbalance found in unmedicated patients. Venables (1975b, 1977, 1979) has discussed the problem of phenothiazine medication in interpreting electrodermal research in schizophrenia, and concluded that, although neuroleptics reduce tonic measures of SCL and NSSCRs, phasic responsivity to simple stimuli does not seem to be altered by medication. However, this latter conclusion on phasic activity may require modification with respect to D- and DL-propanolol, which have been found to reduce significantly both the habituation and occurence of SCRs (Gruzelier *et al.,* 1979).

The recovery limb of the SCR has received considerable attention in schizophrenia, largely because of the initial 'high-risk' research of Mednick and Schulsinger (1968). They reported that the fast recovery of the SCR of children (in the premorbid state) of schizophrenic mothers was predictive of future breakdown in later life. Although the significance of fast SCR recovery is still open to debate, Venables (1974c) has hypothesized that fast recoveries of the SCR may indicate an unusual 'openness' to environmental stimulation and an inability to filter, and that this defective 'gating' capacity may lead to the sensory overload frequently observed in schizophrenia. Reviews by Venables (1977) and Spohn and Patterson (1979) indicate that, although fast recoveries have been observed in adult schizophrenics, variability is evident, and even recoveries that are longer than normal have been documented. Patterson (1976a) observed significant differences in recovery times between schizophrenic fast habituators, habituators and nonhabituators, in that fast habituators exhibited longer recoveries than the other groups, who exhibited equal but very fast recoveries. Concomitant work by Patterson (1976b) with SCRs and pupillometric measures as indicators of adrenergic-cholinergic balance is particularly interesting, in that it has been suggested by Venables (1977) to be supportive of the argument that nonhabituation or hyperresponding and short SCR recovery is a function of hippocampal dysfunction, which could be obtained by cholinergic depletion (Douglas, 1972).

Mednick (1970) has commented that every now and then the scientific community is intrigued by a research project that seems to promise a good deal more than the average. Although Mednick's own continuing high-risk studies in schizophrenia certainly fall into this category, there is a unique ongoing high-risk longitudinal psychophysiological study which, if fruitful, will indeed yield a great more than average. As described by Bell *et al.* (1975), this study differs from the traditional high-risk design in that it will attempt, solely on the basis of psychophysiological indices, to identify children within a normal population who may be at risk for psychiatric breakdown. Preliminary reports and results of the project appear promising: they seem to indicate that abnormal SCRs, as observed in the hyperresponding and hyporesponding dichotomy, appear to be a sensitive discriminator of behaviour patterns even in young (three-year-old) children (Venables, 1977, 1978; Venables *et al.,* 1978). However, the long-term gain of this research — the elucidation of possible causes of schizophrenia — cannot be assessed until the children have reached the age of maximum risk.

Relative to electrodermal studies, data on cardiovascular activity in schizoprenia is not as extensive and, for the most part, it is confined to measures of heart rate, largely because this is a convenient measure and also because of its connection with attentional and arousal processes. Excellent reviews of the literature have been supplied by Venables (1975b) and Spohn and Patterson (1979) and, as with the electrodermal literature, the data has fallen into two categories: studies involving tonic HR and studies involving phasic HR activity. There is fairly unanimous agreement that, relative to normals, schizophrenics exhibit higher tonic HR levels. Although neuroleptics are found to raise tonic activity, elevated HR levels are still apparent in nonmedicated schizophrenics (Venables, 1975c; Zahn, 1979). Individual differences within the schizophrenics are quite apparent, and a number of studies have examined this variability. Gruzelier and Venables (1975) examined the tonic HR and blood-pressure levels of electrodermal responders and nonresponders, and found that responders exhibited higher HR levels and systolic blood-pressure levels than nonresponders, and these differences did not seem to be dependent on drug influence. Tarrier *et al.* (1978) did not replicate these findings.

Although the studies on tonic HR may easily be interpreted within arousal theory, i.e., as meaning that schizophrenics are hyperaroused, Spohn and Patterson (1979) suggested that the elevated HR levels are not a secondary consequence of the disease process on institutionalization and could be interpreted within 'bradycardia-of-attention' theory (Lacey and Lacey, 1970) as reflecting a defensive stance against sensory-information overload — in other words, HR acceleration represents a gating-out of environmental stimuli

In general, studies of phasic HR support the contention that input-attentional dysfunction is a major symptom of schizophrenia. Studies by Dykman *et al.* (1968) and Zahn *et al.* (1968) have indicated that schizophrenics show abnormal HR responsivity to neutral stimuli of moderate intensity, in that they exhibit acceleration while normals exhibit HR deceleration. As stated earlier, HR acceleration is part of a 'defensive' response and, as such, the attentional strategy of the schizophrenic is one of apparent 'closure' to the environment. Gruzelier (1975) examined phasic HR responsivity in skin-conductance responders and nonresponders. All groups, schizophrenics and nonpsychotic patients, exhibited the initial deceleration response to moderate intensity stimuli, but groups were differentiated by the second component of the HR response. Among responders, habituators exhibited deceleration while nonhabituators exhibited acceleration. Nonresponders did not exhibit any regular second component.

Spohn and Patterson (1979) have reviewed studies dealing with HR reactivity to meaningful signal stimuli and have commented that only a few have been guided by the attentional hypothesis of Lacey and Lacey (1970), in which cardiac function is thought to modulate aspects of cognitive-perceptual functioning by a negative-feedback loop from the brain stem to the peripheral cardiovascular system and back to the brain stem via baroreceptor afferent input. Gray (1975) examined HR responsivity in nonmedicated schizophrenics, normal controls and prison controls, using a standard foreperiod reaction-time paradigm similar to Lacey's. Heart-rate reactivity was examined with respect both to acceleration to the warning stimulus and to anticipatory deceleration and subsequent acceleration to the imperative 'go' stimulus of low and high intensity. Whereas increased signal intensity produced increased tonic levels and amplitude of anticipatory responding in control groups, schizophrenics exhibited the reverse pattern of decreased tonic levels and anticipatory deceleration. Results in schizophrenics were interpreted as indicating the presence of a learned anticipatory set that serves to reduce the impact of intense stimuli. Waddington *et al.* (1978) reported similar findings using a modified version of Lacey's reaction-time paradigm. Here, the characteristic pattern of anticipatory HR deceleration was observed in normal controls in response to aversive stimulation, but both deceleration and subsequent HR acceleration components were absent in both medicated and nonmedicated schizophrenics. As the phasic HR response has been shown to be altered by neuroleptics (Goldstein *et al.*, 1966), the investigators caution against interpretation, but do suggest that the diminished HR response could represent a deficit central to the disease process.

Finally, Zahn (1979) reported similar findings of attenuated anticipatory deceleration in nonmedicated schizophrenics during a standard reaction-time task. A potential application of this abnormal responsivity was suggested by Zahn's finding that it discriminated patients with poor treatment outcome from patients who remitted within a three-month period: patients with poor outcome showed the deficit in anticipatory HR responding, while the remitted patients exhibited responding similar to controls. These two groups did not differ in rated psychopathology at the time of testing. Zahn suggested that, despite the marked clinical improvement in the remitted group, the lack of change in anticipatory HR responding over time may indicate that the measure reflects a 'trait' rather than a 'state' interpretation.

In the compass available, this brief review has done no more than select a few salient facts; an overall summary statement on the electrodermal and cardiovascular psychophysiology of schizophrenia is obviously impossible. If anything, this section has indicated the necessity of attending to subgrouping and, since further knowledge of the mechanisms and significance of these noninvasive measures is crucial to an understanding of the psychophysiology of schizophrenia, the request by Spohn and Patterson (1979) for multivariable-measure studies would seem mandatory.

B. In Anxiety

The syndrome of anxiety produces a complex of psychic and somatic symptoms. As such, anxiety is a part of everyday life. It is when anxiety dominates an individual's existence as a relatively fixed and persistent mood resulting in impairment of wellbeing and efficiency that it becomes pathological. This state of anxiety may become fixed, and is referred to as a generalized anxiety disorder.

It may, on the other hand, become a style of reacting to stressors as an habitual predisposition. It is then referred to as an anxiety trait (Rees, 1973).

The physiological accompaniments of an anxiety state suggest that a state of hyperarousal of the autonomic nervous system is occurring. No single physiological measure is unequivocably an index of anxiety since, by definition, the basic symptom is psychic. Nevertheless, adaptation to this state of hyperarousal is impaired in anxious patients. There seems to be a self-feeding cycle between increased arousal and decreased adaptation to it, resulting in self-perpetuating mechanisms. Subsequent removal of anxiogenic stimuli then no longer relieves the state of anxiety or hyperarousal, and the anxiety becomes free-floating, non-situational and generalized (Marks and Lader, 1973).

The physiological variables which have been investigated in anxiety have focused on baseline levels of activity and on changes in this level of activity in response to external and internal stimuli. The stimulation procedures have been either neutral or artificially stressful, to be experienced as stressful to the subject (Lader, 1971).

Electrodermal activity has been extensively studied in anxiety. In normal subjects, the resting levels of skin conductance do not correlate with ratings of anxiety as they do in anxiety neurosis (Stern and Jones, 1973). The electrodermal activity of chronically anxious patients has been found to differ in several respects from normals. Anxiety-state patients have exhibited higher basal skin-conductance levels, exhibited smaller responses to initial stimulus presentations and had much slower rates of habituation with repeated presentation of stimulus (Lader and Wing, 1966). Nonspecific fluctuations unrelated to external stimulation are also greater in anxiety states. Although these findings have been consistently demonstrated in a number of investigations (Marmo *et al.,* 1948; Lader, 1967; Stern *et al.,* 1964; Raskin, 1975), inconsistencies in the magnitude and rate of habituation measures have been observed (Hart, 1974).

By definition, anxiety state is generally considered in the here and now, whereas anxiety trait is a personality characteristic, possibly acquired as an anxiety process but also having a basic genetic predisposition (Endler *et al.,* 1976). In anxiety trait, the physiological accompaniments of arousal are more difficult to define. Attempts at correlating these with levels of anxiety have been directed both towards basal skin conductance and towards nonspecific and specific skin-conductance responses to stimuli. Skin-conductance levels have been found to correlate with the anxiety levels in these subjects. The degree of nonspecific fluctuations in skin conductance is related to variables of the Eysenck personality inventor (EPI). Nielsen and Petersen (1976) observed that nonspecific fluctuations correlated more with levels of anxiety and less with the degree of neuroticism on this personality scale. On the other hand, the number of spontaneous orienting responses correlated negatively with the extraversion score as measured by the EPI. Investigating the concept of trait anxiety further, in an attempt to clarify the physiological basis of anxiety responses, Watts (1975) observed that skin-conductance levels were not much different in these patients, whether they were stimulated in a threatening or a nonthreatening situation. In fact, the skin-conductance response may have been slightly less in threatening situations. Adaptation or habituation to these stimuli improved with time rather than with changes in the experimental condition.

Other personality variables influence electrodermal responses. For example, the orienting response on first stimulation with a noval stimulus was more pronounced in patients classified as 'high sensation seekers.' Their responses subsequently habituated similarly to controls, in spite of initially higher responsivity. This orienting response is different in patients with high levels of anxiety. The expected orienting response pattern may change into that of a defensive reflex (DR), suggesting that perception and integration of stimulation is drastically modified in severe anxiety (Katkin and McCubbin, 1969). Therefore, after reaching a certain level, anxiety reverses the response which was initially attributable to it. Another situation where this interaction of personality or pathological state and physiological response disassociate is in anxious-hysteric patients (Lader and Sartorius, 1968).

Acute anxiety may be observed both in experimental stress situations and in naturalistic conditions. Individuals predisposed to anxiety tend to produce greater nonspecific electrodermal reactions in artifically stressful situations than subjects with low prestress anxiety (Rappaport and Katkin, 1972). This suggests that anxiety trait is accompanied by a physiological predisposition to greater autonomic reactivity. Similarly, experimentally induced anxiety produces a different profile of skin-conductance habituation related to the degree of anxiety manifested in these subjects (Epstein, 1971).

Naturalistic events may also lead to acute anxiety reactions. Such a situation occurs, for example, in patients with acute coronary heart disease. In these patients, skin potentials (SP) were found to be greater in the acute phase. This finding is not clearly related to anxiety levels, because of the difficulty in eliciting higher levels of anxiety on rating scales (Froese *et al.*, 1975). Another example may be seen in preoperative surgical patients. In these patients, greater amounts of intravenous thiopenthal were required to suppress spontaneous skin-conductance responsivity. The dosages required correlated significantly with anxiety ratings. These findings suggest that the persistence of increased spontaneous electrodermal fluctuations reflects greater initial levels of anxiety (Williams *et al.*, 1969).

Finally, the anxiety syndrome of neurotics was compared to the anxiety syndrome of thyrotoxicosis. The electrodermal-activity profile was not differentiable in these two generalized anxiety states, suggesting that they had a similar autonomic basis (Greer *et al.*, 1973).

Vasomotor responses have also been investigated as correlates of anxiety. Heart rate is a rather nonspecific physiological variable subject to numerous influences, of which emotional state is one. Anxiety, viewed as a hyperarousal state, is expected to accelerate heart rate. However, fluctuations in heart rate are not necessarily unidirectional. In experimentally induced anxiety states, it was the change scores from baseline levels which correlated most closely with anxiety levels (Epstein, 1971). Heart rate has also been associated with cognitive activity during the orienting reflex. During the attentional perceptual phase, there is a normal deceleration of heart rate, which then returns to normal baseline. In anxiety, the return to prestimulus levels is slower (McGuinness, 1973). Hart (1974) made a second-by-second comparison of the heart-rate responses of normal and anxious subjects in response to three intensities of signal and non-signal tones. Relative to normal subjects, anxious subjects showed a deficit in orienting responsivity and a greater tendency to respond defensively.

Forearm blood is another parameter studied in anxiety states. The observations of Kelly and Walter (1969) suggest that in chronic anxiety states the basal forearm blood flow is greater in depression or in normal controls. Measurements of digital blood-volume pulse indicated that anxious patients have a greater sympathetic dominance, resulting in a greater peripheral arteriolar constriction. This constriction of finger pulse correlates with anxiety levels (Van der Meuve, 1948, 1950). Therapeutic studies of benzodiazepines in anxiety also suggest a relationship between levels of anxiety and the degree of peripheral vasoconstriction (Lapierre, 1975).

In summary, vasomotor reactivity is modified in anxiety. However, the variability of this correlate of anxiety is quite pronounced which makes significant correlations with anxiety levels in groups of subjects difficult, but it may be of value in assessing changes within a subject.

C. In Depression

Studies of depression are fraught with difficulties in the standardization of diagnoses and in obtaining a homogeneous subject population.

The depressive syndrome has undergone numerous attempts at classification, which invariably reflect the orientation of the investigator and the state of development of technology at the time. These classifications generally follow varying levels of abstraction, and are based on mechanisms ranging from those at the molecular level to the more organismic. A molecular classification is based on a biochemical hypothesis which considers depression to be the result of deficiencies of catecholamines and/or indoleamines in critical areas of the central nervous system (Schildkraut, 1974).

Depression can also be considered on a syndromal basis, and classified as unipolar or bipolar, i.e., manic-depressive illness (Spitzer and Endicott, 1978).

The endogenous-exogenous (reactive) classification has heuristic, therapeutic and etiological implications, as does the neurotic-psychotic dichotomy. Finally, the phenomenological classification into agitated and retarded depression has been favored by investigators of the psychophysiological aspects of depression.

Perceptions of stimuli are altered in each of these types of depression. These alterations are towards increasing variability in the perception of external stimuli and in their integration in the central nervous system (Borge, 1973). In addition to these changes in perception, there is evidence for decreased abilities of learning and of cognition in depression (Friedman, 1964; Hemsi and Whitehead, 1968).

Electrodermal activity is also altered in depression. The initial observations of Alexander (1959) suggested that the electrodermal responses of depressed subjects were reduced in conditional-reflex studies. Subsequent studies indicate that patients who were depressed also had higher basal skin-resistance levels (Richter, 1928). These observations were independent of the classification of depression, and were based solely on the symptom of depression. More recent studies by Lader and coworkers focused on skin conductance. For most, skin-conductance levels correlated highly with severity of depression. Agitated patients tended to have more fluctuations in skin conductance than the retarded depressive. After treatment with ECT, the two subgroups were no longer differentiated (Lader and Wing, 1969; Noble and Lader, 1971).

Skin-conductance levels were also higher in agitated depressives than in normal controls, and lower in retarded depressives than in the latter. Spontaneous fluctuations of skin conductance were also greater in agitated depressives and lower in retarded depressives compared to normal controls. Agitated depressives are also slower to habituate to external stimuli than are normal controls. Similar findings were reported by Lapierre and Butter (1980), who succeeded in differentiating depressives from controls by means of their higher basal skin-resistance levels. Also, a greater number of spontaneous fluctuations of electrodermal activity was found in agitated depressives than in retarded depressive and controls. The latency to an orienting response was also confirmed to be longer in depressed subjects.

The distinction between neurotic and psychotic depressives is also possible by the use of electrodermal correlates. Neurotic depressives generally have a greater amplitude of electrodermal response than do controls and psychotic depressives. The frequency of skin-resistance responses is also greater in neurotic depressives than in controls and psychotic depressives. The habituation rate of these three groups also differs, suggesting that neurotic depressives have a profile more similar to anxiety, whereas psychotic depressives are generally at the other pole of reactivity (Byrne, 1975).

Inhibition of the skin-resistance response using barbiturates is obtained with a lower dosage of sedative in psychotic depression than in normals. Correlation of these changes with levels of anxiety is not as reliable in psychotic depressives, but is more significant in neurotic depression (Byrne and Tottman, 1974).

Treatment with ECT usually does not produce significant changes in skin-conductance levels nor in the phasic responses after recovery from depression, suggesting that depression may be associated with a chronic underlying disorder in which the electrodermal correlates changes are a reflection of the basic physiological defect (Dawson *et al.*, 1977).

Clinically, the differentiation of anxiety and depression may present difficulties. The solution may be facilitated by the pattern of the time-response curve of the electrodermal response, which tends to be somewhat more flattened in depression than in anxiety.

Additional observations on skin-conductance orienting activity of depressives suggests that a bilateral asymmetry exists in depressives, as it does in schizophrenics. Skin-conductance levels are found to be consistently higher on the left in depressives, whereas in schizophrenics they were observed to be more variable from side to side. This persistence of left-sided dominance of skin-conductance responses was independent of the degree of response (Gruzelier and Venables, 1974).

Cardiovascular studies of depression are still in the preliminary stages and have generally been limited to central cardiovascular activity.

Blood pressure is generally more elevated in the acute phase of depression, and normalizes progressively during the course of treatment. In general, heart rate is more elevated in the early stages of treatment than for normal controls (Lapierre and Butter, 1978). The systolic blood pressure changes are generally more elevated than are the diastolic and differential measurements (Escobar *et al.*, 1977).

In summary, electrodermal-activity observations of depressives indicate that skin-conductance levels are generally elevated in agitated depressives and lower in retarded depressives. Response to an orienting stimulus is slower in both types of depression. The spontaneous fluctuations of skin conductance are greater in agitated depressives than in controls, and lower in retarded depressives than controls. The inhibition of orienting reflex response is more difficult in psychotic depressives than in neurotic depressives. The changes in cardiovascular activity of depressives usually result in increased heart rate and systolic blood pressure. These correlates normalize during the course of treatment.

V. SUMMARY AND CONCLUSIONS: LIMITATIONS OF PSYCHOPHYSIO-LOGICAL ASSESSMENT OF MENTAL DISORDERS

Attempts to use psychophysiological measures to assess psychopathology are fraught with difficulties. Alexander (1972) has discussed a number of the problems encountered and, although several of them are not unique to psychophysiology, they do prevent direct study of fundamental pathology. One of the first, and too often foremost, is the problem of reliable psychiatric diagnosis, and an additional related problem is subdiagnosis. In schizophrenia, where variability is manifested in virtually every aspect of psychophysiological functioning, investigations have repeatedly commented on the need to subtype for study purposes. Despite the availability of standardized diagnostic tools (Endicott and Spitzer, 1972; Wing *et al.*, 1974), relatively few studies have employed these techniques.

Another major problem area which has arisen in the past twenty years is the influence of patient medication. Venables (1975b) has discussed several promising techniques of dealing with this, including: persuading the clinician to put all the patients in the study on equal dosages of a particular drug; using a drug-index method such as that of Spohn *et al.* (1971), where the drug dosage made equivalent over different types of drugs may be entered as a factor for statistical control; showing statistically that dosages and types of drugs do not distinguish the patient subpopulations divided on the variable of interest. Other techniques, of course, include withdrawing medication or studying patients who are non-medicated due to noneffectiveness of medication.

As medication prevents an isolated study of the basic pathology, so do secondary aspects of the disease, and on this basis Mednick and McNeil (1968) have strongly recommended the use of 'high-risk' methodology to overcome these stumbling blocks and to get at the pathological process *per se*. There are numerous other problems facing the psychophysiologist dealing with psychiatric patients; these include the effects of hospitalization and institutionalization and the very definite possibility that psychiatric patients may respond differentially to the laboratory situation, demand characteristics of the experiment and inter-personal relations with the experimenter.

The use of psychophysiological measures to assess mental disorders also involves the problems relevant to any area of psychophysiology; in essence, they center around measurement and interpretation. Each psychophysiological measure has characteristic sources of error from electrode, transducer and amplifier to recorder; naivité and blind trust in the efficacy of applying electrodes, taking a couple of 'simple' measures and making unfounded interpretations about central activity can only lead to false conclusions. Psychophysiological measurements are not simple, and considerable skill is required in selecting and properly using appropriate electrodes, electrolyte, transducers and recording equipment in order to carry out a productive, valid psychophysiological study. There is an extensive literature on methods and instrumentation for psycho-physiology, giving detailed accounts of the various types of equipment required for various measures, as well as their construction and care (Brown, 1967, 1972; Venables and Martin, 1967; Thompson and Patterson, 1974a, b).

For the most part, the use of noninvasive autonomic techniques in psycho-pathology is an attempt to examine central events. Knowledge of the extent to which autonomic activity is a reflection of CNS functioning and of how far each mutually modifies each other is continually increasing in complexity. Rickles (1972) has reviewed the literature on central control of the cardiovascular and electrodermal systems, and it is quite apparent that extrapolation from peripheral measures to central events should be carried out with caution. For example, central control of electrodermal activity is suprasegmental (Wang, 1957, 1958) and reflects activity at spinal, reticular, limbic and cortical levels, as well as circulating steroid hormones, and psychophysiologists should be aware of this multitude of factors which vary electrodermal activity. Psychophysiologists are also becoming increasingly aware that excitatory and inhibitory processes coexist, and response systems interact. Lacey (1967) has discussed the neuro-physiology of excitatory and inhibitory aspects of the cardiovascular system and its interaction with the CNS and also how this is reflected in psychophysiological variables, and Sokolov (1963) has discussed speculation of the role of autonomic activity in altering receptor sensitivity and system response by indirect action on the CNS. These few examples serve to demonstrate the complexity of interpreting psychophysiology, especially if interpretation is based on single dependent measures.

The experimental design, the method of response measurement and the type of statistical analysis influence the conclusions reached by the experimenter in the study of psychopathology. Johnson and Lubin (1972) have provided an extensive discussion of these problem areas, together with procedures for controlling subject and environmental factors, and the psychophysiologist who is not aware of the potential of each decision on design, measurement and analysis, as well as the possible alternatives available, may be imposing severe restrictions on the experimental results. One decision and area of controversy which frequently rises in psychophysiological studies is the law of initial values (LIV), which states that the size of a response is related to the level from which it starts (Wilder, 1950). The statistical necessity of eliminating a relationship between response amplitude and tonic levels, so that responses of subjects having different levels may be compared, will depend, of course, on the response system involved. The cardiovascular system is clearly constrained within limits by dual innervation from both branches of the ANS, and the LIV will likely be operative. Procedures for dealing with this are presented by Lacey (1956). In the case of electrodermal activity there is no evidence of antagonistic double innervation at the periphery, but there exists the possibility of ceiling effects due to a limited number of sweat glands, for example. Each experimenter must examine his data to investigate level-response relationships.

These problems are by no means insurmountable and, with careful planning and attention to design, adequate sample sizes, diagnostic subcategories, adequate representation of physiological measurements and proper statistical analysis, psychophysiological studies show promise of yielding significant contributions to the study of mental disorders.

REFERENCES

Alexander, A. (1972). In *Handbook of Psychophysiology*, ed. N. Greenfield and R. Sternbach (Holt, Rinehart and Winston, New York) p. 925

Alexander, L. (1959). *Biological Psychiatry* (Grune and Stratton, New York) Chap. 13, p. 154

Baer, P. and Hines, M. (1977). In *Phenomenology and Treatment of Schizophrenia*, ed. W. Fann, I. Kararan, A. Pokorny and R. Williams (Spectrum, New York) p. 101

Bell, B., Mednick, S., Raman, A., Schulsinger, F., Sutton-Smith, B. and Venables, P. (1975). *Develop. Med. Child Neurol., 17*, 320

Bernstein, A. (1964). *Psychonom. Sci., 1*, 392

Borge, G.F. (1973). *Arch. Gen. Psychiat., 29*, 760

Brener, J. (1967). In *A Manual of Psychophysiological Methods*, ed. P. Venables and J. Martin (North-Holland, Amsterdam) p. 103

Brown, C. (1967). In *Methods in Psychophysiology*, ed. C. Brown (Williams and Wilkins, Baltimore) p. 54

Brown, C. (1972). In *Handbook of Psychophysiology*, ed. N. Greenfield and R. Sternbach (Holt, Rinehart and Winston, New York) p. 159

Byrne, D.G. (1975). *Austral. and NZ J. Psychiat., 9,* 181
Byrne, D.G. and Tottman, U. (1974). *Austral. and NZ J. Psychiat., 8,* 261
Connor, W. and Lang, P. (1969). *J. Exp. Psych., 82,* 310
Dawson, M.E., Schell, A.M. and Catania, J.J. (1977). *Psychophysiology 14,* 569
Depue, R. and Fowles, D. (1973). *Psych. Bull., 79,* 233
Douglas, R. (1972). In *Inhibition and Learning,* ed. R. Boakes and M. Halliday (Academic Press, New York) p. 529
Duffy, E. (1972). In *Handbook of Psychophysiology,* ed. N. Greenfield and R. Sternbach (Holt, Rinehart and Winston, New York) p. 577
Duncan-Johnson, C. (1979). Paper presented at the Society for Psychophysiological Research, Nineteenth Annual Meeting (Cincinatti, Ohio)
Dykman, R., Reese, W., Goldbrecht, C., Ackerman, P. and Sunderman, R. (1968). *Ann. NY Acad. Sci., 47,* 237
Endicott, J. and Spitzer, R. (1972). *Arch. Gen. Psychiat., 27,* 678
Endler, N.S., Magnusson, D., Ekehammar, B. and Okakada, M. (1976). *Scand. J. Psychol., 17,* 81
Engel, B. (1972). In *Handbook of Psychophysiology,* ed. N. Greenfield and R. Sternbach (Holt, Rinehart and Winston, New York) p. 571
Epstein, S. (1971). *Psychophysiology 8,* 319
Epstein, S. and Coleman, M. (1970). *Psychosom. Med., 32,* 113
Escobar, J., Gomez, O. and Tuason, V. (1977). *Dis. Nerv. Syst., 38,* 76
Fowles, D. (1975). *Clinical Applications of Psychobiology* (Columbia University Press, New York)
Friedman, A.S. (1964). *J. Abnorm. Psychol., 69,* 237
Froese, A.P., Cassem, N.H., Hackett, T.P. and Silverberg, E.L. (1975). *J. Psychosom. Res., 19,* 1
Goldstein, M., Acker, C., Crockett, J. and Riddle, J. (1966). *J. Ab. Psychol., 71,* 335
Graham, F. and Clifton, R. (1966). *Psych. Bull., 65,* 305
Gray, A. (1975). *J. Ab. Psychol., 84,* 189
Greer, S., Ramsay, I. and Bagley, C. (1973). *Brit. J. Psych., 122,* 549
Gruzelier, J. (1973). *Biol. Psychol., 1,* 21
Gruzelier, J. (1975). *Biol. Psychiat., 3,* 143
Gruzelier, J. (1979). Paper presented at the Society for Psychophysiological Research, Nineteenth Annual Meeting (Cincinatti, Ohio)
Gruzelier, J. and Venables, P. (1972). *J. Nerv. Ment. Dis., 155,* 277
Gruzelier, J. and Venables, P. (1973). *Neuropsychology 11,* 221
Gruzelier, J. and Venables, P. (1974). *Biol. Psychiat., 8,* 55
Gruzelier, J. and Venables, P. (1975). *Psychophysiology 12,* 66
Gruzelier, J. and Hammond, N. (1977). *Stud. Psychol., 19,* 40
Gruzelier, J., Lykken, D. and Venables, P. (1972). *Arch. Gen. Psychiat., 26,* 427
Gruzelier, J., Hirsch, S., Weller, M. and Murphy, C. (1979). *Acta Psychiat. Scand., 60,* 241
Hart, J. (1974). *Psychophysiology 11,* 443
Hemsi, L.K. and Whitehead, A. (1968). *J. Psychosom. Res., 12,* 145
Johnson, L. (1974). *Int. J. Psychiat. Med., 5,* 565
Johnson, L. and Lubin, A. (1972). In *Handbook of Psychophysiology,* ed. N. Greenfield and R. Sternbach (Holt, Rinehart and Winston, New York) p. 125
Jordan, L. (1974). *Psych. Bull., 81,* 85
Katkin, E.S. and McCubbin, R.J. (1969). *J. Ab. Psychol., 74,* 54
Kelly, D. and Walter, C.J.S. (1969). *Brit. J. Psych., 115,* 401
Knight, M.L. and Borden, R.J. (1979). *Psychophysiology 16,* 209
Lacey, J. (1956). *Ann. NY Acad. Sci., 67,* 123
Lacey, J. (1967). In *Psychological Stress: Issues in Research,* ed. M. Appley and R. Trumbull (Appleton-Century-Crofts, New York) p. 14
Lacey, J. and Lacey, B. (1970). In *Physiological Correlates of Emotion,* ed. P. Black (Academic Press, New York) p. 205
Lader, M. (1967a). In *A Manual of Methods in Psychophysiology,* ed. P. Venables and I. Martin (North-Holland, Amsterdam) p. 159
Lader, M.H. (1967b). *J. Psychosom. Res., 11,* 271

Lader, M.H. (1971). *Sci. Basis Med., Ann. Rev.,* 279
Lader, M. (1975). *The Psychophysiology of Mental Illness* (Routledge, London)
Lader, M.H. and Wing, L. (1966). *Institute of Psychiatry Monograph* (London)
Lader, M. and Sartorius, N. (1968). *J. Neurol. Neurosurg. Psychiat., 31,* 490
Lader, M.H. and Wing, L. (1969). *J. Psych. Res., 7,* 89
Lang, P. and Buss, A. (1965). *J. Ab. Psych., 70,* 77
Lang, P., Rice, D. and Sternbach, R. (1972). In *Handbook of Psychophysiology,* ed. N. Greenfield and R. Sternbach (Holt, Rinehart and Winston, New York) p. 623
Lapierre, Y.D. (1975). *Int. J. Clin. Pharmacol., 11,* 315
Lapierre, Y.D. and Butter, H.J. (1978). *Progress Neuropsychopharm., 2,* 207
Lapierre, Y.D. and Butter, H.J. (1980). *Neuropsychobiology 6,* 217
Lykken, D. and Venables, P. (1971). *Psychophysiolopy 8,* 656
Lynn, R. (1966). *Attention, Arousal and the Orientation Reaction* (Pergamon Press, New York)
McGuinness, D. (1973). *Biol. Psychol., 1,* 115
Mackworth, J. (1969). *Vigilance and Habituation* (Penguin Books, Middlesex, England)
Marks, J. and Lader, M. (1973). *J. Nerv. Ment. Dis., 156,* 3
Marmo, R., Shagass, C., Davis, J., Cleghorn, R., Graham, B. and Goodman, A. (1948). *Science 108,* 509
Mednick, S. (1970). *Ment. Hyg., 54,* 50
Mednick, S. and McNeil, T. (1968). *Psych. Bull., 70,* 681
Mednick, S. and Schulsinger, F. (1968). In *Transmission of Schizophrenia,* ed. D. Rosenthal and S. Kety (Pergamon Press, London) p. 267
Neary, R.A. and Zuckerman, M. (1976). *Psychophysiology 13,* 205
Nielsen, T.C. and Petersen, K.E. (1976). *Scand. J. Psychol., 17,* 73
Noble, P. and Lader, M. (1971). *J. Psychiat. Res., 9,* 61
Obrist, P., Black, A., Brener, J. and DiCara, L. (1974). *Cardiovascular Psychophysiology: Current Issues in Response Mechanisms, Biofeedback and Methodology* (Aldine, Chicago)
Patterson, T. (1976a). *Psychophysiology 13,* 189
Patterson, T. (1976b). *J. Nerv. Ment. Dis., 163,* 200
Patterson, T. and Venables, P. (1978). *Psychophysiology 15,* 556
Pavlov, I. (1927). *Conditioned Reflexes: An Investigation of the Physiological Activity of the Cerebral Cortex,* ed. and trans. G.V. Anrep (Oxford University Press, New York)
Rappaport, H. and Katkin, E.S. (1972). *J. Consult. Clin. Psychol., 38,* 219
Raskin, M. (1975). *Biological Psychol., 2,* 309
Rees, W.L. (1973). In *Comprehensive Patient Care,* ed. W.L. Rees, International Congress Series No. 305, (Excerpta Medica, Amsterdam)
Richter, C.P. (1928). *Arch. Neurol. Psych., 19,* 488
Rickles, W. (1972). In *Handbook of Psychophysiology,* ed. N. Greenfield and R. Sternbach (Holt, Rinehart and Winston, New York) p. 93
Roessler, R. (1977). In *Psychiatric Medicine,* ed. G. Usdin (Brumer/Mazel, New York) p. 95
Schachter, J., Kerr, J., Lachin, J. and Faer, M. (1975). *Psychophysiology 12,* 483
Schildkraut, J.J. (1974). *Ann. Rev. Med., 25,* 333
Schneiderman, N., Dauth, G. and Van Dercar, D. (1974). In *Bioelectric Recording Techniques,* Ed. R. Thompson and M. Patterson (Academic Press, New York) p. 165
Serebro, B. (1971). *S-A Mediese Tydskrif,* 1273
Sokolov, E. (1963). *Perception and the Conditioned Reflex* (Macmillan, New York)
Spitzer, R.L. and Endicott, J. (1978). In *Critical Issues in Psychiatric Diagnosis,* ed. R.L. Spitzer and D.F. Klein (Raven Press, New York) p. 15
Spohn, H. and Patterson, T. (1979). *Schiz. Bull., 5,* 581
Spohn, H., Thetford, P. and Cancro, R. (1971). *J. Nerv. Ment. Dis., 152,* 129
Stern, G.A., Jones, C.L., Lader, M.H. and Wing, L. (1964). *J. Neurol. Neurosurg. Psychiat., 27,* 210
Stern, J.A. and Jones, C.L. (1973). In *Personality and Psychopathology in Electrodermal Activity in Psychosocial Research,* ed. W.F. Caucasy, and D.C. Raskin (Academic Press, New York) Chap. 6
Sternbach, R. (1966). *Principles of Psychophysiology* (Academic Press, New York)
Straube, E. (1979a). *J. Nerv. Ment. Dis., 167,* 601
Straube, E. (1979b). Paper presented at the Society for Psychophysiological, Nineteenth

Annual Meeting (Cincinatti, Ohio)
Tarrier, M., Cooke, E. and Lader, M. (1978). *Acta. Psychiat. Scand., 57,* 369
Thompson, R. and Patterson, M. (1974a). *Bioelectric Recording Techniques,* Vol. 1-B (Academic Press, New York)
Thompson, R. and Patterson, M. (1974b). *Bioelectric Recording Techniques,* Vol. 1-C (Academic Press, New York)
Van der Meuve, A.B. (1948). *Psychosom. Med., 10,* 347
Van der Meuve, A.B. (1950). *Proc. S. Afr. Psychol. Assoc., 1,* 10
Venables, P. (1966). *Brit. J. Med. Psychol., 39,* 289
Venables, P. (1971). *Psychol. Med., 1,* 185
Venables, P. (1973). In *Psychopathology: Contributions from the Social, Behavioural and Biological Sciences,* ed. M. Hammer, K. Salzinger and S. Sutton (Wiley, New York) p. 261
Venables, P. (1974a) In *Neuropsychopharmacology – Proceedings of the Ninth Congress of the Collegium Internationale Neuropsychopharmacologicum,* International Congress Series No. 359 (Excerpta Medica, Amsterdam) p. 159
Venables, P. (1974b). In *Assessment of Pharmacodynamic Effects in Human Pharmacology. Part I: Psychopharmacological Screening Tests* (F.K. Schattauer Verlag, Stuttgart-New York) p. 61
Venables, P. (1974c). In *Genetics, Environment and Psychopathology,* ed. S. Mednick, F. Schulsinger, J. Higgins and B. Bell (North-Holland, Amsterdam) p. 117
Venables, P. (1975a). In *Research in Psychophysiology,* ed. P. Venables and M. Christie (Wiley, London) p. 418
Venables, P. (1975b). In *Research in Psychophysiology,* ed. P. Venables and M. Christie (Wiley, London). p. 282
Venables, P. (1975c). In *Clinical Applications of Psychophysiology,* ed. D. Fowles (Columbia University Press, New York) p. 106
Venables, P. (1977). *Schiz. Bull., 3,* 28
Venables, P. (1978). *Psychophysiology 15,* 302
Venables, P. (1979). In *Current Themes in Psychiatry. II,* ed. R. Gaind and B. Hudson (Macmillen, London) p. 239
Venables, P. and Martin, I. (1967). *A Manual of Psychophysiological Methods* (North-Holland, Amsterdam)
Venables, P. and Christie, M. (1973). In *Electrodermal Activity in Psychological Research,* ed. W. Prokasy and D. Raskin (Academic Press, New York) p. 1
Waddington, J., Maccolloch, M., Schalken, M. and Sambrooks, J. (1978). *Psychol. Med., 8,* 157
Wang, G. (1957). *Am. J. Phys. Med., 36,* 295
Wang, G. (1958). *Am. J. Phys. Med., 37,* 35
Watts, J.M.J. (1975). *Psychophysiology 12,* 596
Weinman, J. (1967). In *A Manual of Psychophysiological Methods,* ed. P. Vanables and I. Martin (North-Holland, Amsterdam) p. 185
Wilder, T. (1950). *Psychosom. Med., 12,* 392
Williams, J.G.L., Jones, J.R. and Williams, B. (1969). *Psychosom. Med., 31,* 522
Wing, J., Cooper, J. and Sartorius, N. (1974). *The Measurement and Classification of Psychiatric Symptoms* (Cambridge University Press, London)
Zahn, T. (1976). *J. Nerv. Ment. Dis., 162,* 195
Zahn, T. (1979). Paper presented at the Society for Psychophysiological Research Nineteenth Annual Meeting (Cincinatti, Ohio)
Zahn, T., Rosenthal, D. and Lawlor, W. (1968). *J. Psychiat. Res., 6,* 117

16 HORMONAL MARKERS IN SCHIZOPHRENIA AND DEPRESSION

Gregory M. Brown, John M. Cleghorn, Prakash G. Ettigi and
Peter Brown

TABLE OF CONTENTS

I	INTRODUCTION	341
II	THE DOPAMINE HYPOTHESIS OF SCHIZOPHRENIA: RELEVANCE OF NEUROENDOCRINE STUDIES	341
III	OVERVIEW OF STUDIES ON ENDOCRINE FUNCTION IN SCHIZOPHRENICS	343
	A. Growth-hormone	343
	(a) Neural Regulation	343
	(b) Comparison of Schizophrenics and Controls	343
	B. Prolactin	344
	(a) Neural Regulation	344
	(b) Comparison of Schizophrenics and Controls	344
	C. TRH	346
	D. LH and FSH	346
	E. Endocrine Changes in Chronic Hospitalized Patients	346
	F. Endogenous Opiates (Endorphins/Enkephalins)	346
IV	PROBLEMS AND PROSPECTS IN SCHIZOPHRENIA	348
V	HORMONAL MARKERS IN DEPRESSION	349
	A. Cortisol	350
	B. Growth Hormone	353
	C. Prolactin	354
	D. Thyroid-stimulating Hormone	355
	E. Luteinizing Hormone	356
VI	SUMMARY AND PROSPECTS IN DEPRESSION	356
	REFERENCES	358

I. INTRODUCTION

A variety of neuroendocrine strategies which are useful for the examination of neural mechanisms in patients, have been described, and widely applied in the study of depression and schizophrenia.

The examination of resting levels of hormones, and particularly the 24-hour pattern, permits an assessment of the physiologic control mechanisms. Resting levels of most hormones demonstrate diurnal variation, episodic secretion and sleep-related changes. These techniques have proved extremely valuable in the study of depressed patients. They have been of less use, to date, in studies on schizophrenics. A variety of challenge tests have been devised. Challenge tests of pituitary function can be done using stimulation by thyrotropin-releasing hormone or luteinizing-hormone-releasing hormone, or by examining prolactin responses to the administration of dopamine (DA) blockers or agonists. Challenges which act primarily at the neural level include insulin-induced hypoglycemia, the dexamethasone suppression test and administration of neuropharmacologic agents which act at those sites where a particular neurotransmitter is present.

These techniques are based on an understanding of the mechanisms by which the brain controls the pituitary. Secretion of hormones from the anterior pituitary is influenced by hypothalamic hormones which travel from the median eminence via the pituitary portal system. Nerve terminals are present in the median eminence, from which are released those hormones that control the pituitary. The neurons which terminate in the median eminence originate in areas throughout the base of the hypothalamus and form the final neural link in the network of pathways which originate widely throughout the brain. Thus, many brain structures, as well as a wide variety of neurotransmitters, may be involved in endocrine regulation.

II. THE DOPAMINE HYPOTHESIS OF SCHIZOPHRENIA: RELEVANCE OF NEUROENDOCRINE STUDIES

In humans, the relative clinical potency of each neuroleptic is directly related to the ability to block postsynaptic DA receptors (Carlsson, 1978), while the DA precursor, L-DOPA, can aggravate schizophrenic symptoms (Angrist *et al.*, 1973). However, untreated schizophrenics do not have elevations of the DA metabolite, homovanillic acid (HVA), in their cerebrospinal fluid (Bowers, 1974). The possibility exists that a disorder in DA is not distributed uniformly throughout the CNS dopamine systems, but is restricted to a specific DA system, so that significant elevations in cerebrospinal-fluid HVA levels might not occur.

The search for an anatomic locus for schizophrenia began with the work of Alzheimer in the last century. While examination of the cortex proved of little value, the discovery of the importance of subcortical structures in the regulation of internal homeostasis, modulation of movement, affect and complex behaviour patterns has led to the concept that these structures are involved in schizophrenia. In their review of the literature regarding schizophrenic-like syndromes associated with organic illness, Davidson and Bagley (1969) conclude that patients with a number of focal CNS disorders frequently demonstrated schizophrenic symptoms. The correlation is highest in those patients with lesions of the limbic system or areas of the immediately adjacent temporal lobe and diencephalon.

Led by the work of Stevens (1973), the beginning of a neuroanatomic basis for schizophrenia has been elaborated.

Histofluorescent techniques demonstrate three main DA systems in the brain (Carlsson, 1978).

(1) *The nigrostriatal tract* from the nigra to the caudate and the putamen. DA depletion of this system has been implicated in Parkinson's disease. Treatment with various neuroleptics results in changes in DA turnover in this region which are in rank order with their ability to induce extrapyramidal symptoms (Crow *et al.,* 1977).

(2) *The mesolimbic system* from the ventral tegmental nucleus to: (a) the nucleus accumbens; (b) the stria terminalis; and (c) the olfactory tubercle. This system has been implicated by Stevens (1979) in the pathophysiology of schizophrenia. The terminals of this system are in the region investigated by Heath (1966). In studies using chronically implanted electrodes, Heath demonstrated focal spike activity in this area in the EEG of acute schizophrenics. Changes in HVA levels in this region following treatment with various neuroleptics occur in rank order with their ability to diminish antipsychotic symptoms (Crow *et al.,* 1977).

(3) *The tuberoinfundibular system* from the median eminence to the pituitary stalk. DA release in the tuberoinfunibular system has an important role in regulation of pituitary hormonal activity (Stevens, 1979). Thus the possibility arises of more closely examining systems directly involved in schizophrenia by studying the associated endocrine changes. Specifically, this strategy allows for closer testing of hypotheses of differences in the number or sensitivity of DA receptors *in toto* or within a particular system as a factor in the illness.

The advent of modern endocrine techniques renewed interest in the psychoendocrine characterization of schizophrenia, which was first reviewed by Kraepelin in 1892. Gradually, hormones were more clearly seen as an integral part of organismic self-regulation. The role of the steroid hormones in coordinating widespread metabolic response to physical or emotional changes was demonstrated by the work of Mason (1975) and others. Sachar *et al.,* (1963) demonstrated elevations of the steroid metabolite 17-hydroxycorticosteroid in acutely ill schizophrenics which correlated with the severity of the psychosis. While later work demonstrated that this steroid elevation is a response to an overwhelming stress rather than part of the illness process *per se*, this study marked the beginning of the most recent phase of neuroendocrine research in schizophrenia. The identification first of the pituitary hormones and then of the hypothalamic factors which regulate them has been a most important contribution, demonstrating more specifically the intimate relationship between brain and endocrine function in the effects that they have upon one another (Schally, 1978). This latest stage in the endocrine characterization of schizophrenia coincided with the first explorations of the DA theory. To date, the DA theory has provided both an organizing hypothesis and the impetus for much of the current work.

III. OVERVIEW OF STUDIES ON ENDOCRINE FUNCTION IN SCHIZO-PHRENICS

A. Growth Hormone

(a) Neural Regulation. Growth-hormone (GH) release involves a number of neurotransmitters. Dopaminergic, serotoninergic and α-adrenergic stimulation enhance release and β-adrenergic function exerts an inhibiting effect (Martin, 1973). DA agonists such as L-DOPA, bromocryptine and apomorphine, serotoninergic agonists such as 5-hydroxytryptophan and L-tryptophan, α-adrenergic agonists (e.g., clonidine) and β-adrenergic antagonists (e.g., propranolol) all produce a rise in GH levels (Blackard and Hubble, 1970). The GH response to hypoglycemia can be reduced by serotonin antagonists such as cyproheptadine and methysergide (Bivens *et al.*, 1973). The GH response to apomorphine can be abolished by DA antagonists such as pimozide and haloperidol, but not by the serotonin antagonist methysergide (Lal *et al.*, 1977). The role of various neurotransmitters in modulating GH release is not yet well understood, but different neurotransmitters appear to underlie the responses to different types of stimuli.

(b) Comparison of Schizophrenics and Controls. No differences have been found in baseline GH levels between schizophrenic patients and controls (Brambilla *et al.*, 1979; Rotrosen *et al.*, 1979). Apomorphine, which has been shown to decrease psychotic symptoms in schizophrenic patients (Tamminga *et al.*, 1978), produced a variable GH response in schizophrenic patients. While there are two reports (Meltzer *et al.*, 1978; Busch *et al.*, 1979) showing no differences in GH responses, four separate investigations (Ettigi *et al.*, 1976; Pandey *et al.*, 1977; Tamminga *et al.*, 1977; Rotrosen *et al.*, 1978) demonstrated exaggerated GH response in acute and a small number of chronic shcizophrenics. Most, but not all, chronic schizophrenics show a blunting of response; the degree of GH-response blunting appears to be related to the duration of prior neuroleptic treatment (Ettigi *et al.*, 1976; Rotrosen *et al.*, 1978). Elevations in GH levels have been reported in patients who demonstrate acute transient dyskinesias on withdrawal from neuroleptics but not in patients with chronic dyskinesia (Brown and

Laughren, 1980). The presence of tardive dyskinesia did not correlate with the degree of GH response (Smith *et al.*, 1977). This suggests that, if GH response and tardive dyskinesia are both results of the effects of long-term neuroleptic treatment on DA systems, then separate systems which respond differently to the same drug may be involved. Further evidence for this differential response will be presented below; see Section IV. Rotrosen *et al.* (1979) suggest that the blunted GH response to apomorphine found in some chronic patients can be explained as a result of prolonged neuroleptic treatment, while the exaggerated GH response seen in many acute and some chronic patients is a reflection of relative dopaminergic overactivity, which is part of the illness process. Preliminary reports which suggested that response to neuroleptics could be predicted on the basis of the GH response to apomorphine have not been confirmed by later studies (Rotrosen *et al.*, 1979). It is possible, however, that real differences between groups were obscured by the large variation in response seen between subjects. Brambilla *et al.* (1979) demonstrated a normal GH response to L-DOPA in chronic schizophrenics when neuroleptic therapy had been withdrawn for at least ten days. After a 30-day course of treatment with a standardized dose of haloperidol, a repeat of the L-DOPA challenge divided the patients into two groups: those with a normal GH response to L-DOPA had responded favourably to haloperidol; those who had little or no clinical response to haloperidol had an increased GH response. The patients were also typed genetically for the presence of the leukocyte antigen HLA-AI, which has the capacity to bind DA and the neuroleptics (Scorza-Smeraldi *et al.*, 1975). The two groups could be distinguished by the presence or absence of the antigen; those who had a normal GH response and favourable drug response were positive for the antigen; in those who had an accentuated GH response and unfavourable drug response, the antigen was absent.

B. Prolactin

(a) Neural Regulation. A variety of pharmacologic agents have been used to characterize the neural regulation of prolactin (PRL). Dopamine exerts an inhibitory influence on PRL either by release of PIF (prolactin-inhibitory factor) or by direct action on the pituitary. DA agonists such as apomorphine and L-DOPA result in suppression of PRL in man, while DA blockers such as pimozide or haloperidol produce marked elevation in PRL (MacLeod, 1976). Serotonin causes PRL release and is assumed to be responsible for elevations of PRL in response to stress (Martin *et al.*, 1977). Administration of the serotonin precursor 5-hydroxytryptophan results in elevation of PRL (Kato *et al.*, 1974), while the serotonin antagonist methysergide causes a decrease in PRL levels. Finally, TRH, in addition to stimulating TSH release, also stimulates the release of PRL.

(b) Comparisons of Schizophrenics and Controls. Comparisons of baseline PRL in schizophrenics and controls have been attempted. Two studies (Gruen *et al.*, 1978b; Rotrosen *et al.*, 1978) have demonstrated such a variability across

individuals that it is impossible to differentiate between the two populations, though the patient levels are generally in the low normal range. However, repeated tests of individuals permits the control of interindividual variation (Langer *et al.,* 1977), suggesting that with the appropriate techniques the issue is testable. In unmedicated chronic patients, there is evidence that severity of formal thought disorder is inversely related to PRL levels; this is the opposite of what might be expected if PRL levels were simply elevated in response to stress (Johnstone *et al.,* 1977). Administration of chlorpromazine to both schizophrenics and controls results, after 72 hours of administration, in elevations of PRL to levels three times normal (Meltzer and Fang, 1976). PRL elevations with neuroleptics persist throughout the course of treatment, with no evidence of habituation of response (Chouinard and Jones, 1980). After three months of continuous administration, a return to normal levels was noted within 48-96 hours of cessation of treatment (Gruen *et al.,* 1978b).

It is well known clinically that individuals vary markedly in their response to neuroleptic drugs. Even when such factors as body weight, sex or age are allowed for, dosage requirements vary markedly amongst individuals. Similarly, there is poor correlation between dosage and serum chlorpromazine levels when comparing subjects (Kolakowska *et al.,* 1979). Serum chlorpromazine levels and PRL vary widely with large interindividual differences but correlate well for a particular individual (Kolakowska *et al.,* 1979). Two studies (Meltzer and Fang, 1976; Siris *et al.,* 1978) found a correlation between increase in PRL and clinical response to neuroleptics, while three other studies (Wiles *et al.,* 1976; Gruen *et al.,* 1978a; Kolakowska *et al.,* 1979) did not. While Kolakowska *et al.* (1979) did not find a correlation between PRL levels and response, there was a correlation between PRL and the development of extrapyramidal symptoms. A similar study, comparing neuroleptic responders with nonresponders in a chronic schizophrenic population, demonstrated a linear relationship between neuroleptic dosage and PRL for both groups (Cleghorn *et al.,* 1980). In sum, PRL does not appear to be a useful predictor of neuroleptic response in either acute or chronic schizophrenic patients. A complicating factor is that maximal individual PRL response to neuroleptics occurs at doses below those necessary for clinical response (Gruen *et al.,* 1978a), so that real differences in response sensitivity may well be obscured by the use of clinically effective doses. At lower doses, PRL responses to neuroleptics do correlate with their DA blocking potency (Langer *et al.,* 1977).

When compared to controls, a subgroup of chronic schizophrenic patients show a blunted PRL response to apomorphine (Rotrosen *et al.,* 1979). This diminished suppression of PRL with apomorphine is a less consistently reproducible effect than the exaggerated GH response in schizophrenics, possibly because of the wide interindividual variations, or because of differences in GH and PRL regulating systems. Busch *et al.* (1979) reported that a correlation exists between severity of schizophrenic symptoms and PRL suppression by apormorphine. However, no difference was found in PRL response to L-DOPA in a group of chronic schizophrenics when compared to normals either before or after treatment with haloperidol (Brambilla *et al.*, 1979). A lowered PRL response to TRH in acute schizophrenics compared to controls has also been reported (Brambilla *et al.*, 1976). The significance of this finding is uncertain, since a similar diminished response has been reported in unipolar depressives (Linnoila *et al.*, 1979).

C. TRH

In a double-bind study TRH, was found to significantly reduce symptomatology in schizophrenics (Prange *et al.*, 1979). Lowered baseline levels of T_3, increased T_3 response to TRH, increased free T_4 and decrease in T_4 binding sites were noted in these patients. Three other routes of administration, dosages and rating scales showed either no change (Clark *et al.*, 1975), slight worsening (Lindström *et al.*, 1977) or substantial improvement (Inanaga *et al.*, 1975), making interpretation of these findings difficult.

D. LH and FSH

Brambilla et al. (1974) found diminished levels of LH, FSH and testosterone, with increased LH and FSH response to LHRH in untreated acute schizophrenics. These changes were all reversed by treatment with neuroleptics. In another study, one half of a sample of unmedicated chronic schizophrenics also showed diminished levels of LH (Johnstone *et al.*, 1977).

E. Endocrine Changes in Chronic Hospitalized Patients

One of the authors (GMB) has found loss of diurnal variation of cortisol and diminished GH, PRL and cortisol response to insulin-induced hypoglycemia in a substantial subgroup of treated, chronically hospitalized schizophrenics. However, these changes were also found in a matched population of chronically hospitalized patients with other psychiatric diagnoses and were therefore not considered to be specific for schizophrenia. These changes are possibly indicators of chronic illness or the result of prolonged treatment rather than markers of a specific disease process.

F. Endogenous Opiates (Endorphins/Enkephalins)

Considerable attention has recently been focused on the possible role played by endogenous opiates in schizophrenia (Davis *et al.*, 1979). Small peptides (enkephalins) have been demonstrated in both the brain and the pituitary (Kosterlitz and Hughes, 1977), and shown to have endocrine as well as behavioural effects. Administration of an enkephalin analog to normal subjects resulted in elevations of serum PRL, GH and TSH and decreases in serum LH, FSH, cortisol and ACTH (Besser *et al.*, 1978). Prior administration of the opiate antagonist naloxone attenuated the hormonal responses. Administration of naloxone alone decreases PRL levels (Gold *et al.*, 1979a, b). Marked decreases in serum PRL are also seen in opiate withdrawal, while administration of morphine and methadone has an opposite effect (Gold *et al.*, 1978). Although naloxone decreases GH in animals, administration of naloxone to manic patients produces an elevation of GH and diminishes euphoria and arousal (Janowsky *et al.,* 1978).

ACTH and β-endorphin share a common precursor. This precursor, although lacking a known biologic activity itself, appears to function as a prohormone to a number of cleavage products, including β-lipotropin and ACTH. In turn, β-lipotropin is a prohormone for β-endorphin (Guillemin, 1978). In man, β-endorphin is released simultaneously with ACTH in response to acute stress (Guillemin, 1978). In most instances studied, elevation of ACTH parallels rises in β-endorphin (Guillemin, 1978).

The effect of endorphins on endocrine measures may be mediated in part by DA. In animal studies, administration of β-endorphin resulted in diminished turnover of hypothalamic DA (Van Loon *et al.,* 1980). Moreover, administration of endorphins to AMPT-pretreated rats profoundly decreases nigrostriatal DA (Versteeg *et al.,* 1979). Interest in the possible role of endorphins in schizophrenia has been pursued via three broad hypotheses (Veregey *et al.,* 1978):

(a) *The hypothesis that there is an excess of endorphins in schizophrenics.* Double-blind administration of opiate antagonists to schizophrenics has resulted in some temporary improvement (Emrich *et al.,* 1977; Gunne *et al.,* 1977), however, others (Janowsky *et al.,* 1977; Davis *et al.,* 1979) have been unable to replicate these results. Differences in methodology have made these ostensibly contradictory data somewhat difficult to interpret. There have been reports of elevated CSF endorphin levels in both schizophrenic and manic patients which decreased to normal levels with neuroleptic treatment (Lindström *et al.,* 1978). A report of beneficial effects of hemodialysis in schizophrenics (Wagemaker and Cade, 1977) was followed by identification of β-endorphin in the dialysate (Palmour *et al.,* 1978). A further study, however, could not confirm this finding (Bloom, 1978). A small single-blind study (Lehmann *et al.,* 1979) demonstrated pronounced improvement in some, but not all, schizophrenics treated with naloxone. The greatest improvement was demonstrated in the patient who demonstrated the widest variations in diurnal plasma ACTH and the least improvement was shown in a patient with the smallest variations of diurnal ACTH.

(b) *The hypothesis that there is a deficiency of endorphins in schizophrenia.* Animal studies indicate a similarity between the effects on behaviour of endorphins and neuroleptics (Jacquet and Marks, 1976). Double-blind administration of γ-endorphin to schizophrenic patients resulted in clear-cut transient improvement in eight out of 13 patients (van Ree *et al.,* 1978), suggesting a possible therapeutic value of this peptide.

(c) *The hypothesis that there is a structural abnormality in the endorphins of schizophrenics.* Endorphins and other peptides share a number of common steps in their synthesis. Recent interest has focused on possible abnormal peptides produced by errors in cleavage. The identification of the various types of endogenous opiates and the sequence involved in their synthesis would allow this hypothesis to be formally tested. Elaboration of the role of endogenous opiates in the etiology of schizophrenia and its possible treatment will await further study.

IV. PROBLEMS AND PROSPECTS IN SCHIZOPHRENIA

The study of neuroendocrine changes appears to be one of the most promising avenues in an area in which there has been a paucity of definite findings. If this promise is to materialize, a number of issues must be resolved.

Firstly, in most of the studies reviewed in this section, the patients had been drug-free only for periods from three months to as little as one week. Clearly, future studies must include populations of both acute untreated patients and chronic schizophrenics on drug holidays of six months or more. An additional strategy would be to include longitudinal examination of neuroendocrine measures in a cohort of patients, both while acutely ill and during remission. Until data from these populations are available, it may prove impossible to separate definitively changes related to the illness and those due to administration of neuroleptics.

A record point to be taken into account is that differences between DA systems in the brain, as recently reviewed by Mishra and Cleghorn (1979), suggest that current models do not allow us to generalize from findings in one system to another. For example, tardive dyskinesia, implying nigrostriatal DA supersensitivity, is found concurrently in some patients with blunting of GH and PRL responses, which implies hyposensitivity of the tuberoinfundibular DA system (Tamminga *et al.*, 1977; Smith *et al.*, 1977; Ettigi *et al.*, 1976). Indeed, this potential difficulty may be turned to advantage by using agents which have a selective action on the various DA systems. One of the authors (GMB) has recently found that domperidone, a peripheral DA blocker, does not eliminate the GH response to apomorphine. Furthermore, administration of carbidopa, an agent which blocks peripheral DA synthesis, increases PRL levels but does not affect GH (Brown *et al.*, 1976a, b). These findings suggest that the DA receptors controlling PRL are peripheral (located in the pituitary and/or median eminence), while the DA receptors principally controlling GH are central. Comparison of GH responses to apomorphine in domperidone-pretreated schizophrenic subjects with the GH responses to apomorphine alone could conceivably demonstrate a nonpituitary location in the brain for differences in DA sensitivity between normals and schizophrenics.

There is also increasing evidence of differences in the responses of the various DA systems to chronic neuroleptic administration. Recent animal data demonstrate that rats pretreated with haloperiodol show decreased pituitary but increased striatal binding of both apomorphine and haloperidol, indicating not only that long-term effects exist but that they may be qualitatively different (Friend *et al.*, 1978). In humans, chronic neuroleptic treatment has been associated with an increase in DA-receptor sites both in the striatum and in the mesolimbic system (Owen *et al.*, 1979). It has been suggested that the former is associated with the development of tardive dyskinesia, while the latter has been recently implicated in the supersensitivity-psychosis syndrome described by Chouinard *et al.* (1978). In contrast, there has been no evidence to demonstrate similar DA supersensitivity in the tuberoinfundibular system (Chouinard and Jones, 1980). This may well account for the persistence of elevated PRL levels found in medicated chronic patients regardless of the duration of treatment.

The pharmacological and endocrine profiles of some of the most recent neuroleptics contribute to our understanding of the difference in the activity of the various DA systems. For example, clozapine, a dibenzodiazepine derivative, is a potent DA blocker (Miller and Iversen, 1974; Karobath and Leitich, 1975) and a clinically effective antipsychotic which produces relatively few extrapyramidal symptoms (Matz *et al.*, 1974; Nair *et al.*, 1977). Animal studies provide supporting evidence indicating that clozapine has less effect on the nigrostriatal DA system than do other neuroleptics (Bartholini *et al.*, 1972; Sedval and Nyback, 1973). Moreover, clozapine does not raise PRL levels, although it does blunt GH response to apomorphine profoundly (Meltzer *et al.*, 1979; Nair *et al;.* 1979). In animals, in contrast, clozapine does induce a marked increase in turnover of DA in the mesolimbic system (Anden and Stock, 1973). This suggests that its antipsychotic properties may be related to a more specific effect on the mesolimbic DA system. While clozapine may prove an important investigative tool, its current therapeutic value is limited by concern, resulting from reports of agranulocytosis associated with prolonged use (Amsler *et al.*, 1977).

A third problem that remains to be resolved is that there is no compelling evidence to suggest that schizophrenia is a single biological entity, and substantial evidence to the contrary. Indeed, in medicine, biological and therefore biochemical heterogeneity is the rule rather than the exception (Buchsbaum and Rieder, 1979). Differences in neuroendocrine profiles of schizophrenic patients may well reflect different underlying pathophysiological mechanisms. Until this issue is clarified, all data must be interpreted with caution, and careful consideration must be given to identification of possible biologic subgroups of schizophrenics.

Recent animal studies demonstrate that the pituitary hormones, including GH and PRL, are regulated in part by a number of peptides (Vijayan and McCann, 1980). The progressive accumulation of information on the role of peptides in the CNS suggests the opening of a broad new vista for the neuroendocrine characterization of schizophrenia.

V. HORMONAL MARKERS IN DEPRESSION

Major affective disorders are generally considered to be associated with an abnormality in CNS neurotransmitter activity (Schildkraut, 1974). The evidence for neurotransmitter regulation of the hypothalamo-pituitary axis is rapidly increasing in both animal and human studies (Lissak, 1973; Martin *et al.*, 1977; Brown *et al.*, 1979). Several clinical investigations have revealed the existence of definite neuroendocrine abnormalities in patients with depression. These abnormal endocrine changes in depression may well reflect the hypothesized abnormal neurotransmitter activity in the brain. Demonstration of consistent endocrine abnormalities in carefully selected subgroups of affective disorder may lead to their being considered as biological markers, and may have value in the classification of affective disorder. The value of psychoneuroendocrine research in the diagnosis, management and prognosis of depressive subtypes is evident in some studies (Furlong *et al.*, 1976; Ettigi *et al.*, 1979; Gold *et al.*, 1979a, b; Schlesser *et al.*, 1979). A possible relationship between behavioural, physiological and neuroendocrine changes and altered neurotransmitter activity in the central nervous system is outlined in Figure 16.1. Changes in brain centers, such as cortex, limbic system and reticular activating system, may result in altered neurotransmitter activity. Such changes may simultaneously alter behaviour and hypothalamo-pituitary function, leading to the coexistence of behavioural and neuroendocrine abnormalities in some patients. The major neuroendocrine abnormalities reported in affective disorder are in the secretion of cortisol, growth hormone and thyroid-stimulating hormone. Prolactin and luteinizing hormone (LH) have also been studied with conflicting results and require further investigation.

A. Cortisol

Plasma cortisol and the hypothalamo-pituitary-adrenal (HPA) axis have been extensively studied, and hypersecretion of cortisol in a significant proportion of depressed patients is well established (Sachar, 1975a, b; Carroll, 1976). Earlier studies measuring 17-OH corticosteroid (17-OHCS) excretion in depressed patients demonstrated the occurrence of high corticosteroid levels in depression (Mason, 1968). Urinary corticosteroid measurement in manic-depressive patients revealed low levels during the manic phase and higher levels in the depressed phase of the illness (Bryson and Martin, 1954; Bunney *et al.*, 1965; Rubin *et al.*, 1968; Stancer *et al.*, 1969). However, in these earlier studies, 17-OHCS in urine correlated poorly with severity of depression, which may be partially explained by evidence that 17-OHCS and 17-KS do not correlate well with the actual cortisol secretion rate (Cope and Pearson, 1965). Further, it has been suggested that urine 17-OHCS is not an accurate estimate of HPA activity, particularly in stressful situations. Urine free cortisol (UFC) has been shown to be a more reliable measure of the HPA axis, and studies in depressed patients show a high UFC excretion compared to psychiatric inpatient controls or to acutely psychotic schizophrenic patients (Carroll, 1976).

This elevation of cortisol secretion in depressed patients was initially considered to be the result of nonspecific factors such as stress, effects of hospitalization, first sampling effect, depressive decompensation or anxiety (Sachar, 1967). Controlled studies taking these factors into account still showed evidence

Figure 16.1: A defect in neurotransmitter can lead to behavioural, endocrine and visceral changes.

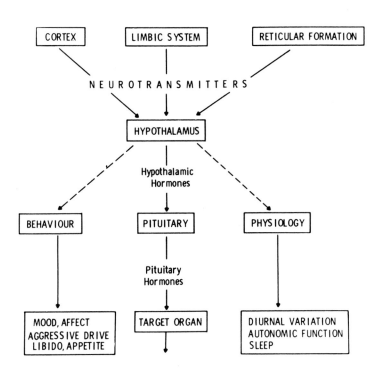

Source: Ettigi and Brown (1979).

of cortisol hypersecretion (Carroll, 1972). Hypersecretion of cortisol was evident in apathetic depressed patients and while the patients were asleep, thus refuting the concept that hypersecretion is simply a response to stress (Sachar, *et al.*, 1973b).

Plasma cortisol evidences a normal diurnal variation and secretion occurs in bursts which are more frequent between 2.00 a.m. and 9.00 a.m. (Weitzman *et al.*, 1971). These cortisol bursts reflect episodic bursts of adrenocorticotrophic hormone (ACTH) from the pituitary, which in turn is regulated by release of corticotrophin-releasing factor (CRF) from the hypothalamus (Weitzman *et al.*, 1971). Measuring plasma cortisol levels once or twice during the day may reflect secretory bursts occurring at the time of plasma collection, or may miss them entirely. This problem could explain the discrepancy noted in the results of some earlier studies in which plasma cortisol was measured at only one point of time. The most comprehensive measurement of HPA activity is the measurement of plasma cortisol at frequent intervals over a 24-hour period. Studies measuring plasma cortisol in depressed patients every 20 minutes for 24 hours have revealed an increase in both the number and the magnitude of secretory episodes, as well as a pervasive increase in plasma cortisol levels not only during the night, but also during the day (when control subjects have very low levels; Sachar *et al.*, 1973b).

Cortisol hypersecretion has also been investigated by means of the dexamethasone-suppression test. Dexamethasone is a potent synthetic corticosteroid which, when given in 2-mg dosage to normal subjects, produces complete suppression of plasma cortisol and 17-OHCS excretion for 24 hours (Carroll, 1976), secondary to suppression of pituitary ACTH secretion. Depressed patients with elevated cortisol secretion fail to show normal suppression with dexamethasone, suggesting that there is continued increaaed ACTH secretion in these subjects (Carroll, 1976). There have been some studies which report a normal dexamethasone suppression in depressed patients (Carpenter and Bunney, 1971; Shopsin and Gershon, 1971). The conflicting results observed in these studies may be due to heterogeneity in the diagnostic criteria. Carroll and coworkers find greater correlation of abnormal dexamethasone suppression test with endogenous than nonendogenous depressions. Schlesser *et al.* (1979) studied primary unipolar depressed patients subcategorized according to Winokur's genetic classification (Winokur, 1974), and found that 82% of patients with familial pure depressive disease demonstrated a lack of suppression by dexamethasone, while only 4% of patients with depressive spectrum disease failed to show cortisol suppression with dexamethasone. Moreover, only 37% of the group labelled sporadic depressive disease failed to show cortisol suppression after dexamethasone. Based on these results, the authors suggest that unipolar depression has several subtypes, each with its own mode of inheritance, neurochemistry, pathophysiology, clinical course and treatment outcome. These findings may indicate that this abnormality in cortisol secretion may be a biological marker differentiating genetically determined subtypes of primary depression. Further research in carefully selected homogenous subgroups of depressed patients will be necessary in order to replicate the study and strengthen this hormonal marker as a useful tool in diagnosis of depression.

Plasma cortisol, as mentioned previously, is closely related to the secretion of ACTH from the pituitary, which in turn is regulated by corticotrophin-releasing factor (CRF) elaborated in the hypothalamus. The abnormality in the hypothalamo-pituitary axis observed in depressed patients is considered to be related to a disturbance in central neurotransmitter activity. Studies in both animals and human subjects have shown that norepinephrine activity in the brain has a tonic inhibitory effect on ACTH, which is presumably due to reduced CRF release from the hypothalamus. Serotoninergic and cholinergic mechanisms have been noted to increase ACTH secretion and seem to be involved in the circadian rhythm of ACTH secretion (Krieger, 1973). These very same neurotransmitters have been implicated in the biogenic-amine hypotheses of depression. The hypotheses suggest that depression is associated with a relative deficiency of catecholamines, particularly norepinephrine, at functionally important adrenergic-receptor sites of the brain (Schildkraut, 1965) or a diminished activity of serotonin (Coppen *et al.*, 1972). It could therefore be speculated that the abnormality in cortisol secretion could be a function of such a biogenic-amine defect in depressed patients. A disturbance in acetylcholine activity in the CNS has also been proposed in depression. Carroll *et al.* (1978) have speculated that the escape of the hypothalamo-pituitary axis from dexamethasone suppression may occur by a muscarinic cholinergic hyperfunction in the limbic system of depressed patients. This was based on studies of control subjects wherein escape from dexamethasone suppression was induced by physostigmine but was prevented by the prior administration of atropine. Cortisol hypersecretion and lack of its suppression by dexamethasone is thus a hormonal marker which may be of value not only in the diagnostic classification of primary unipolar depression but also in understanding the biological etiology of depression.

B. Growth Hormone

The secretion of growth hormone (GH) and its response to a variety of pharmacological agents have also been studied in depression. The reported abnormalities in GH release need to be carefully examined because a variety of factors — stress, sleep, prolonged fasting, protein intake, plasma glucose levels, plasma fatty-acid levels, sex of the patient, oral-contraceptive medication and phase of menstrual cycle — have an influence on GH release (Brown and Reichlin, 1972; Martin, 1973; Ettigi *et al.*, 1975; Brown *et al.*, 1978). Growth-hormone response to insulin hypoglycemia has been shown to be abnormal in depressed patients (Sachar *et al.*, 1971; Müller *et al.*, 1972; Garver *et al.*, 1975; Gruen *et al.*, 1975). Gruen *et al.* (1975) showed a clear reduction in GH in response to insulin hypoglycemia in a group of depressed postmenopausal women compared to normal postmenopausal women. In our own study (Ettigi *et al.*, 1979) we were able to show that GH response to insulin hypoglycemia was reduced in depressed male patients. It was also evident that primary unipolar depressed patients had a significantly lower GH response to insulin hypoglycemia when compared to secondary depressed patients and normal controls. L-DOPA, an amino-acid precursor of catecholamines, causes a rise in plasma GH in normal subjects (Boyd *et al.*, 1970). It has been shown that 5-hydroxytryptophan (5-HTP), the amino-acid precursor of serotonin, which also causes a rise in plasma GH levels in normal subjects, produces an inadequate response in depressed patients (Takahashi *et al.*, 1974a). These abnormalities in GH secretion may well reflect an abnormality in biogenic-amine activity in the brain. Maeda *et al.* (1975) first reported that depressed patients show an abnormally increased GH response to thyrotropin-releasing hormone (TRH). Eight out of their 13 depressed patients showed a significant increase in GH levels, while control subjects showed no response to 500 μg of TRH infusion. In our sample of primary unipolar depressed and secondary depressed patients and control subjects, we could find no significant difference between the three groups in their GH response to intravenous administration of 500 μg of TRH (Ettigi and Brown, 1979). Though the levels of GH in depressed patients during the postinfusion period were higher than those in control subjects, the difference between preinfusion and postinfusion GH levels was not significant. Our group of depressed patients had slightly higher basal GH levels compared to controls. It is therefore important to determine baseline GH levels in depressed patients when studying responses to TRH. Further carefully controlled studies are required to confirm these abnormalities in GH regulation in depression.

Dopamine, norepinephrine and serotonin exert a positive control on GH release (Frohman and Stachura, 1975; Brown *et al.*, 1979). Amphetamines have been known to produce an increase in GH; this response is less marked in depressed patients (Langer *et al.*, 1976). Pimozide, a dopamine-blocking agent, does not block the GH response to amphetamine, suggesting that dopaminergic mechanisms may not be influential in GH response to amphetamine. Garver *et al.* (1975) have shown that a linear correlation exists between GH response to insulin and the urinary excretion of 3-methoxy-4-hydroxphenylglycol. This evidence suggests that the GH response to insulin hypoglycemia is likely to be mediated by the noradrenergic system. The GH abnormalities observed in depression in response to amphetamine and to hypoglycemia are therefore both consistent with the hypothesis that there is a central catecholamine deficiency, particularly that of norepinephrine, in some depressions.

C. Prolactin

Sachar *et al.* (1973a) have reported elevated basal prolactin levels in unipolar and bipolar depressed patients when compared to normal subjects, but noticed no difference in prolactin suppression when L-DOPA was given to the two groups of subjects. Prolactin secretion is regulated by the hypothalamus, which releases prolactin-inhibiting factor (PIF). L-DOPA, which increases central catecholamine activity, and apomorphine, a dopamine-receptor agonist, both decrease prolactin levels, presumably by causing PIF release (Friesen and Hwang, 1973). Direct infusion of dopamine into the portal system also inhibits prolactin release, suggesting that the postulated PIF may well be dopamine itself (Takahara *et al.*, 1974). This concept is supported by the finding that the pituitary contains stereospecific dopamine receptors (Brown *et al.*, 1976b). No strong evidence exists for an abnormality in dopaminergic activity in the brain in depression, though the antidepressant nomifensine (Hoffmann, 1977; Brogden *et al.*, 1979) has a preferential effect on dopamine reuptake. Systematic studies on prolactin regulation may be of importance in studying the dopaminergic system in certain types of depression. Elevated prolactin levels reported in depression have been thought to reflect central dopaminergic depletion. However, Arana *et al.* (1977) reported no change in prolactin levels in a large group of mildly depressed patients compared to matched controls.

The prolactin response to thyrotropin-releasing hormone (TRH) has also been studied in depressed patients. TRH stimulates the release of prolactin in normal subjects, presumably by acting on the pituitary gland. Ehrensing *et al.* (1974) found a diminished prolactin response to TRH, while Maeda *et al.* (1975) found an increased prolactin response to TRH in depressed patients compared to normal controls. In our own study involving primary unipolar depressed patients, secondary depressed patients and normal controls, we were unable to find any differences in the prolactin response to TRH in the three groups of subjects (Ettigi and Brown, 1979).

Dopamine mechanisms have been shown to have an inhibiting effect on prolactin secretion, while serotoninergic mechanisms tend to stimulate the release of prolactin. 5-HTP has been shown to increase prolactin levels, presumably either by decreasing PIF or, more likely, by stimulating the release of prolactin-releasing factor (PRF) from the hypothalamus (Valverde *et al.*, 1972; Kato *et al.*, 1974). These results suggest that prolactin secretion from the pituitary can be stimulated by reduction of catecholamine activity and by an increase in serotoninergic activity. Further research on prolactin secretion in depressed patients is clearly warranted.

D. Thyroid-stimulating Hormone

Interest in thyroid-stimulating-hormone (TSH) secretion and depression has been related to the availability of synthetic TRH (Guillemin and Burgus, 1972; Schally *et al.*, 1973) and the antidepressant-like action that was reported by many investigators conducting clinical trials of TRH in depressed patients (Prange and Wilson, 1975). The reported transient antidepressant effect of TRH has not been reproduced by other investigators (Ehrensing *et al.*, 1974; Hollister *et al.*, 1974; Takahashi *et al.*, 1974b; Maeda *et al.*, 1975) and the psychological changes, if any, are thought to be a placebo effect. Despite the lack of antidepressant effect, the TSH response to TRH was found to be subnormal in many depressed patients (Kastin *et al.*, 1972; Prange *et al.*, 1973; Ehrensing *et al.*, 1974; Takahashi *et al.*, 1974b; Maeda *et al.*, 1975). It is not clear from these studies whether the TSH response to TRH could be of value in identifying subgroups of depressive illness. Gold *et al.* (1979a, b) in a recent study of a small sample of biopolar and unipolar depressives, reported an augmented TSH response to TRH in patients with bipolar depression, while patients with unipolar depression had an abnormally low response. The authors suggest that this test may be clinically useful in diagnosis and choice of drug treatment.

The abnormal (lowered or elevated) response to TRH could be a function of thyroid state (hyper- or hypofunction). However, the depressed patients in the above studies have been reported as being euthyroid. A convincing explanation of the subnormal TSH response in depressed patients has yet to be found. It could be a reflection of inappropriate TRH production or deficiency in hypothalamic TRH production. Research into neurotransmitter regulation of TRH reveals that central noradrenergic mechanisms facilitate release of TRH (Reichlin *et al.*, 1974; Frohman and Stachura, 1975), while cholinergic systems do not alter

TRH secretion. One can therefore speculate that the TSH abnormality observed in the depressed patient could be a reflection of central biogenic-amine dysfunction.

E. Luteinizing Hormone

Plasma luteinizing hormone (LH) has been reported to be low in depressed postmenopausal women compared to normal postmenopausal women (Altman *et al.*, 1975). Basal LH concentration was measured in 13 normal women and twelve women with unipolar depression. The mean LH concentration in the depressed group was 33% lower than in the control group; 50% of the depressed patients had LH levels lower than the lowest values found in the control group. The authors mention the possibility that this reflects a diminished hypothalamic noradrenergic activity in such patients. The only other study measuring LH in depressed patients was the one by a group (Ettigi *et al.*, 1979), wherein a significant difference between LH response to LHRH was reported in a group of secondary depressed patients compared to primary unipolar depression and normal controls. Baseline LH levels in all three groups were not different. All subjects were males and received 100 mg of LHRH at 8.0 a.m. following an overnight fast. The group of secondary depressed patients had a significantly greater LH response at 30, 45, 60 and 90 minutes following the LHRH infusion compared to those of primary unipolar depression and normal controls (Figure 16.2). Our results suggest that the difference between the LH responses to LHRH may be useful in distinguishing these two subgroups of depression. The role of central biogenic amines in the regulation of LH release in humans is still unclear but LH release seems to be closely related to the feedback regulatory influence of estrogen and testosterone on the pituitary and the hypothalamus. It should be noted that testosterone levels in the above three groups of subjects were not different. Thus, the abnormality in LH secretion does not seem to be a defect in the feedback regulation by testosterone, but could be the result of a central neurotransmitter abnormality, and remains to be studied in larger numbers of patients.

It has been shown that norepinephrine turnover is increased in rats whose gonads have been removed (Anton-Tay and Wurtman, 1968). This finding implies that estrogen deficiency causes an increase in catecholamines that results in increased production of LHRH, which in turn enhances LH secretion. It has been shown in animals that noradrenergic mechanisms stimulate the release of LH (Ojeda and McCann, 1973). However, in human subjects, administration of such agents as apomorphine (Lal *et al.*, 1973) or clonidine (Lal *et al.*, 1974) does not have any effect on LH secretion. Evidence for a role of neurotransmitters in the regulation of LH, although apparent from animal studies, is lacking in humans, and further research is required for its clarification.

VI. SUMMARY AND PROSPECTS IN DEPRESSION

The hormonal abnormalities which exist in certain groups of depression provide

Figure 16.2: Mean luteinizing-hormone responses (±SEN) to TRH and LHRH in depressed patients as compared to matched controls. Responses in secondary depressed patients are significantly greater than other groups ($p < 0.05$; ANOVA and multiple range test).

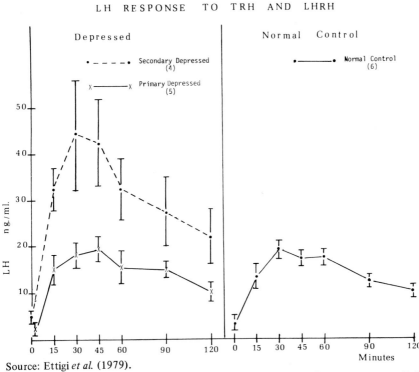

LH RESPONSE TO TRH AND LHRH

Source: Ettigi *et al.* (1979).

additional biological evidence in support of specific subtypes of depression. Failure of cortisol suppression by dexamethasone has been shown to be a useful marker in differentiating subtypes of primary unipolar depression based on family history and genetic background. GH and LH responses may also be useful in identifying different subgroups of affective disorders. Some, if not all, hormonal markers studied to date seem to support the biogenic-amine hypotheses of depression. As the determination of a variety of hypothalamic hormones becomes available in the future, it may be possible to use neuroendocrine parameters as indicators of a common disturbance in brain biogenic-amine function. The enkephalins and related substances have also been found to influence neuroendocrine regulation. The long-acting enkephalin analog DAMME has been reported to raise prolactin and GH levels and lower LH, FSH, cortisol and ACTH levels in man (Stubbs *et al.*, 1978). The authors suggest that enkephalins and other peptides with opiate-like activity may provide an important link between perception, behaviour and neuroendocrine regulation, and have therefore a considerable potential as diagnostic and perhaps therapeutic agents. However, a word of caution is necessary in interpreting neuroendocrine dysfunction in affective disorders. Several factors need to be taken into account and controlled for in neuroendocrine investigations: physical and psychological variables which alter neuroendocrine secretion, extrapolation from animal studies to humans, indirect interpretation of neurotransmitter activity in the central nervous system and difficulty in adequately defining the subgroups of affective disorder. Nevertheless, on the basis of a careful survey of the existing studies, there appears to be enough evidence to indicate that specific neuroendocrine abnormalities do exist in affective disorders, and that neuroendocrine techniques constitute a powerful approach useful in the assessment of brain biogenic-amine function, which promises to increase our understanding of the biology of the disorder.

REFERENCES

Altman, N., Sachar, E.J., Gruen, P.H. Halpern, F.S. and Eto, S. (1975). *Psychosom. Med., 37,* 274.

Amsler, H.A., Teerenhovi, L., Barth, E., Karjula, K. and Vuopic, P. (1977). *Acta Psychiat. Scand., 56,* 241

Anden, N.E. and Stock, G. (1973). *J. Pharm. Pharmacol., 25,* 346

Angrist, B., Sathananthan, G. and Gershon, S. (1973). *Psychopharmacology, 31,* 1

Anton-Tay, F. and Wurtman, R.J. (1968). *Science, 159,* 1245

Arana, G., Boyd, A.E. III, Reichlin, S. and Lipsitt, D. (1977). *Psychosom. Med., 39,* 193

Bartholini, G., Haefely, W., Jalfre, M., Keller, H. and Pletscher, A. (1972). *Br. J. Pharmacol., 46,* 737

Besser, G., Bloom, S. and Alberti, K. (1978). *Lancet ii,* 1225

Bivens, C.H., Lebovitz, H.E. and Feldman, J.M. (1973). *New Engl. J. Med., 289,* 236

Blackard, W.G. and Hubble, G.J. (1970). *Metabolism, 19,* 547

Bloom, F.E. (1978). In *Cell Receptor Disorders,* ed. T. Melnechuk (Western Behavioural Sciences Institute, La Jolla). p. 134

Bowers, M.B. (1974). *Arch. Gen. Psychiat., 31,* 50

Boyd, A.E. III, Lebovitz, H.E. and Pfeiffer, J.B. (1970). *New Engl. J. Med., 283,* 1425

Brambilla, F., Guerrini, A., Riggi, F. and Ricciardi, F. (1974). *Dis. Nerv. Syst., 35,* 362

Brambilla, F., Guerrini, A., Guastalla, A., Rovere, C. and Riggi, F. (1975). *Psychopharmacologia* (Berlin), *44,* 17

Brambilla, F., Guastalla, A., Guerrini, A., Rovere, C., Legnani, G., Sarno, M. and Riggi, F. (1976). *Acta Psychiat. Scand., 54,* 275

Brambilla, F., Bellodi, L., Negri, F., Smeraldi, E. and Malagoli, G. (1979). *Psychoneuroendocrinology, 4,* 329

Brogden, R.N., Heel, R.C., Speight, T.M. and Avery, G.S. (1979). *Drugs, 18,* 1

Brown, G.M. and Reichlin, S. (1972). *Psychosom. Med., 34,* 45

Brown, G.M., Garfinkel, P.E., Warsh, J.J. and Stancer, H.C. (1976a). *J. Clin. Endocrinol. Metab., 43,* 236

Brown, G.M., Seeman, P. and Lee, T. (1976b). *Endocrinology, 99,* 1407

Brown, G.M., Seggie, J.A., Chambers, J.W. and Ettigi, P.G. (1978). *Psychoneuroendocrinology 3,* 131

Brown, G.M., Friend, W.C. and Chambers, J.W. (1979). In *Clinical Neuroendocrinological Approach,* ed. G. Tolis, F. Labrie, J.B. Martin and F. Naftolin (Raven Press, New York). p. 47

Brown, W.A. and Laughren, T.P. (1980). *Lancet i,* 259

Bryson, R.W. and Martin, D.F. (1954). *Lancet ii,* 365

Buchsbaum, M.S. and Rieder, R.O. (1979). *Arch. Gen. Psychiat., 36,* 1163

Bunney, W.E., Jr., Hartmann, E.L. and Mason, J.W. (1965). *Arch. Gen. Psychiat., 12,* 619

Busch, D.A., Meltzer, H.Y., Young, M. and Fang, V.S. (1979). Scientific Proceedings, American Psychiatric Association

Carlsson, A. (1978). *Am. J. Psychiat., 135,* 164

Carpenter, W.T. and Bunney, W.E. (1971). *Am. J. Psychiat., 128,* 31

Carroll, B.J. (1972). In *Depressive Illness. Some Research Studies* ed. B. Davies, B.J. Carroll and R.M. Mowbroy (Charles C. Thomas, Springfield, Illinois). p. 69

Carroll, B.J. (1976). In *Modern Trends in Psychosomatic Medicine,* ed. O.W. Hill (Butterworth, London). p. 121

Carroll, B.J., Greden, J.F., Rubin, R.J., Haskett, P., Feinberg, M. and Schteingart, D. (1978). *Acta Endocr., 89,* (5220), 14

Chouinard, G. and Jones, B.D. (1980). *Am. J. Psychiat., 137,* 16

Chouinard, G., Jones, B. and Annable, L. (1978). *Am. J. Psychiat., 135,* 1409

Clark, M.L., Paredes, A., Costiloe, J.P. and Wood, F. (1975). *Psychopharmacol. Commun., 1,* 191

Cleghorn, J., MacCrimmon, D. and Brown, G. (1980). *Proceedings of the Third Annual Meeting of the Canadian College of Neuropsychopharmacology* (Edmonton, May 12-13).

Cope, C.L., and Pearson, J. (1965). *J. Clin. Pathol., 18,* 82

Coppen, A., Prange, A.J., Whybrow, P.C. and Noguera, R. (1972). *Arch. Gen. Psychiat., 26,* 474

Crow, T., Deakin, J. and Longden, A. (1977). *Psychosom. Med., 7,* 213

Davidson, K. and Bagley, C.R. (1969). In *Current Problems in Neuropsychiatry,* ed. R.N. Herrington (Headley, UK). p. 113

Davis, G.C., Buchsbaum, M.S. and Bunney, W.E. (1979). *Schizophrenia Bull., 5,* 244

Ehrensing, R.H., Kastin, A.J., Schalch, D.S. Friesen, H.G., Vargas, J.F. and Schally, A.V. (1974). *Am. J. Psychiat., 131,* 714

Emrich, H.M., Cording, C., Piree, S., Kolling, A., Zerssen, D.V. and Herz, A. (1977). *Pharmakopsychiatrie, 10,* 265

Ettigi, P. and Brown, G.M. (1978). *Psychosom. Med., 41,* 1

Ettigi, P. and Brown, G.M. (1979). In *Neuroendocrine Correlates in Neurology and Psychiatry,* ed. E.E. Müller and A. Agnoli (Elsevier/North-Holland Biomedical Press, Amsterdam). p. 225

Ettigi, P., Lal, S., Martin, J.B. and Friesen, H.G. (1975). *J. Clin. Endocrinol. Metab., 40,* 1094

Ettigi, P., Nair, N.P.V., Lal, S., Cervantes, P. and Guyda, H. (1976). *J. Neurol. Neurosurg. Psychiat., 39,* 870

Ettigi, P., Brown, G.M. and Seggie, J. (1979). *Psychosom. Med., 41,* 203

Friend, W.C., Brown, G.M., Jawahir, G., Lee, T. and Seeman, P. (1978). *Am. J. Psychiat., 135,* 839

Friesen, H.G. and Hwang, P. (1973). *Ann. Rev. Med., 24,* 251

Frohman, L.A. and Stachura, M.E. (1975). *Metabolism, 24,* 211

Furlong, F.W., Borwn, G.M. and Beeching, M.F. (1976). *Am. J. Psychiat., 133,* 1187

Garver, D.L., Pandey, G.N., Dekirmenjian, H. and Deleon-Jones, F. (1975). *Am. J. Psychiat., 132,* 1149

Gold, M.S., Redmond, E. Jr., Donabedian, R.K., Goodwin, F.K. and Extein, I. (1978). *Am. J. Psychiat., 135,* 1415

Gold, M.S., Pottash, A.L.C., Davies, R.K., Ryan, N., Sweeney, D.R. and Martin, D.M. (1979a). *Lancet ii,* 411

Gold, M.S., Sweeney, D.R., Pottash, A.L.C. and Kleber, H.D. (1979b). *Am. J. Psychiat., 136,* 849

Gruen, P.H., Sachar, E.J., Altman, N. and Sassin, J. (1975). *Arch. Gen. Psychiat., 32,* 31

Gruen, P.H., Sachar, E.J., Altman, N., Lnager, G., Tabrizei, M.A. and Halpern, F.S. (1978a). *Arch. Gen. Psychiat., 35,* 1220

Gruen, P.H., Sachar, E.J., Langer, G., Altman, N., Leifer, M., Frantz, A. and Halpern, F.S. (1978b). *Arch. Gen. Psychiat., 35,* 108

Guillemin, R. (1978). *Res. Publ. Assoc. Res. Nerv. Ment. Dis., 56,* 155

Guillemin, R. and Burgus, R. (1972). *Scientific American, 227* (5), 24

Gunne, L.M., Lindstrom, L. and Terenius, L. (1977). *J. Neural Transm., 40,* 13

Heath, R.G. (1966). *Int. J. Neuropsychiat., 2,* 597

Hoffmann, I. (1977). *Br. J. Clin. Pharmacol., 4,* 69S
Hollister, L.E., Berger, P., Ogle, F.L., Arnold, R.C. and Johnson, A. (1974). *Arch. Gen. Psychiat., 31,* 468
Inanaga, K., Nakano, T. and Nagat, T. (1975). *Kurume Med. J., 22,* 159
Jacquet, Y.F. and Marks, N. (1976). *Science, 194,* 632
Janowsky, D., Judd, L., Huey, L., Roitman, N., Parker, D. and Segal, D. (1978). *Lancet ii,* 320.
Janowsky, D., Segal, D., Bloom, F., Abrams, A. and Guillemin, R. (1977). *Am. J. Psychiat., 134,* 926
Johnstone, E.C., Crow, T. and Mashiter, K. (1977). *Psychol. Med., 7,* 223
Karobath, M.I. and Leitich, H. (1975). *Proc. Nat. Acad. Sci. USA, 71,* 2915
Kastin, A.J., Ehrensing, R.H., Schalch, D.S. and Anderson, M.S. (1972). *Lancet ii,* 740
Kato, Y., Nakai, Y., Imura, H., Chihara, K. and Ohgo, S. (1974). *J. Clin. Endocrinol. Metab., 38,* 695
Kolakowska, T., Orr, M., Gelder, M., Heggie, M., Wiles, D. and Franklin, M. (1979). *Brit. J. Psychiat., 135,* 352
Kosterlitz, H. and Hughes, J. (1977). *Brit. J. Psychiat., 130,* 298
Krieger, D.T. (1973). *Mt. Sinai J. Med., 40,* 302
Lal, S., de la Vega, C.E., Sourkes, T.L. and Friesen, H.G. (1973). *J. Clin. Endocrinol. Metab., 37,* 719
Lal, S., Ettigi, P., Martin, J.B., Tollis, G., Brown, G.M., Guyda, H, and Friesen, H.G. (1974). *Clin. Res., 22,* 732A
Lal, S., Guyda, H. and Bedadoroff, S. (1977). *J. Clin. Endocrinol. Metab., 44,* 766
Langer, G., Heinze, G., Reim, B and Matussek, N. (1976). *Arch. Gen. Psychiat., 33,* 1471
Langer, G., Sachar, E.J., Gruen, P.H. and Halpern, F.S. (1977). *Nature, 266,* 639
Lehmann, H., Nair, V. and Kline, N.S. (1976). *Am. J. Psychiat., 136,* 762
Lindström, L., Gunne, L. and Oest, L. (1977). *Acta Psychiat. Scand., 55,* 74
Lindström, L.H., Widerlov, E. and Gunne, L.M. (1978). *Acta Psychiat. Scand., 57,* 153
Linnoila, M., Lamberg, B., Rosberg, G., Karonene, S. and Welin, G. (1979). *Acta Psychiat. Scand., 59,* 536
Lissak, K. (1973). *Hormones and Brain Function,* (Pleum Press, New York).
MacCrimmon, D.J., Cleghorn, J.M., Brown, G.M., Blackall, M.H. and Brown, P.J. (1980). In *Proceedings of the Third Annual Meeting of the Canadian College of Neuropsychopharmacology* (Edmonton, May 12-13)
MacLeod, R.M. (1976). In *Frontiers in Neuroendocrinology,* ed. L. Martini and W.F. Ganong (Raven Press, New York). p. 169
Maeda, K., Kato, Y., Ohgo, S., Chihara, K., Yoshimoto, Y., Yamaguchi, N., Kuromaru and Imura, H. (1975). *J. Clin. Endocrinol. Metab., 40,* 501
Martin, J.B. (1973). *New Engl. J. Med., 288,* 1384
Martin, J.B., Reichlin, S., and Brown, G.M. (1977). *Clinical Neuroendocrinology,* Contemporary Neurology Series No. 14 (F.A. Davis Co., Philadelphia)
Mason, J.W. (1968). *Psychosom. Med., 30,* 576
Mason, J.W. *(1975). In Emotions — Their Parameters and Measurement,* ed. L. Levi (Raven Press, New York). p. 143
Matz, R., Rick, W., Thompson, H. and Gershon, S. (1974). *Curr. Ther. Res., 16,* 687
Meltzer, H.Y. and Fang, V.S. (1976). *Arch. Gen. Psychiat., 33,* 279
Meltzer, H., Simonovic, M., Fang, V., Pijalalamala, S. and Young, M. (1978). *Life Sci., 23,* 605
Meltzer, H.Y., Goode, D.J., Schyve, P.M., Young, M. and Fang, V.S. (1979). *Am. J. Psychiat., 136,* 1550
Miller, R.J. and Iversen, L.L. (1974). *Trans. Biochem. Soc.,* (544th Meeting, London), *2,* 256
Mishra, R.M. and Cleghorn, J.M. (1979). In *Recent Advances in Canadian Neuropsychopharmacology,* ed. P. Grof and B. Saxena (S. Karger, Basel, in press)
Müller, P.S., Henninger, G.R. and MacDonald, R.K. (1972). In *Recent Advances in the Psychobiology of Depressive Illnesses,* ed. T.A. Williams, M.M. Katz and J.A. Schields. Department of Health, Education and Welfare, Publication No. 70-9053, (Washington, DC)
Nair, N.P., Zicherman, V. and Schwartz, G. (1977). *Can. Psychiat. Ass. J., 22,* 285

Nair, N.P., Lal, S., Cervantes, P., Yassa, R. and Guyda, H. (1979). *Neuropsychobiology, 5,* 136

Ojeda, S.R. and McCann, S.M. (1973). *Neuroendocrinology, 12,* 295

Owen, F., Cross, A. and Crow, T. (1979). *Lancet ii,* 223

Palmour, R.M., Ervin, F.R., Wagemaker, M. and Cade, R. (1978). In *Endorphins in Mental Health Research,* ed. E. Usdin, W.E. Bunney, Jr. and N.S. Kline (MacMillan, New York). p. 581

Pandey, G.M., Garver, D.L., Tamminga, C., Ericksen, S., Ali, S.I. and Davis, J.M. (1977). *Am. J. Psychiat., 134,* 518

Prange, A.J. and Wilson, I. (1975). *Psychopharmacol. Bull., 11,* 22

Prange, A.J., Wilson, I.C., Lara, P.P., Wilber, J.K., Breese, G.B., Alltop, L.B., Lipton, M.A. and Hill, C. (1973). *Arch. Gen. Psychiat., 29,* 28

Prange, A., Loosen, P., Wilson, I., Meltzer, H. and Fang, V. (1979). *Arch. Gen. Psychiat., 36,* 1086

van Ree, J.M., Verhoeven, W.M.A., van Pragg, H.M. and de Wied, D. (1978). In *Characteristics and Function of Opioids,* ed. J.M. van Ree and S. Terenius (Elsevier/North-Holland Biomedical Press, Amsterdam). p. 62

Reichlin, S., Jackson, I.M.D., Seyler, L.E. and Grimm-Jorgenson, Y. (1974). In *Frontiers in Neurology and Neurosience Research,* ed. P. Seeman and G.M. Brown (Neuroscience Institute, University of Toronto). p. 48

Rotrosen, J., Angrist, B. and Paquin, J. (1978). *Psychopharmacol. Bull. 14,* 14

Rotrosen, J., Angrist, B., Gershon, S., Paquin, J., Branchey, L., Oleshansky, M., Halpern, F. and Sachar, E.J. (1979). *Brit. J. Psychiat., 135,* 444

Rubin, R.T., Young, W.M. and Clark, B.R. (1968). *Psychosom. Med., 30,* 162

Sachar, E.J. (1967). *Arch. Gen. Psychiat., 17,* 544

Sachar, E.J. (1975a). *Hosp. Practice, 10,* 49

Sachar, E.J. (1975b). *Prog. Brain Res., 42,* 81

Sachar, E.J., Mason, J.W., Kolmer, H.S. and Artess, K.L. (1963). *Psychosom. Med., 25,* 510

Sachar, E.J., Finkelstein, J. and Hellman, L. (1971). *Arch. Gen. Psychiat., 25,* 263

Sachar, E.J., Frantz, A.G., Altman, N. and Sassin, J. (1973a). *Am. J. Psychiat., 130,* 1362

Sachar, E.J., Hellman, L., Roffwarg, H.P., Halpern, F.S., Fukushima, D.K. and Gallagher, T.F. (1973b). *Arch. Gen. Psychiat., 28,* 19

Schally, A.V. (1978). *Science, 202,* 18

Schally, A.V., Arimura, A. and Kastin, A.J. (1973). *Science, 179,* 341

Schildkraut, J.J. (1965). *Am. J. Psychiat., 122,* 509

Schildkraut, J.J. (1974). *Ann. Rev. Med., 25,* 333

Schlesser, M.A., Winokur, G. and Sherman, B.M. (1979). *Lancet i,* 739

Scorza-Smeraldi, R., Smeraldi, E., Fabio, G., Bellodi, L., Sacchetti, E. and Rugarli, C. (1975). *Tissue Antigens, 9,* 163

Sedval, G. and Nyback, H. (1973). *Isr. J. Med. Sci., 9,* (suppl. x), 24

Shopsin, B. and Gershon, S. (1971). *Arch. Gen. Psychiat., 24,* 320

Shopsin, B., Wilk, S., Sathanathan, G., Gershon, S. and Kavis, K. (1974). *J. Nerv. Ment. Dis., 158,* 369

Siris, S., Van Kammen and de Fraites, E. (1978). *Psychopharmacol. Bull., 14,* 11

Smith, R.C., Tamminga, C., Haraszti, J., Pandey, G. and Davis, J. (1977). *Am. J. Psychiat., 134,* 763

Stancer, H.C., Quarrington, B., Cookson, B.A., Brown, G.M., Bonkalo, A. and Lyall, W.A.L. (1969). *Arch. Gen. Psychiat., 20,* 290

Stevens, J. (1973). *Arch. Gen. Psychiat., 29,* 177

Stevens, J. (1979). *Trends Neurosci., 2,* 102

Stokes, P.E. (1972). In *Recent Advances in the Psychobilogy of Depressive Illness,* ed. T.A. Williams, M.M. Katz and J.A. Schield, US DHEW Publication Nos. 70-9053, (Washington, DC). p. 189

Stubbs, W.A., Jones, A., Edwards, C.R.W., Delitala, G., Jeffcoate, W.J., Ratter, S.J., Besser, G.M., Bloom, S.R. and Alberti, K.G.M.M. (1978). *Lancet ii,* 1225

Takahara, J., Arimura, A. and Schally, A.V. (1974). *Endocrinology, 95,* 464

Takahashi, S., Kondo, H., Yoshimura, M. and Ochi, Y. (1974a). In *Psychoneuroendocrinology,* ed. N. Hatolani (S. Karger, Basel). p. 32

Takahashi, S., Kondo, H., Yoshimura, M. and Ochi, Y. (1974b). *Fol. Psychiat. Neurol., 28,* 355

Tamminga, C.A., Smith, R.C., Pandey, G., Frohman, L.A. and Davis, J.M. (1977). *Arch. Gen. Psychiat., 34,* 1199

Tamminga, C.A., Schaffer, M.H., Smith, R.C. and Davis, J.M. (1978). *Science, 200,* 567

Valverde, R.C., Chieffo, V. and Reichlin, S. (1972). *Endocrinology, 91,* 982

Van Loon, G.R., Ho, D. and Kim, C. (1980). *Endocrinology, 106,* 76

Verebey, K., Volavka, J. and Clouet, D. (1978). *Arch. Gen. Psychiat., 35,* 877

Versteeg, D.H., Dekleot, E.R. and de Wied, D. (1979). *Brain Res., 179,* 85

Vijayan, E., and McCann, S. (1980). *Life Sci., 26,* 321

Wagemaker, H. and Cade, R. (1977). *Am. J. Psychiat., 134,* 684

Weitzman, E.D., Fukushima, D., Nogeire, C., Roffwarg, H.P., Gallaghar, T.F. and Hellman, L. (1971). *J. Clin. Endocrinol., 33,* 14

Wiles, D.H., Kolakowska, T., McNeilly, A.S., Mandelbrote, B.M. and Gelder, M.G. (1976). *Psychol. Med., 6,* 407

Winokur, G. (1974). *Int. Pharmacopsychiat., 9,* 5

17 CONTEMPORARY NEUROENDOCRINE RESEARCH STRATEGIES AND METHODOLOGIES IN PSYCHIATRY

Russell E. Poland and Robert T. Rubin

TABLE OF CONTENTS

I.   INTRODUCTION TO NEUROENDOCRINE REGULATION 365

II. METHODOLOGIES FOR NEUROENDOCRINE INVESTIGATION 366

III. STATISTICAL ANALYSIS 370
     A. Repeated-measures data 370
     B. Statistical-power analysis 372

IV. APPLICATION OF RESEARCH STRATEGIES AND METHODOLOGIES TO CURRENT PSYCHIATRIC RESEARCH 373

V.   SUMMARY AND CONCLUSIONS 376

REFERENCES 378

I. INTRODUCTION TO NEUROENDOCRINE REGULATION

As a consequence of the pioneering work of Scharrer and Scharrer (1940) and Harris (1960), it is now appreciated that the secretion of anterior pituitary hormones is prominently regulated by the central nervous system (CNS). Small polypeptide and catecholamine hormones are secreted into the portal vessels of the pituitary stalk from neurosecretory cells in the external layer of the median eminence of the basal hypothalamus and carried to the anterior pituitary to stimulate or inhibit the synthesis of pituitary hormones and their release into the systemic circulation (Guillemin, 1978; Schally, 1978). Various putative neurotransmitters such as dopamine, norepinephrine, serotonin, acetylcholine, GABA and histamine (Weiner and Ganong, 1978), as well as some of the recently described CNS peptides such as endorphins, bombesin, substance P and vaso-active intestinal peptide, appear to be involved in the regulation of hypothalamic neuropeptide release (Hökfelt *et al.,* 1978). These neurotransmitters and/or neuromodulators may act singly to increase or decrease the secretion of a specific pituitary hormone, but more likely a number of different transmitters are involved in a complex web of interactions, with the final response of the pituitary being determined by their net influence on the secretion of hypothalamic releasing and inhibiting factors into the hypophysial portal blood.

The secretion rates of anterior pituitary hormones depend on a balance between open-loop CNS driving mechanisms and closed-loop negative feedback to the pituitary and brain by the pituitary hormones and their target-organ hormones (Rubin, 1977). Since pituitary-hormone secretion is regulated by the CNS in a major way, alterations in the secretory patterns of pituitary hormones can reflect changes in CNS activity. Thus, changes in the dynamics of pituitary-hormone secretion, as determined by the measurement of blood hormone levels and secretion patterns, may be used as a neuroendocrine window to reflect alterations of CNS function.

For example, because both the transition between wakefulness and sleep and the shifts between sleep stages are periodic and easily determined by electro-physiological recording techniques, and because they represent major changes in the functional activity of the CNS, sleep has been used as a paradigm for the study of CNS open-loop regulation of endocrine function. Numerous studies have shown that some hormones have important temporal secretory relation-ships to the sleep-wake cycle. Growth hormone (GH) appears to be the only pituitary hormone which is tightly linked to a specific sleep stage, being secreted only after the onset of slow-wave (stage 3-4) sleep. The secretion of several other pituitary hormones such as prolactin (PRL), luteinizing hormone (LH) during

365

puberty and thyroid-stimulating hormone (TSH) bears a close relationship to sleep onset, but not to any specific sleep stage. Other pituitary and target-gland hormones (corticotropin, cortisol, and aldosterone) show clear circadian rhythms, but these rhythms bear little specific relationship to the mechanism of sleep. Still other hormones (LH in adults and follicle-stimulating hormone, FSH) appear to have no circadian rhythm; nor do their secretory episodes bear any relationship to sleep staging (Boyar, 1978).

Current psychoendocrine research has focused on the endocrine alterations associated with diseases such as schizophrenia, anorexia nervosa and endogenous depression. It is possible that other changes in CNS activity may produce alterations in pituitary-hormone secretion both during the day and during nocturnal sleep. The endocrine disturbances associated with some types of mental illness are being elucidated for diagnostic purposes, such as subtyping endogenous depression, which might aid in differential diagnosis and possibly in the prediction of treatment response (Rubin *et al.*, 1979). Similarly, the PRL secretory response to neuroleptic drugs is being investigated as an *in vivo* bioassay for the dopamine-blocking activity of these drugs, for the possible prediction of clinical drug dosages necessary to achieve optimal therapeutic response (see Section IV).

II. METHODOLOGIES FOR NEUROENDOCRINE INVESTIGATION

The two classical techniques for investigating neuroendocrine function are measurements of basal hormone secretion and hormone responses to perturbation tests. These two complementary approaches are used to gain a better understanding of neuroendocrine regulation under conditions such as waking, sleep, endocrine, neurological and psychiatric illness and various stress situations. A basic methodology for many neuroendocrine studies is the collection of blood samples for measurement of circulating hormone levels. Serial blood sampling may often continue for 24-hours or longer. However, certain inherent characteristics of neuroendocrine systems require that special attention be paid to blood-drawing procedures. One characteristic is that some hypothalamic-pituitary-end-organ axes are very responsive to stress, so that the secretion patterns of several pituitary hormones may change dramatically as a consequence of the anticipation and insertion of an intravenous catheter, a procedure frequently used for neuroendocrine studies. Thus, depending upon the hormone system studied, several hours should elapse after catheter insertion prior to initiation of blood sampling. For example, PRL, a stress-responsive hormone, has a metabolic halflife of 20-30 minutes. Based upon first-order metabolism kinetics, it takes from one and a half to two hours (four to five halflives) for PRL levels to return to baseline after a single episode of hormone secretion (Adler *et al.*, 1975). Similarly, cortisol has a metabolic halflife of 60-90 minutes, so that serum levels of cortisol will not return to baseline until three to five hours after adrenal cortical activation occurs. Many laboratory animals, such as mice and rats, are also particularly responsive to minor stresses and external stimuli. In fact, handling an animal or the mere entrance of the investigator into the animal room can cause rapid and dramatic changes in circulating hormone levels (Krulich *et al.,* 1974; Döhler *et al.,* 1977; Poland *et al.,* 1979). Thus, stress effects in both human and animal studies must be controlled.

In addition to the need for an appropriate adaptation period, as discussed above, additional methodological considerations include the frequency and duration of blood sampling. While the specifics depend on the individual experimental protocols, there are certain fundamental aspects to be considered in all studies. As a consequence of the episodic nature of hormone secretion, as well as the short metabolic halflives of many pituitary hormones, it is important that sampling be frequent enough to achieve a precise estimate of average circulating hormone levels for baseline and for response purposes (Santen and Bardin, 1973; Goldzieher *et al.,* 1976). Similarly, the duration of sampling must be long enough to adequately characterize basal hormone levels as well as the entire hormone response to a perturbation stimulus. Figure 17.1 shows the serum PRL changes from baseline in individual subjects after intravenous injections of 0.5 mg of haloperidol and 25 or 500 μg of thyrotropin-releasing hormone (TRH), two different stimuli to PRL secretion (Hays and Rubin, 1979). While the serum PRL responses to the two doses of TRH return to baseline within two to three hours, the PRL levels after haloperidol administration remain elevated for as long as seven hours in some of the subjects. We have recently determined that sampling PRL for three hours after haloperidol administration correlates well with the full seven-hour PRL profile ($r = +0.9$; Rubin and Forster, unpublished observations). It is first necessary to determine the entire profile of hormone response and then establish satisfactory minimum sampling periods. However, it should be emphasized that delineation of the pattern of hormone response still requires sampling for a full seven to eight hours. A constraint on the frequency and duration of sampling is that the amount of blood that can be withdrawn should be kept below 300 ml per 24 hours, to avoid hormone changes secondary to blood loss. Therefore, there must be a compromise among the length of the sampling period, the frequency of sampling within that period and the amount of blood withdrawn in each sample.

Recently, investigators have begun measuring steroid-hormone levels in saliva as a noninvasive reflection of serum steroid-hormone levels. Saliva can be considered an ultrafiltrate of plasma, and certain saliva steroid-hormone levels (cortisol, progesterone, testosterone) have been shown to adequately reflect serum-hormone concentrations (Walker *et al.,* 1978, 1979; Türkes *et al.,* 1979). Many steroid hormones in blood circulate in both protein-bound and free forms, and it is only the free fraction which is presumed to be physiologically active. Saliva levels may reflect the free, physiologically relevant hormone levels, and thus may actually provide more information than the measurement of total blood hormone concentrations. For example, 24-hour saliva cortisol levels

Figure 17.1: Serum PRL changes from baseline in five individual subjects after intravenous injections of 0.5 mg of haloperidol (left) and 25 or 500 μg or TRH (right).

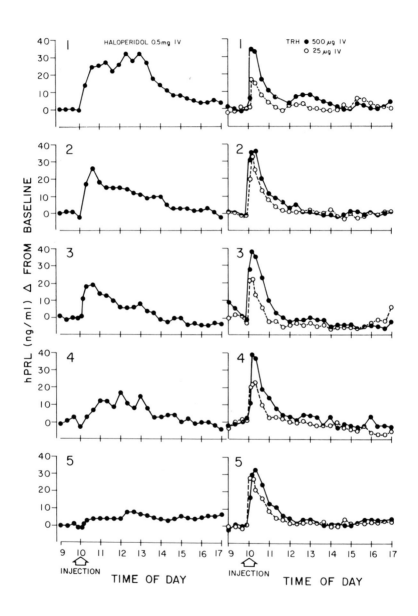

Source: Hays and Rubin (1979).

show a marked circadian variation paralleling blood cortisol levels, the saliva cortisol levels being similar to the free cortisol levels in serum, which are 5-10% of the
total serum cortisol concentration (Walker *et al.*, 1978). While free hormone levels
in blood can be determined, the methodologies for these determinations are cum-
bersome and time-consuming, particularly for studies in which multiple samples
are collected.

   The final methodologic variable is the time-of-day effect. Because many hormones show prominent circadian (and in some instances ultradian) rhythms, baseline hormone values can vary several-fold depending upon the time of day the
samples are obtained. Circulating cortisol, testosterone, TSH and PRL concentrations all have circadian rhythms. Sachar *et al.* (1973) provided one example of
the importance of considering the circadian variation of serum hormone levels by
obtaining a complete 24-hour profile of hormone concentration. These investigators showed that six psychotically depressed patients had substantially greater
serum levels of cortisol during their illness than after their recovery. Blood samples
were taken every 20 minutes for 24 hours; it was apparent that the greatest difference in cortisol levels between illness and recovery occurred during the daytime
and the early evening hours, when cortisol secretion normally was minimal. During
the early morning period of maximal cortisol secretion, differences in the circulating hormone levels between illness and recovery were less apparent, and sampling
during these hours only might not have yielded discernible differences. Figure 17.2
shows similar data from our laboratory (Rubin *et al.*, 1980a). The circadian rhythm
of serum cortisol was elucidated in 15 endogenously depressed patients and eight
normal age, sex and endocrine-matched controls. While the circadian variation of
serum cortisol is present in both groups, the circulating cortisol levels are consistently higher in the depressed patients. However, if the cortisol samples had been
obtained only during the early morning hours, significant differences might not
have been observed.

   In addition to the rhythmic nature of basal hormone secretion, there are examples in the literature indicating that a number of hormone perturbation tests
employed as diagnostic tools to assess abnormalities in endocrine function are susceptible to circadian variation. These tests include intravenous administration of
various doses of TRH, gonadotrophin-releasing hormone (GnRH), arginine, L-
DOPA, insulin and oral dexamethasone, followed by blood sampling to assess
hormonal responses to these challenges. There are circadian variations in dexamethasone suppression of the serum cortisol rhythm (Nichols *et al.*, 1965; Krieger
et al., 1971), the GH response to insulin-induced hypoglycemia (Takebe *et al.*,
1969; Nakagawa, 1971; Gibson *et al.*, 1975), and the LH, FSH and testosterone
responses to GnRH (Schwarzstein *et al.*, 1975). In addition, results from our
laboratory indicate that TSH and PRL responses to TRH during sleep are much
greater than the responses during any other time of the day when subjects are
awake (Peters *et al.*, 1981). Thus, chronoendocrinological apsects of experimental protocols are important, not only because peaks and troughs of serum
hormone levels affect the baseline control values against which experimental data

Figure 17.2: Serum cortisol levels in 15 endogenously depressed patients and eight normal matched controls studied for 48 hours. Dexamethasone (1.0 mg, orally) was given at 11.0 p.m. on the second night, and blood samples were taken eight, 16 and 24 hours later.

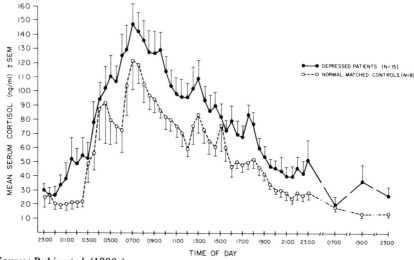

Source: Rubin *et al.* (1980a).

are compared, but also because hormonal responses to various perturbation tests may change at different times of the day and night. Investigators have only recently begun to appreciate that time of day can have a significant effect on the magnitude and duration of hormonal responses to various perturbation tests.

III. STATISTICAL ANALYSIS

A. Repeated-measures Data

As noted above, the design of many neuroendocrine experiments incorporates the use of the subject or patient more than once, and often entails the collection of more than one blood sample in each experimental testing session. The statistical analysis of data obtained by this approach requires certain precautions. For example, Figure 17.3 shows the mean PRL responses of five normal subjects, studied on three separate occasions, to haloperidol (0.25 and 0.5 mg i.m.) and saline. On each occasion 28 blood samples were taken from each subject, at approximately 30-minute intervals, over a nine-hour period (four samples prior to drug or saline injection and 24 samples after injection). Thus, the protocol is both a repeated-measures design across subjects (each subject was studied more than once) and a repeated-measures design across time (more than one blood sample was taken from each subject at each session) (Rubin *et al.*, 1976). The statistical analysis of such correlated repeated-measures data often requires multivariate analytic techniques that employ sophisticated computations (Rubin and Poland, 1976). However,

Figure 17.3: Daytime plasma **PRL** levels, averaged across five subjects, for the three experimental conditions.

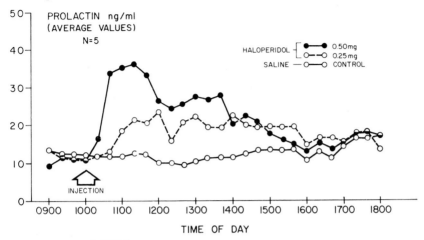

Source: Rubin *et al.* (1976).

this may not always be necessary if multivariate data can be reduced to univariate data. For example, the integrated PRL responses to haloperidol and saline treatment were calculated by subtracting the average of the four preinjection PRL values from the average of the 24 postinjection values, yielding a single PRL response value for each subject during each treatment and eliminating the effect of repeated (and thus correlated) samples. Many investigators then would proceed to analyze such data by a two-way analysis of variance (ANOVA) for independent measures. This is not correct, because the repeated testing sessions on the same subject (treatments) are still correlated, and thus violate the assumptions underlying the usual independent-measures ANOVA. A downward adjustment of the degrees of freedom for the *F* ratio is required, the most conservative being the Greenhouse-Geisser technique, in which the degrees of freedom are reduced to one and *K*-one, *K* being the number of independently selected subjects (Winer, 1971). This correction assumes the greatest possible interdependence among the correlated measures.

A reasonable scheme for statistical analysis for repeated measures data is to evaluate the *F* ratio for treatment effects with the usual degrees of freedom for independent measures; if it is nonsignificant, one can stop here. However, if the *F* ratio for treatment is significant, it should be reevaluated with the Greenhouse-Geisser correction for the degrees of freedom. If the *F* ratio is still significant with this conservative test, one again may stop. However, if it is nonsignificant, further evaluation must be done; Lana and Lubin (1963) suggest the use of Hotelling's multivariate analysis of variance, in which the actual correlation matrix is computed. It should be emphasized that a significant overall ANOVA *F* ratio does not give any information about where in the data the significant differences lie. Further investigation into the location of significant differences requires appropriate multiple-comparison techniques (Dunn and Clark, 1974).

B. Statistical-power Analysis

By setting the alpha (α) probability level at $p = 0.05$ or 0.01, most investigators are extremely cautious about reporting a significant effect when significance really does not exist (a type-I statistical error). In contrast, investigators are generally much less cautious about committing a type-II statistical error, i.e., not finding a significant effect when one really does exist. The use of statistical-power analysis allows the investigator to control the probability of committing a type-II error (i.e., erroneously failing to reject the null hypothesis) (Cohen, 1977). There are four parameters involved in this analysis: (1) power (the probability of finding a true significant difference); (2) the alpha (α) significance level (usually $p < 0.05$ by convention); (3) a dimensionless estimate of the magnitude of the measured effect; and (4) sample size. Knowledge of any three of these four variables permits determination of the fourth (Cohen, 1977). A power of 0.8 for example, means an 80% chance of rejecting the null hypothesis, given a certain effect size, alpha level and sample size.

Cohen (1962) concluded that, in many experiments reported in the *Journal of Abnormal and Social Psychology,* the number of observations was so small that tests of significance had relatively little power, even when the differences between groups were moderately large. Similar conclusions were reached on negative results from randomized clinical trials (Freiman *et al.,* 1978) and on the probability of finding a circadian rhythm of various anterior pituitary hormones in rats (Poland *et al.,* 1980). Thus, because of small sample sizes, many study designs are initially biased against appropriately rejecting the null hypothesis.

Of the four parameters involved in power analysis, sample size is the one most frequently altered. An increase in sample size increases the power of statistical tests, and thus decreases the chance of a type-II statistical error. Similarly, an increase in the magnitude of the measured effect or an increase in the acceptable alpha significance level (for example from $p = 0.05$ to $p = 0.10$) also increases the power of statistical testing. However, changes in effect magnitude are usually outside the investigator's control, and increasing the acceptable probability value enhances the chances of a type-I error. Therefore, increasing sample size is the only practical maneuver.

We have applied these principles of statistical-power analysis to determine the sample size necessary to consistently discern hormone rhythms in adult male rats (Poland *et al.,* 1980). Figure 17.4 shows trough and peak serum TSH values of individual animals and their means and standard errors. As mentioned earlier, pituitary hormones are secreted in a pulsatile, episodic fashion, and they have relatively short metabolic halflives. Therefore, a high degree of variability among animals in circulating hormone levels is usually found. In this example, the range of serum TSH values at 7.0 a.m. was 112-1190 ng/ml and at 3.0 p.m. it was 100-1404 ng/ml; the mean peak hormone value (721 ng) was 90% higher than the

Figure 17.4: Trough and peak serum TSH levels obtained from individually sacrificed adult male rats. Mean trough and peak values were significantly different at the $p = 0.0025$ level. Each time-point represents the mean (±SEM) of 34-36 hormone values.

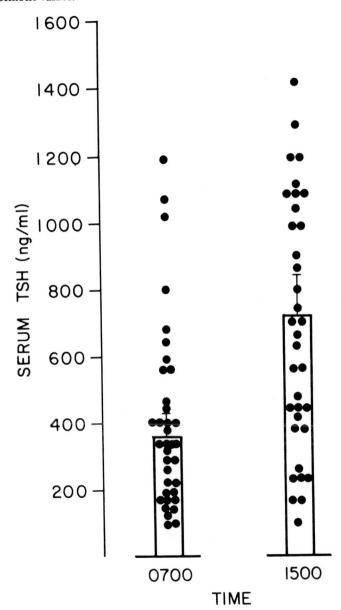

mean trough (376 ng) value. This large within-group variability made it difficult to delineate small between-group changes in circulating hormone levels. However, power analysis enabled us to determine the minimum number of animals necessary at each time-point to consistently reproduce our initial significant findings (Cohen, 1977). Because of the skewed distributions of the hormone values (Figure 17.4), the data were log-transformed in order to achieve Gaussian distributions and to equalize the variances. Using the formulas and tables provided by Cohen (1977), the power analysis indicated that with a preselected power level of 0.8 (80% chance of correctly rejecting the null hypothesis) and an alpha significance level of 0.05, approximately 20 animals would be needed at each of the two time-points to detect a statistically significant time-of-day difference in replication experiments. Both the magnitude of the mean TSH difference and the variability of the individual circulating hormone values in this experiment are similar to those previously reported by other investigators (Männistö *et al.*, 1978; Rookh *et al.*, 1979). However, in many of the previous experiments, only five to ten animals were sampled at each of the time-points, so that a much smaller chance existed for finding a statistically significant difference in circulating TSH levels (weaker statistical power).

The same principles of statistical-power analysis can be applied to other experimental protocols, including repeated-measures designs. With a small amount of preliminary data regarding the variability usually found for values of the particular hormone under consideration, and the expected magnitude of the effect, the appropriate sample size for prospective experiments of many kinds can be determined, thus importantly minimizing the chances of a type-II statistical error (Cohen, 1977).

IV. APPLICATION OF RESEARCH STRATEGIES AND METHODOLOGIES TO CURRENT PSYCHIATRIC RESEARCH

Our laboratory is presently involved in two major areas of psychoneuroendocrine research which utilize the previously discussed research strategies and methodologies. These areas of research are: (1) the investigation of the neuroendocrine abnormalities in endogenomorphic depression and the relationship of these endocrine changes to the severity of the disease and to treatment response; and (2) the usefulness of the PRL response to low-dose neuroleptic administration as a predictor of potential clinical responsivity to these drugs. One of the most frequently described abnormalities in endogenously depressed patients is the alteration of the hypothalamic-pituitary-adrenal axis (Carroll *et al.*, 1976). We have studied this endocrine system using the two classic approaches: measurement of basal circulating hormone levels and hormone response to perturbation tests. Figure 17.2 indicates that, under basal conditions, the average circulating levels of serum cortisol are consistently higher in endogenously depressed patients compared to matched controls. The circadian rhythm of serum cortisol is maintained in the depressed patients, but the values, particularly during the night and late afternoon, are higher.

The patients were administered an overnight dexamethasone suppression test. At 11.0 p.m. on the second night of the study, each patient (and control) was given dexamethasone (1.0 mg, orally). Blood samples were taken eight, 16 and 24 hours later. As can be seen in Figure 17.2, the cortisol levels were suppressed in both groups at 7.0 a.m. (eight hours after dexamethasone), but by 3.0 p.m. the depressed patients escaped from the dexamethasone suppression, while the control subjects showed suppressed cortisol values at both 3.0 p.m. and 11.0 p.m. The cortisol escape from dexamethasone suppression does not appear to be a consequence of altered dexamethasone metabolic halflife (Rubin *et al.*, 1980a), suggesting that there are differences in the CNS and/or pituitary sensitivity to dexamethasone in the depressed patients. We are currently investigating other endocrine axes such as the pituitary-thyroid and the pituitary-gonadal systems, and examining the interrelationships among these hormone systems in depressive illness.

The second major area of our research is the investigation of the PRL response to low-dose neuroleptic administration as a potential indicator of a subject's CNS sensitivity to these drugs. Because the regulation of PRL secretion is primarily under inhibitory dopaminergic influence (Rubin and Hays, 1980), and because it is believed that some forms of schizophrenia also involve alterations in central dopaminergic function, possibly in the mesolimbic dopamine system (Stevens, 1973), it is conceivable that the sensitivity of the PRL secretion response after low-dose neuroleptic administration may reflect CNS sensitivity to these drugs. PRL secretion may thus be an *in vivo* bioassay to provide information for a given patient's potential responsiveness to these psychopharmacological agents.

Because clinically effective doses of neuroleptics often stimulate PRL secretion maximally, we (Rubin and Hays, 1980) and others (Langer *et al.*, 1977; Gruen *et al.*, 1978; Busch *et al.*, 1979), have attempted to study neuroleptic sensitivity using subclinical low-dose paradigms. Figure 17.1 shows the serum PRL changes from baseline in individual subjects after intravenous injections of 0.5 mg of haloperidol and 25 or 500 μg of TRH. While the PRL response to TRH (peak level and integrated response) was dose-related and similar across subjects particicuarly for the high dose of TRH, the PRL response to haloperidol was highly variable across subjects and did not correlate with either the low or the high TSH response to TRH. Subject 1 responded to the i.v. haloperidol injection almost immediately, showing a somewhat biphasic PRL response with values reaching a peak change from baseline of approximately 40 ng/ml and returning to baseline five to six hours later. In contrast, subject 5 barely showed any PRL response at all, although the actual baseline values were similar in all subjects, as were serum haloperidol pharmacokinetics, as measured by radioimmunoassay (Hays and Rubin, 1979; Rubin *et al.*, 1980b).

Figure 17.5 shows the change of PRL from baseline in three subjects treated with haloperidol (0.5 mg, i.v.) on two separate occasions at least one week apart. These results show the good reproducibility of the PRL response within each subject. The within-subject variability of PRL secretion (Figure 17.5) was much less than the PRL response to haloperidol observed between subjects (Figure 17.1), suggesting true interindividual differences rather than test-retest variability (Hays and Rubin, 1979). Therefore, our present working hypothesis is that there are interindividual differences in PRL-secreting sensitivity to haloperidol. Since these differences in sensitivity do not appear to be a consequence of pharmaco-kinetic parameters or general differences in the capacity of the pituitary to secrete PRL, as judged by the TRH response, the different PRL responses to haloperidol may reflect real differences in receptor sensitivity to this drug. If this is true, it may be feasible to use this difference in PRL sensitivity as an *in vivo* bioassay to indirectly measure CNS sensitivity to neuroleptic drugs. We then might predict the dosage regimen of neuroleptic medication for each patient and possibly even choose the appropriate type of neuroleptic for maximal therapeutic effect. Such studies are presently in progress.

V. SUMMARY AND CONCLUSIONS

Because the CNS appears to be a predominant controller of neuroendocrine systems and the hypothalamic-pituitary axes reflect changes in CNS function, the secretory patterns of pituitary hormones may serve as a marker for various aspects of CNS dysfunction. The performance of neuroendocrine studies requires that certain methodological aspects be controlled. The aspects include: minimiz-ing stress effects, choosing adequate sampling intervals and durations, perform-ing appropriate data analysis, particularly for experiments with repeated measures designs and using sample sizes sufficiently large to optimize the chance of discerning statistically significant differences. Regarding the statistical analysis, it should be emphasized that the relation between statistical significance and physiological significance is variable. Large statistically significant differences may have minimal physiological importance; conversely, small statistically significant differences may have very important physiological or pathological consequences. Thus, statistical analysis is a tool used to discern effects above the background noise of random variation, but the results of the analysis do not place any value on the importance of the observed effects. What is important from a methodological point of view is that type-I and type-II statistical errors should be minimized, so that the physiological interpretation of the results is based upon accurate data.

Figure 17.5: Changes in serum PRL concentrations from baseline and serum haloperidol concentrations after the intravenous injection of haloperidol (0.5 mg) on two separate occasions in each of three subjects.

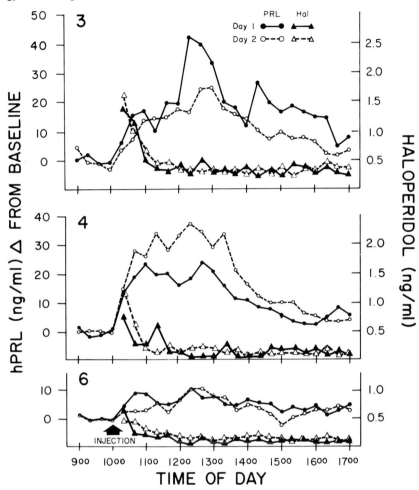

Source: Hays and Rubin (1979).

ACKNOWLEDGEMENTS

This work was supported in part by NIMH Grants MH 29491, MH 28380 and MH 34471, NIMH Research Scientist Development Award MH 47363 (to RTR), NICHD Grant HD 09277, Office of Naval Research Contract N00014-77-C-0245 and NIH Clinical Research Center Grant RR 00425.

REFERENCES

Adler, R.A., Noel, G.L., Wartofsky, L. and Frantz, A.G. (1975). *J. Clin. Endocrinol. Metab.*, *41*, 383

Boyar, R.M. (1978). In *The Hypothalamus*, ed. S. Reichlin, R.J. Baldessarini and J.B. Martin (Raven Press, New York). p. 373

Busch, D.A., Fang, V.S. and Meltzer, H.Y. (1979). *Psychiat. Res.*, *1*, 153

Carroll, B.J., Curtis, G.C. and Mendels, J. (1976). *Arch. Gen. Psychiat.*, *33*, 1039

Cohen, J. (1962). *J. Abnormal Social Psychol.*, *65*, 145

Cohen, J. (1977). *Statistical Power Analysis for the Behavioural Sciences.* (Academic Press, New York).

Döhler, K.D., Gärtner, K., von zur Mühler, A. and Döhler, U. (1977). *Acta Endocrinol.*, *86*, 489

Dunn, O.J. and Clark, V.A. (1974). *Applied Statistics: Analysis of Variance and Regression.* (Wiley, New York).

Freiman, J.A., Chalmers, T.C., Smith, H. Jr. and Kuebler, R.R. (1978). *N. Engl. J. Med.*, *299*, 690

Gibson, T., Stimmler, L., Jarret, R.J. Ruthland, P. and Shiu, M. (1975). *Diabetologia*, *11*, 83

Goldzieher, J.W., Dozier, T.S., Smith, K.D. and Steinberger, E. (1976). *J. Clin. Endocrinol. Metab.*, *43*, 824

Gruen, P.H., Sachar, E.J., Langer, G., Altman, N., Leifer, M., Frantz, A. and Halpern, F. (1978). *Arch. Gen. Psychiat.*, *35*, 108

Guillemin, R. (1978). *Science*, *202*, 18

Harris, G.W. (1960). In *Handbook of Physiology, Section 1, Neurophysiology*, Vol. 2, ed. H.W. Magoun (American Physiological Society, Washington, DC)

Hays, S.E. and Rubin, R.T. (1979). *Psychopharmacology*, *61*, 17

Hökfelt, T., Elde, R., Fuxe, K., Johansson, O., Ljuangdahl, H., Goldstein, M., Lift, R., Efendic, S., Nilsson, G., Terenius, L., Ganten, D., Jeffcoate, S.L., Rehfeld, J., Said, S., Perez de la Mora, M., Possani, L., Tapia, R., Teran, L. and Palacios, R. (1978). In *The Hypothalamus*, ed. R. Reichlin, R.J. Baldessarini and J.B. Martin (Raven Press, New York). p. 69

Krieger, D.T., Allen, W., Rizzo, F. and Krieger, H. (1971). *J. Clin. Endocrinol. Metab.*, *32*, 266

Krulich, L., Hefco, E., Illner, P. and Read, B. (1974). *Neuroendocrinology*, *16*, 293

Lana, R.E. and Lubin, A. (1963). *Educ. Psychol. Measur.*, *23*, 729

Langer, G., Sachar, E.J., Halpern, F.S., Gruen, P.H. and Solomon, M. (1977). *J. Clin. Endocrinol. Metab.*, *45*, 996

Männistö, P.T., Pakkanen, J., Ranta, T., Koivusalo, F. and Leppäluoto, J. (1978). *Life Sci.*, *23*, 1343

Nakagawa, K. (1971). *J. Clin. Endocrinol. Metab.*, *33*, 854

Nichols, T., Nugent, C.A. and Tyler, F.H. (1965). *J. Clin. Endocrinol. Metab.*, *25*, 343

Peters, J.A., Santa-Cruz, F., Tower, B.B. and Rubin, R.T. (1981). *J. Clin. Endocrinol. Metab.* (in press)

Poland, R.E., Weichsel, M.E., Jr. and Rubin, R.T. (1979). *Horm. Metab. Res.*, *11*, 222

Poland, R.E., Rubin, R.T. and Weichsel, M.E. Jr. (1980). *Psychoneuroendocrinology*, *5*, 209

Rookh, H.V., Azukizawa, M., DiStefano, J.J., Ogihara, T. and Hershman, J.M.(1979).*Endocrinology, 104,* 851

Rubin, R.T. (1977). In *Neuroregulators and Psychiatric Disorders*, ed. E. Usdin, D.A. Hamburg and J.D. Barchas (Oxford University Press, New York). p. 233

Rubin, R.T. and Poland, R.E. (1976). *Psychoneuroendocrinology, 1,* 281

Rubin, R.T. and Forster, B. (1980). *Commun. Psychopharmacol., 4,* 41

Rubin, R.T. and Hays, S.E. (1980). *Psychoneuroendocrinology, 5,* 121

Rubin, R.T., Poland, R.E., O'Connor, D., Gouin, P.R. and Tower, B.B. (1976). *Psychopharmacology, 47,* 135

Rubin, R.T., Poland, R.E. and Hays, S.E. (1979). In *Biological Psychiatry Today*, ed. J. Obiols, C. Ballús, E. Gonzáles and J. Pujol (Elsevier/North-Holland Biomedical Press, Amsterdam). p. 684

Rubin, R.T., Poland, R.E., Blodgett, A.L.N., Winston, R.A., Forster, B. and Carroll, B.J. (1980a). In *Progress in Psychoneuroendocrinology*, eds F. Brambilla, G. Racagni and D. de Wied (Elsevier/North-Holland Biomedical Press, Amsterdam). p. 223

Rubin, R.T., Tower, B.B., Hays, S.E. and Poland R.E. (1980b). In *Methods of Enzymology*, Vol. B, *Immunochemical Techniques*, (Academic Press, New York, in press)

Sachar, E.J., Hellman, L., Roffwarg, H.P., Halpern, F.S., Fukushima, D.K. and Gallagher, T.F. (1973). *Arch. Gen. Psychiat., 28,* 19

Santen, R.J. and Bardin, C.W. (1973). *J. Clin. Invest., 52,* 2617

Schally, A.V. (1978). *Science, 202,* 390

Scharrer, E. and Scharrer, B. (1940). In *The Hypothalamus*, Vol. 20 (Hafner, New York). p. 170

Schwarzstein, L., Laborde, N.P., Aparicio, N.J., Turner, D., Mirkin, A., Rodriguez, A., Lhullier, F.R. and Rosner, J.M. (1975). *J. Clin. Endocrinol. Metab., 40,* 313

Stevens, J.D. (1973). *Arch. Gen. Psychiat., 29,* 177

Takebe, K., Kunita, H., Sawano, S., Horiuchi, Y. and Mashimo, K. (1969). *J. Clin. Endocrinol. Metab., 29,* 1630

Türkes, A.O., Türkes, A., Read, G.F. and Fahmy, D.R. (1979). *J. Endocrinol., 83,* 31P

Walker, R.F., Riad-Fahmy, D. and Read, G.F. (1978). *Clin. Chem., 24,* 1460

Walker, R.F., Read, G.F. and Riad-Fahmy, D. (1979). *Clin. Chem., 25,* 2030

Weiner, R.T. and Ganong, W.F. (1978). *Physiol. Rev., 58,* 905

Winer, B.J. (1971). In *Statistical Principles in Experimental Design* wnd edn. (McGraw-Hill, New York). pp. 261, 518

18 RELEVANCE OF PLASMA AND URINE AMINE METABOLITES AS DIAGNOSTIC AND PHARMACODYNAMIC INDICATORS IN PSYCHIATRIC DISORDERS

Jerry J. Warsh, Harvey C. Stancer and Peter P. Li.

TABLE OF CONTENTS

I. INTRODUCTION   383

II. LIMITATIONS OF METABOLITES AS INDICES OF CNS
    MONOAMINE FUNCTION   383

III. UTILITY OF NORADRENERGIC METABOLITES   384
    A. CNS metabolism of NE   384
    B. Clearance of NE metabolites from the CNS   385
    C. Neuronal activity and NE metabolite production   387
    D. Validation studies of the origin of peripheral MHPG   388
    E. Clinical significance of peripheral noradrenergic metabolites 391
    F. Pharmacodynamic utility of peripheral MHPG measurements   392

IV. UTILITY OF DOPAMINERGIC METABOLITES   393

V. UTILITY OF SEROTONERGIC METABOLITES   396

VI. SUMMARY AND CONCLUSIONS   397

REFERENCES   398

I. INTRODUCTION

Hypothesized dysfunction of central-nervous-system (CNS) neurotransmitter systems (for reviews, see Goodwin and Potter, 1978; Van Praag, 1978) stimulated numerous clinical investigations entailing measures of monoamine metabolites in readily accessible body fluids, such as urine, blood and cerebrospinal fluid (CSF). To the present time, data derived through the use of such measures have provided only tenuous support to these hypotheses. This may indicate deficiencies in our concepts of the nature of these disorders, but could also reflect the insensitivity of peripheral monoamine metabolites to detect CNS change. In this chapter we critically reevaluate the utility of plasma and urine monoamine-metabolite measures as indicators either of the presumed aberrant CNS neurotransmitter function in psychiatric disorders or of the pharmacodynamic effects of psychotropic drugs.

Investigations of the norepinephrine (NE) metabolite, 3-methoxy-4-hydroxy-phenylethylene glycol (MHPG) have dominated the field, and its examination will form a major part of the ensuing discussion to illustrate the conceptual approaches used. However, the present chapter is not intended to be a selective review of MHPG, as this has been accomplished by others (Maas, 1975; DeMet and Halaris, 1979). The recent advent of suitable assays of homovanillic acid (HVA) and 3, 4-dihydroxyphenylacetic acid (DOPAC) by gas chromatography-mass spectrometry has stimulated renewed interest in the possible utility of plasma dopamine (DA) metabolites. Finally, the limited work evaluating the significance of 5-hydroxyindoleacetic acid (5HIAA) in plasma and urine will also be briefly considered.

II. LIMITATIONS OF METABOLITES AS INDICES OF CNS MONOAMINE FUNCTION

Brain monoamine metabolites such as MHPG, HVA or 5-HIAA may arise from intraneuronal as well as extraneuronal pools. However, the bulk of metabolite formation is intraneuronal, since selective manipulation of neuronal function markedly affects net brain monoamine-metabolite concentrations (Sheard and Aghajanian, 1968; Korf et al., 1973, 1976). Neuronal metabolite production and transport from the CNS is a complex function of dynamic processes influencing synthesis, release, reuptake and degradation of the respective neurotransmitters. Moreover, a variety of data demonstrate the dependence of monoamine-metabolite formation on presynaptic neuronal impulse flow. The latter is influenced by a confluence of inputs from other interacting neuronal systems through postsynaptic neuronal feedback loops, recurrent collateral innervations and autoreceptor mechanisms (Aghajanian, 1978). Finally, monoamine-metabolite transport from CNS into blood is a function of the total end-product outflow from a number of CNS regions. Thus, monoamine metabolites measured in blood or urine would only detect cumulative changes in CNS presynaptic neuronal function, and would not permit conclusions about specific factors affecting impulse flow or release.

For peripherally measured monoamine metabolites to provide valid indicators of CNS monoamine function, certain additional conditions should be satisfied. Firstly, metabolic clearance of CNS neuronal monoamines should occur preferentially through production and outflow of the major metabolite. Where significant multiple metabolites are formed, such as for DA and NE, it may be necessary to measure each one. A second condition already alluded to is that metabolite formation and turnover should bear a direct relationship to neuronal activity (i.e., firing rate and release). Thirdly, the fraction of the peripherally measured monoamine metabolite of CNS origin should be sufficiently large that net changes in CNS outflow of the metabolite can be resolved in the presence of contributions from peripheral pools. Finally, the peripheral disposition and elimination of the metabolite should be relatively free of potentially confounding metabolic, vascular or renal dynamic effects. Such effects might include, for example, interconversion of metabolites in peripheral pools, variations in metabolite excretion as a result of changes in vascular blood volume and factors affecting the renal filtration or secretion of the metabolites.

The above considerations establish a framework upon which the various monoamine metabolites may be critically examined with respect to their potential utility as indicators of CNS monoaminergic function and the action of psychotropic drugs.

III. UTILITY OF NORADRENERGIC METABOLITES

A. CNS Metabolism of NE

It is well established that CNS NE is metabolized principally through oxidative deamination and enzyamatic reduction to the glycol metabolites, MHPG and 3, 4-dihydroxyphenylethylene glycol (DHPG). The biochemical basis for this preferential metabolism lies in the higher affinity of brain aldehyde reductase compared to aldehyde dehydrogenase for the intermediate phenylglycolaldehyde derived from β-hydroxylated monoamines (Duncan and Sourkes 1974; Tipton et al., 1977). Thus, the concentrations and rate of vanillylmandelic-acid (VMA) formation from NE are markedly lower compared to the glycol metabolites and unresponsive to conditions which alter NE neuronal impulse flow in the CNS (Karoum et al., 1976; Ader et al., 1978).

While MHPG is a major CNS metabolite of NE in all animal species studied, including humans, the relative importance of DHPG formation in the metabolic clearance of NE remains to be clarified. Radioisotopic studies and recent work employing direct chemical assay of DHPG clearly suggest that DHPG formation may account for a significant fraction of NE metabolism in the brain of some animal species such as the rat and mouse (Sharman, 1969; Braestrup *et al.*, 1974; Karasawa *et al.*, 1978). Recent mass spectrometric studies in this center have demonstrated the occurrence of significant concentrations of DHPG in the cerebrospinal fluid of humans (Table 18.1), thus supporting the notion that DHPG formation may be important in CNS NE metabolism in humans as well.

Table 18.1: Occurrence of Total MHPG and DHPG in Rat, Mouse and Humans*

		Total DHPG	Total MHPG
Rat	Whole brain	99.2 ± 4.11	86.0 ± 3.70
	Cerebral cortex	67.9 ± 4.62	75.3 ± 4.57
	Hypothalamus	290 ± 19	100 ± 6.2
	Midbrain	155 ± 7.35	117 ± 6.75
	Hindbrain	151 ± 8.71	129 ± 8.37
	Cerebellum	44.7 ± 1.70	54.5 ± 2.18
	Spinal cord	68.0 ± 3.97	69.2 ± 1.95
	Plasma	2.01 ± 0.26	11.5 ± 0.47
	Urine	14.0 ± 0.76	56.5 ± 2.81
Mouse	Whole brain	17.3 ± 1.23	42.5 ± 2.06
Human	CSF	2.17 ± 0.57	9.92 ± 1.15
	Urine	0.36 ± 0.05	1.43 ± 0.20

* Values are expressed as the means ± SEM for at least six determinations (μg/g for brain tissue, ng/ml for plasma and CSF, μg per 24 hours for urine). Data were obtained by multiple ion mass fragmentography.

Although normetanephrine (NMN) is known to occur in the CNS, its importance in the metabolic clearance of NE remains obscure. It has been suggested that during conditions of increased neuronal impulse flow, released NE is preferentially metabolized to NMN and subsequently to MHPG (DeMet and Halaris, 1979). Although brain NMN concentrations are quite low compared to the glycol metabolites, this need not signify that NMN is of lesser importance in CNS NE metabolism, since low NMN levels could result from rapid turnover.

The routes of catecholamine metabolism are summarized in Figure 18.1.

B. Clearance of NE Metabolites from the CNS

Clearance of MHPG (and presumably DHPG) from the CNS is mediated by passive diffusion (Kessler *et al.*, 1976) or, after conjugation by brain phenolsulphotransferase (PST), by the probenecid-sensitive active efflux process (Meek and Neff,

Figure 18.1: Routes of catecholamine metabolism: Abbreviations used: AD, aldehyde dehydrogenase; AR, aldehyde reductase; COMT, catechol-O-methyltransferase; DA, dopamine; DBH, dopamine-β-hydroxylase; DHPG, 3, 4-dihydroxyphenylethyleneglycol; DHPGA, 3, 4-dihydroxyphenylglycol aldehyde; DOPET, 3, 4-dihydroxyphenylethanol; DOPAC, 3, 4-dihydroxyphenylacetic acid; DPA, 3, 4-dihydroxyphenylacetaldehyde; HVA, 3-methoxy-4-hydroxyphenylacetic acid; MHPG, 3-methoxy-4-hydroxyphenylethylene glycol; MHPGA, 3-methoxy-4-hydroxyphenylglycolaldehyde; MAO, monoamine oxidase; MOPET, 3-methoxy-4-hydroxyphenylethanol; 3-MT, 3-methoxytyramine; MPA, 3-methoxy-4-hydroxyphenylacetaldehyde; NE, norepinephrine; NMN, normetanephrine; VMA, vanillylmandelic acid.

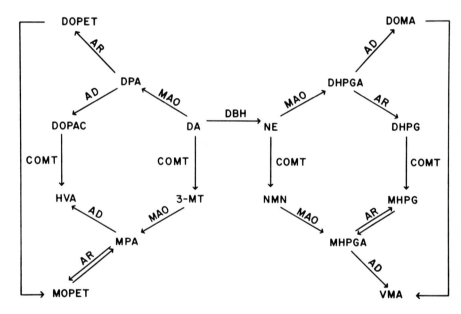

1972a). The functional significance of conjugation is not clear, although detoxification or facilitation of transport from CNS has been suggested (DeMet and Halaris, 1979). Marked species differences exist in the degree of conjugation of CNS glycol metabolites. In primates and humans, brain MHPG is almost totally free (Maas *et al.,* 1976; Karoum *et al.,* 1977a) while in rats MHPG and DHPG are principally conjugated as the sulphate esters (Meek and Neff, 1972b; Karasawa *et al.,* 1978). The extent of conjugation appears to be related to brain PST activity which in rat, for example, is ten to 20 times that in mouse, a species having little conjugated brain MHPG and DHPG (Foldes and Meek, 1974). In plasma and urine, MHPG occurs as both the sulphate and glucuronide conjugates (Karoum *et al.,* 1977a). However, MHPG-SO$_4$ is the primary CNS conjugate of species exhibiting significant conjugation of monoamine metabolites. This suggests that formation of glucuronide conjugates occurs mainly in peripheral tissues.

In clinical studies the question has been raised as to which form of peripheral MHPG best reflects CNS production. Since little conjugation of MHPG occurs in human brain, free MHPG might be the most relevant fraction to measure clinically in plasma or urine. Only 3% of MHPG in human urine exists in free form, the remaining 97% consisting of almost equal amounts of sulphate and glucuronide conjugates (Karoum *et al.*, 1977a). However, as approximately 26% of plasma MHPG exists in free form (Karoum *et al.*, 1977a), the latter may bear a closer relationship to CNS MHPG formation and outflow.

C. Neuronal Activity and NE Metabolite Production

Procedures which increase central NE activity in rats, such as electrical stimulation of the locus coeruleus (Korf *et al.*, 1973) or treatment with phenoxybenzamine (Meek and Neff 1973; Nielson and Braestrup, 1976), elevate brain MHPG or, where measured, brain DHPG. Similarly, procedures which decrease central NE activity, including the electrothermic lesion of locus coeruleus (Ader *et al.*, 1978) or treatment with clonidine (Tang *et al.*, 1978a), desipramine (Nielsen and Braestrup, 1977) and 6-hydroxydopamine (Helmeste *et al.*, 1979) reduce brain MHPG or, where measured, DHPG. Thus, brain MHPG and DHPG appear to reflect the integrity and functional activity of central NE neurons, satisfying a necessary condition for clinical application.

The results of *in vitro* brain-slice perfusion studies suggest that DHPG is preferentially formed under resting conditions as well as during electrically stimulated release of NE (Farah *et al.*, 1977; Taube *et al.*, 1977). These findings indicate that released NE is recaptured and metabolized presynaptically. However, the relationship between the disposition of DHPG and MHPG in intraneuronal pools is not clear. It has been argued that during increased noradrenergic neuronal activity, metabolism of released NE to NMN and subsequently MHPG is the principal pathway of metabolic flow (DeMet and Halaris, 1979). Under conditions of resting or low neuronal activity, however, DHPG formation may be relatively more important than the production of MHPG. Since DHPG must diffuse to the extraneuronal location of COMT for conversion to MHPG, the possibility arises that this relatively polar metabolite may also diffuse into the vascular system without further metabolism. The presence of significant DHPG concentrations in human CSF supports this idea. In species having high brain PST activity, such as the rat, DHPG formation and efflux would be expected to be of substantial significance since sulphation of DHPG would prevent subsequent O-methylation to MHPG. The DHPG-SO_4 would probably be excreted by an active transport process, as is the case for MHPG-SO_4 (Meek and Neff, 1972a). The greater concentration of total DHPG compared to total MHPG in rat brain (Table 18.1) supports these notions. Accordingly, in some species DHPG formation may be of considerable importance in the net metabolic clearance of NE from the CNS. This may have implications for studies evaluating the pools of origin of NE metabolites in peripheral fluids.

D. Validation Studies of the Origin of Peripheral MHPG

Essentially, four different approaches have been employed in animal or human studies to obtain estimates of the fraction of urinary or plasma MHPG derived from the CNS. The earliest approach involved a double-isotope impulse-labelling technique in which ^{3}H-NE and ^{14}C-NE were simultaneously infused intraventricularly and intravensously, respectively, into dogs, following which the isotopic ^{14}C/^{3}H ratios for urinary NE metabolites were determined (Maas and Landis, 1968). These authors estimated that about 27-30% of urinary MHPG originated from the CNS in the dogs, while less than 10% of urinary NMN was of central origin. This was in agreement with the earlier estimate of Mannarino *et al.* (1963) for cats.

An alternative radioisotopic technique which appears applicable to human studies (Ebert and Kopin, 1975; Tang *et al.*, 1978b) uses the principle of isotope dilution. The central fraction of MHPG is estimated through the decline of urine specific activities of urinary MHPG and VMA following intravenously infused ^{3}H-NE or ^{3}H-DA. A mathematical relationship based on blood-brain barrier dynamics and the fortuitous finding that CNS NE produces unequal quantities of VMA and MHPG is used; the derived formula provides an estimate of the contribution of CNS to urinary MHPG. Estimates in humans of the central fraction of urinary MHPG obtained with this technique were quite high (Ebert and Kopin, 1975), varying from 67 to 91%, although this range has recently been revised downwards (Kopin, 1978).

Several factors may compromise the validity of the CNS MHPG estimates obtained by the above tracer techniques. Peripheral conversion of MHPG to VMA (Blombery *et al.*, 1979) via the intermediate phenylglycolaldehydes could lead to significant recycling and dilution of MHPG specific activity in tracer studies. Similarly, interconversion of DHPG with MHPG by methylation and demethylation (Gale and Maas, 1977; Tyce, 1977) could also produce phenomena of recycling and dilution of tracer specific activity. The inverted (1:2) ratio of urinary MHPG/VMA concentrations as compared to the ratio (2.5:1) for plasma in humans (Stancer *et al.*, 1980) suggests that conversion of MHPG to VMA is physiologically significant. Further, the kidney may play a significant role in such phenomena since the renal metabolism of catecholamines has been demonstrated (Silva *et al.*, 1979). As a result of such interconversion of metabolites and possible dilution of tracer specific activities, the CNS fraction of urinary MHPG estimated by tracer techniques might be substantially overestimated.

A second strategy, used exclusively in animals, utilizes urinary or plasma MHPG measures following the destruction of CNS NE neurons by 6-hydroxydopamine (6-OHDA). The results are conflicting. Breese *et al.* (1972) depleted rat brain NE by 82% with 6-OHDA and found a 29% decrease in urinary MHPG excretion, whereas a number of other studies (Helmeste *et al.*, 1979) have not been able to replicate this finding. The same disparity is seen in studies using nonhuman primates. Maas *et al.* (1972) found a 32% decrement in urinary MHPG after a 72% decrease in brain NE following 6-OHDA, while Breese *et al.* (1972) found that a 60% depletion of brain NE did not change urinary MHPG excetion. In mice, a 50% depletion of brain MHPG by 6-OHDA was accompanied by a 40% decrease in urinary free MHPG levels, whereas the excretion of urinary MHPG conjugates remained unchanged (Howlett *et al.*, 1975).

The basis for these conflicting findings has not been clarified, although several possibilities have been considered. Tang *et al.* (1978b) and Helmeste *et al.* (1979) suggested that a critical degree of destruction of CNS neurons must be surpassed before urinary MHPG excretion or plasma MHPG would be influenced If such is the case, clearly the use of urinary MHPG clinically would reflect only the most dramatic alterations in CNS NE function. A second possibility, now supported by experimental data, suggests that destruction of central NE neurons by 6-OHDA may elicit concomitant decrements in peripheral sympathetic activity and, as a consequence, NE metabolite production. Spinal projections of the CNS noradrenergic system exert excitatory control on sympathetic presynaptic neurons, whereas the spinal serotonergic projections appear to be inhibitory (Sangdee and Franz, 1979). Destruction of CNS NE neurons with 6-OHDA would be expected to alter peripheral sympathetic activity not only by disturbing direct spinal noradrenergic control but also by disrupting the intercommunication and balance between noradrenergic and serotonergic control. Similarly, increments in brain and spinal-cord MHPG levels induced by locus coeruleus (LC) stimulation may not be the sole or major source leading to increased plasma MHPG levels (Crawley *et al.*, 1979a, b). Rather, spinal-mediated increased peripheral sympathetic activity from stimulatory noradrenergic inputs could increase peripheral NE metabolite formation and hence plasma MHPG levels. This suggestion is strengthened by the findings that systemic administration of 6-OHDA or hexamethonium greatly reduced the magnitude of the plasma MHPG increment after LC stimulation (Crawley *et al.*, 1979a) and resulted in an excellent correlation between plasma and spinal MHPG levels after LC stimulation (Crawley *et al.*, 1979b).

Recently Maas *et al.* (1977, 1979a) have measured the brain venous-arterial (V-A) difference to validate the rate of brain MHPG production and outflow. In stump-tailed monkeys, the difference in plasma MHPG concentrations between internal jugular venous and descending aortic arterial blood was determined to be 1.7 ± 0.6 ng/ml, while in humans the difference between internal jugular venous and radial arterial or pulmonary arterial blood was 0.7 ± 0.1 ng/ml. Using mean values of the cerebral blood flow and brain weight, the cerebral MHPG outflow for monkeys and humans was estimated to be 1.45 and 2.56 µg/100 g brain/hour, respectively. Using the latter estimate of brain MHPG outflow, in humans, it was calculated that about 63% of total body MHPG production arises from the brain (Maas *et al.*, 1979a). In further support of the validity of this technique, it was demonstrated in monkeys that the α-adrenergic antagonist piperoxane elevates the V-A difference in plasma MHPG, whereas the α-agonist clonidine decreased the V-A difference. This agrees with the known effect of these drugs on brain MHPG levels.

The reliability of the V-A-difference technique is open to some question, since marked interindividual differences were obtained for the MHPG difference with values varying by as much as sevenfold. Such variation gives rise to estimates of the CNS contribution to urinary MHPG ranging from 20 to 120%. The V-A-difference technique essentially gives instantaneous estimates of brain MHPG outflow and may not provide an accurate determination of the contribution of brain to urinary MHPG obtained over a prolonged period if substantial intrasubject variation exists. While stress-related influences on the V-A difference in MHPG have been alluded to (Maas *et al.*, 1979a), systematic evaluation of such effects has not been reported as yet. Accordingly, at the present time the estimated contribution of brain to peripheral MHPG pools derived through this technique must also be regarded with caution.

The final validation technique that will be considered is the use of selective peripheral inhibitors of synthesis or degradation of NE. The fact that such agents do not enter the CNS permits an estimate of the CNS contribution to peripheral MHPG production. Two such agents have been employed in this regard: the peripheral decarboxylase inhibitor carbidopa and the peripheral monoamine-oxidase (MAO) inhibitor debrisoquine. In animal and human studies, carbidopa significantly reduces peripheral NE formation and MHPG excretion (Wurtman and Watkins, 1977; Garfinkel *et al.*, 1979). During seven days of carbidopa administration to humans, reductions of 15-55% of urinary MHPG excretion have been obtained, along with similar degrees of decrease in 5-HIAA excretion. A concomitant reduction of 90% in urinary tryptamine (TA) excretion would suggest marked blockade of peripheral decarboxylase, since TA is derived from the direct decarboxylation of tryptophan. This could lead to the conclusion that the residual MHPG excretion during carbidopa reflects CNS produced MHPG. Indeed, the estimates of 45-85% of the brain contribution to urinary MHPG agree with the other estimates cited. However, it is unlikely that carbidopa actually produces the degree of peripheral decarboxylase inhibition, as would be implied by the decrease in urinary TA excretion. The difference in affinities of DOPA (high affinity) and tryptophan (low affinity) for the decarboxylating enzyme (Lovenberg *et al.*, 1962) makes it likely that the competitive inhibition by carbidopa would be less for DOPA than for tryptophan. This being the case, following carbidopa, peripheral inhibition of NE synthesis will be incomplete and the CNS contribution to urinary MHPG will probably be overestimated.

Selective peripheral MAO inhibition by debrisoquine produces significant decrements in urinary and plasma MHPG in rats (Karoum *et al.*, 1974; Helmeste *et al.*, 1979). Perturbation of peripheral ^{3}H-NE metabolism with debrisoquine led to an estimate that in the rat about 30% of urinary MHPG originates from the CNS (Karoum *et al.*, 1974). However, the validity of this latter estimate is also open to question, since there was no attempt to separate and determine the specific activities of the urinary deaminated NE metabolites nor was the considerable formation of DHPG in this species taken into account. More recently,

Helmeste *et al.* (1979) have examined the effect of debrisoquine on plasma MHPG levels prior to, and again after, the destruction of the CNS NE neurons with 6-OHDA. Central NE neuronal destruction did not produce any further decrease in plasma and urinary MHPG, nor did clonidine enhance the depleting effect of debrisoquine. This led to the conclusion that there is a negligible CNS contribution to peripheral MHPG in the rat. Using a modification of the debrisoquine method in normal volunteers, Karoum *et al.* (1978) estimated that 25% of urinary MHPG derives from the CNS. However, in view of the power of debrisoquine to reduce central NE metabolism in primate brain (Maas *et al.*, 1979b), the validity of this technique in humans remains to be substantiated.

In summary, although a variety of techniques have been employed to estimate the CNS contribution to peripheral MHPG in animals and humans, serious questions remain with respect to the validity of the techniques. Estimates have varied from as low as 25% to 85% or more, with a trend towards the value of 60%. However, consistency in estimates between methods may unjustifiably signify accuracy in the face of techniques having potentially serious limitations. A major factor which may account for these high estimates of the CNS contribution to peripheral MHPG is the intricate control which the CNS monoamine systems exert over peripheral sympathetic function. Accordingly, manipulation or dysfunction of CNS monoaminergic systems may be accompanied by changes in peripheral sympathetic activity and consequent MHPG production; such production might confound attempts to determine central MHPG output accurately.

E. Clinical Significance of Peripheral Noradrenergic Metabolites

Despite the problems in unequivocally establishing that a significant fraction of peripheral MHPG orginates from the CNS, there have been a number of reports evaluating the diagnostic and pharmacodynamic utility of urinary and plasma MHPG determinations particularly in affective disorders. The initial work of Maas *et al.* (1968) revealed that the urinary MHPG levels in a heterogenous group of depressed patients tended to be lower compared to a normal control group. Higher MHPG excretion in the manic phase than in the same or other patients in the depressive phase has been reported (Greenspan *et al.*, 1970; Bond *et al.*, 1972). In addition, low urinary MHPG excretion has been found in bipolar depressed patients compared to normal controls or unipolar depressed subjects (Schildkraut, 1973; Garfinkel *et al.*, 1979). However, Hollister *et al.* (1978) reported no consistent pattern of change in urinary MHPG levels in depressed patients. Investigations of the urinary excretion of NE, epinephrine, NMN, metanephrine and VMA have revealed no consistent differences in urinary levels of these compounds between categories of patients with affective disorders or normal controls (Schildkraut *et al.*, 1978; Van Praag, 1978).

Urinary MHPG excretion in schizophrenic patients has been found lower (Subrahmanyam, 1975) or unchanged when compared to normal controls (Joseph *et al.*, 1976). Recently, Deakin *et al.* (1979) have shown that there is an inverse relationship between urinary MHPG excretion and measures of arousal in schizophrenia. Finally, decreased urinary MHPG excretion has been observed in patients with anorexia nervosa (Halmi, 1978) and Gilles de la Tourette syndrome (Sweeny *et al.*, 1978b).

A number of variables, such as age, sex, stage of illness, degree of anxiety, diet and motor activity, may affect noradrenergic function and MHPG excretion (Goodwin and Potter, 1978). For example, discontinuation of a low monoamine diet increased MHPG excretion in depressed patients, but produced no change in normal controls (Muscettola *et al.*, 1977). However, few studies have controlled for possible dietary effects. Sweeney *et al.* (1978a) demonstrated an interrelationship between changes in urinary MHPG excretion and anxiety over short intervals in depressed female patients with diagnoses of unipolar and bipolar depression or schizoaffective illness. This relationship was not evident when baseline urinary MHPG excretion and anxiety state were examined.

While urinary MHPG excretion in normals is relatively stable over periods of a few days, variability in excretion increases substantially over intervals of several weeks (Hollister *et al.*, 1978). Diurnal variation in MHPG excretion has also been found with peak urinary excretion rates between 4.0 and 6.0 p.m. (Hollister *et al.*, 1978). Daily urinary MHPG excretion tends to be lower in females than males, and variations in MHPG during the menstrual cycle have also been reported (Maas, 1975; De Leon-Jones *et al.*, 1978).

Because distrubances in psychomotor activity frequently occur in psychiatric disorders, there has been special interest in the effects of activity on urinary MHPG excretion both in patients and normal individuals. Ebert *et al.* (1972) reported that physical activity may alter urinary MHPG levels, and that reduced MHPG levels could be related to the decreased activity usually observed in these patients. However, these findings were challenged by the data of Goode *et al.* (1973) and Stancer *et al.* (1980), who showed that vigorous exercise by normal individuals was not associated with changes in urinary MHPG levels. Moreover, there are reports indicating that low urinary MHPG levels occur in both agitated and retarded depression (De Leon-Jones *et al.*, 1975) and that MHPG excretion is not altered by either increasing or decreasing the physical activity of depressed patients (Sweeney *et al.*, 1978a). Thus, there is no clear association between activity and urinary MHPG excretion in depressive patients. The same may not apply to plasma MHPG, which was found to be elevated during exercise adjusted to individual fitness (Stancer *et al.*, 1980), contrary to the earlier findings of Goode *et al.* (1973).

F. Pharmacodynamic Utility of Peripheral MHPG Measurements

The observed differences in urinary MHPG excretion in subgroups of depressed patients (Garfinkel *et al.*, 1979) has raised the issue of whether there is a differential response to antidepressant medications in subgroups of patients distinguished by low or high MHPG excretion. It has been postulated that low MHPG excreters have primarily a disorder of CNS NE function (Goodwin and Potter, 1978), and indeed low excreters show a better response to desipramine and imipramine (Maas, 1975), which are relatively selective in affecting NE neuronal reuptake. In contrast, high MHPG excretion found predominantly in unipolar depressed patients is associated with evidence of CNS serotonergic dysfunction (Garfinkel *et al.*, 1979). These patients respond better to anti-depressant drugs which are more selective in affecting CNS serotonergic activity, such as amitriptyline (Schildkraut *et al.*, 1973; Beckmann and Goodwin 1975). However, Coppen *et al.* (1979) were unable to substantiate a relationship be-tween urinary MHPG excretion and response to amitriptyline in male or female patients with a diagnosis of primary affective disorder. Finally, Zis *et al.* (1979) have observed a positive relationship between urinary MHPG levels and the latency in development of mania or hypomania in depressed patients with bipolar affective disorder.

The interpretation of the above clinical findings is difficult. On the one hand, it is tempting to infer that reduced MHPG excretion in depression reflects de-creased CNS NE function in this disorder. Better response to the relatively selective noradrenergic antidepressants would seem to support this contention. On the other hand, recent data suggest changes in urinary or plasma MHPG may result from alterations of peripheral sympathetic activity, which is under CNS NE control, mediated through descending spinal noradrenergic and serotonergic systems (Sangdee and Franz, 1979), as discussed earlier. Moreover, selectivity of antidepressants in facilitating peripheral sympathetic activity through spinal mediated mechanisms has also been reported (Sangdee and Franz, 1979). Thus, changes in urinary or plasma MHPG may indeed reflect alterations in CNS NE function and the effects of antidepressants on CNS NE systems, but not because of any direct relationship of peripheral MHPG to CNS NE metabolite formation. Instead, peripheral MHPG changes may be affected indirectly through the spinal control of the peripheral sympathetic nervous system involving descending stimulatory noradrenergic and descending inhibitory serotonergic inputs. Thus, urinary MHPG excretion would be more likely to reflect the net balance between CNS noradrenergic and serotonergic function. This possibility would explain the occurrence of elevated urinary MHPG excretion in depressed patients with evidence of disturbed serotonin metabolism (Goodwin and Potter, 1978; Garfinkel *et al.*, 1979) as well as the differential response to noradrenergic as against serotonergically acting antidepressants based on urinary MHPG excretion.

IV. UTILITY OF DOPAMINERGIC METABOLITES

As already noted, the relative affinities of aldehyde reductase and dehydrogenase enzymes predispose NE to glycol metabolite formation. For DA, the absence of the β-hydroxyl group leads primarily to acid metabolite formation. Accordingly, DOPAC and HVA are the major metabolites of DA in rat brain, whereas 3-methoxy-tyramine (3-MT) and 3-methoxy-4-hydroxyphenylethanol (MOPET) only occur in negligible amounts (Karoum *et al.*, 1977b; Muskiet *et al.*, 1978;

Westerink, 1979). In the rat, a species with high brain PST activity, HVA and DOPAC exist in both sulfate conjugated (about 40-50%) and free from (Swahn and Wiesel, 1976). Species differences in DA metabolism are also evident. While substantial levels of DOPAC occur in rat brain, in primate brain and CSF, DOPAC concentrations are about 3% and 1%, respectively, of HVA levels (Bacopoulos *et al.,* 1978; Wilk and Stanley, 1978). Moreover, HVA is not conjugated in human or monkey brain as it is in the rat.

The biochemical basis for the differences in brain DA metabolism between rat and primates has not been clarified. Moreover, there has been some dispute about whether DOPAC or HVA represents the major acid metabolite of DA in rat brain (Karoum *et al.,* 1977b; Dedek *et al.,* 1979; Westerink, 1979). In human brain DOPAC may be rapidly O-methylated to HVA in the absence of significant conjugate formation as occurs in rat. An alternative possibility is that HVA formation in human brain might proceed primarily through extraneuronal 3-MT formation that is subsequently deaminated to HVA. In rat brain, 3-MT formation may not be significant in conversion of DA to HVA (Westerink, 1979). However, little is known about this possible metabolic route in primates and, as can be seen from the differences in DOPAC formation between rat and primates, extrapolating from one to another species may not be valid.

As for other acidic or conjugated monoamine metabolites, HVA, DOPAC and their sulphate conjugates are cleared from the CNS by the same probenecid-sensitive active-transport system (Dedek *et al.,* 1979). Conjugation of DOPAC and HVA in rat brain may serve to further facilitate the efflux of these metabolites since higher concentrations of probenecid are required to induce accumulations of the free as compared to conjugated acid metabolites in rat striatum (Dedek *et al.,* 1979).

Levels of HVA and DOPAC in DA-rich regions also bear a close relationship to the functional activity of the central DA system, as has been noted for MHPG and DHPG with respect to the NE system. Pharmacological manipulations that increase impulse flow in central DA neurons such as electrical stimulation of the nigrostriatal and mesolimbic pathways and treatment with neuroleptics haloperidol, pimozide, etc.) elevate DOPAC and HVA levels in neostriatum and olfactory tubercle in rats (Korf *et al.,* 1976; Roth, 1976). In contrast, procedures that reduce central DA activity such as electrothermic lesion of the substantia nigra and treatment with drugs such as γ-butyrolactone, apomorphine and amphetamine decrease striatal levels of DOPAC and HVA. Thus, striatal DOPAC and HVA concentrations appear to reflect neuronal activity-dependent changes in DA metabolism.

Alterations in receptor function are also reflected by changes in levels of dopamine metabolites. Following DA-receptor blockade by neuroleptics, brain levels of DOPAC and HVA are elevated as a result of transsynaptically mediated feedback stimulation of presynaptic neuronal impulse flow and DA release (Sedvall, 1975). In rats withdrawn from chronic haloperidol, for example, the apomorphine-induced decrease in brain HVA is enhanced indicating the development of receptors supersensitivity (Smith *et al.*, 1978). Thus measurements of these acid metabolites before or after perturbation with DA-receptor agonists such as apomorphine or antagonists such as the neuroleptics will provide information with respect to the status of dopaminergic receptor activity.

Determination of CSF HVA has been used extensively to monitor central DA metabolism in psychiatric and neurological disorders (Van Praag, 1978), but little has been done in utilizing plasma and urine measures. This is rather interesting in view of recent work demonstrating a relationship between plasma DA metabolites and nigrostriatal dopaminergic function (Bacopoulos *et al.*, 1979) and estimates indicating that a significant fraction of plasma and urinary HVA derives from CNS (Kopin, 1978; Maas *et al.*, 1979c). In rats, plasma HVA and DOPAC levels change in parallel with brain DA metabolites following electrical stimulation or electrothermal lesion of the nigrostriatal DA system, intraventricular 6-OHDA or haloperidol administration (Bacopoulos *et al.*, 1979). Moreover, rat plasma and striatal HVA but not DOPAC levels were highly correlated after bilateral lesion of the nigrostriatal pathways. In rhesus monkeys and humans, haloperidol administration elevates both plasma and CSF HVA concentrations, with significant correlation between these concentrations (Bacopoulos *et al.*, 1978; Heninger *et al.*, 1979).

A possible relationship may also exist between urinary HVA and DOPAC excretion and brain DA function. Hoeldtke *et al.* (1974) demonstrated that urinary excretion of free HVA decreased by 27% following destruction of CNS DA systems with intraventricular 6-OHDA, but urinary free DOPAC excretion was unchanged. They estimated that approximately one third of urinary HVA in the rat is derived from the CNS. In contrast, Peyrin *et al.* (1978) found a decrease in urinary excretion of conjugated DOPAC and HVA of 26% and 42%, respectively, following unilateral nigral destruction with 6-OHDA in rats. Urinary free DOPAC and HVA levels were unchanged, however. These authors estimated that 88% of conjugated HVA and 50% of conjugated DOPAC in urine derive from striatal HVA and DOPAC. The reason for the difference in findings between these two groups is not clear. Moreover, since HVA and DOPAC are both about 50% conjugated in rat brain and urine, one would have expected both free and conjugated levels of urinary HVA and DOPAC to manifest the same degree of change.

Recently, Kopin and coworkers have employed stable isotope infusion techniques to evaluate the relative contribution of central HVA pools to plasma and urine HVA. Using this technique, it was shown that the kidney does not contribute significantly to urinary HVA (Elchisak *et al.*, 1978). Probenecid administration to inhibit the renal excretion and CNS efflux of HVA led to estimates that 53-61% of total body HVA production derives from the CNS (Kopin, 1978). However, the accuracy of this estimate must be regarded with caution as the infusion of high concentrations of deuterium-labeled tracer (1 mg) would lead to nonsteady-state conditions influencing the disposition and turnover of HVA.

Finally, Maas *et al.* (1979c) have applied the V-A-difference technique to estimate the CNS contribution to peripheral HVA and DOPAC in primates. A highly significant V-A-difference was found for HVA but not DOPAC, and they estimated that between 9 and 30% of urinary HVA derives from brain in the monkeys used. Calculated brain HVA production did not correlate with cerebral blood-flow measures. As yet, these authors have not reported examination of the effect of manipulating brain DA metabolism and function on the V-A difference in plasma HVA.

Despite the promising relationships between plasma DA metabolites and CNS dopaminergic function that have been demonstrated, so little work has been undertaken evaluating the urine and plasma DA metabolites in psychiatric disorders that even tentative conclusions cannot be made. Preliminary findings on plasma HVA determination in schizophrenic patients (Heninger *et al.*, 1979) suggested that this measure may reflect the effects of neuroleptic medication on the CNS dopaminergic system. Clearly, more extensive evaluation of the utility of plasma and urine dopamine metabolites in psychiatric disorders is indicated.

V. UTILITY OF SEROTONERGIC METABOLITES

The major route of serotonin (5-HT) metabolism proceeds through oxidative deamination by MAO to form 5-hydroxyindolealdehyde, which is subsequently oxidized to the acidic metabolite, 5-HIAA, or reduced to the neutral metabolite 5-hydroxytryptophol (5-HTOL). In rat brain, 5-HT is metabolized mainly to 5-HIAA, while formation of 5-HTOL is significant in peripheral 5-HT metabolism (Cheifetz and Warsh, 1980). This preferential formation of 5-HIAA in brain results from the higher affinity of aldehyde dehydrogenase compared to aldehyde reductase for the intermediate acetaldehyde (Duncan and Sourkes 1974). Low 5-HTOL compared to 5-HIAA (about 1%) in human CSF (Swahn *et al.*, 1976; Takahashi *et al.*, 1978) supports the notion that 5-HIAA is the major 5-HT metabolite in human brain as well. Brain 5-HIAA exists primarily in the free form and is removed by a probenecid-sensitive active transport mechanism (Meek and Neff, 1972a).

As for NE and DA metabolites, brain 5-HIAA levels are also dependent upon CNS serotonergic neuronal activity. Activation of rat-brain ascending 5-HT pathways by electrical stimulation of the dorsal and median raphe nuclei (Sheard and Aghajanian 1968), elevates rat-brain 5-HIAA concentrations. On the other hand, procedures that reduce central 5-HT neuronal firing, such as electrolytic lesion of dorsal and median raphe nuclei (Lorens and Guldberg, 1974) or treatment with chlorimipramine (Collard, 1978), reduce brain 5-HIAA levels. Thus, brain 5-HIAA levels provide a global index of brain serotonergic activity.

Little has been done to evaluate directly whether the CNS contributes significantly to blood or urine 5-HIAA pools. Because peripheral organs such as the gastrointestinal tract and lung are known to be high in 5-HT and 5-HIAA (Cheifetz and Warsh, 1980), it has been generally accepted, but poorly substantiated, that urinary 5-HIAA derives almost entirely from peripheral 5-HT metabolism (Van Praag, 1978). Recently, two studies have evaluated the contribution of CNS 5-HIAA to blood and urine in the rat. During peripheral decarboxylase inhibition, the fraction of 5-HIAA in blood derived from CNS in rats was estimated to be no more than 20% (Warsh and Stancer, 1975). Aizenstein and Korf (1979) utilized a ventricular-lumbar cisternal perfusion technique to estimate the output rate of 5-HIAA from rat CNS. Comparison of this output rate to the mean urinary 5-HIAA excretion rate gave the estimate that about 8% of urinary 5-HIAA derives from CNS.

These findings lend direct support to the concept that urinary and blood 5-HIAA derive primarily from peripheral sources at least in the rat. It is noteworthy that Maas *et al.* (1979d) have demonstrated a small but significant V-A difference for plasma 5-HIAA in monkeys. However, these authors have not as yet provided an estimate of the CNS contribution of 5-HIAA to blood or urine in this species, using the V-A-difference technique.

Because urinary 5-HIAA was thought to emanate almost entirely from peripheral tissues, investigators have concentrated instead on CSF 5-HIAA in the biochemical investigations of psychiatric disorders (Van Praag, 1978). However, in one of the few available clinical studies evaluating 5-HIAA excretion, abnormalities have been detected (Garfinkel *et al.*, 1979). Although such findings cannot be related directly to CNS disturbances in serotonergic function, they could reflect disturbances in peripheral utilization of tryptophan, which has been observed in subgroups of patients with affective disorders (Garfinkel *et al.*, 1979).

VI. SUMMARY AND CONCLUSIONS

Although there has been considerable investigation both in animals and clinically, there remain to be resolved major problems that limit the clinical utility of urinary or blood metabolites as indices of CNS monoaminergic function. Only in the case of the CNS NE and DA systems does there appear to be a significant CNS contribution to blood and urinary MHPG and HVA pools, respectively. However, the reliability of the estimates of the CNS fractions of these metabolites in peripheral fluids has not been convincingly substantiated. Moreover, if the major proportion of MHPG or HVA in peripheral body fluids originates from the CNS, these metabolites would probably reflect only generalized changes in CNS presynaptic neuronal function of their respective parent monoamine systems. Also, peripheral formation of metabolites such as MHPG may be subject to indirect CNS control via spinal efferent NE and 5-HT systems. Consequently, the regulation of urinary or blood MHPG levels probably involves both direct and indirect CNS mechanisms, but the relative partitioning of these mechanisms, in so far as they affect peripheral MHPG levels, remains to be established. The relationship of MHPG levels in peripheral body fluids to CNS NE function is, therefore, probably much more complicated than was originally thought. This suggests that the interpretation of clinical findings of reduced or elevated MHPG in groups of psychiatric disorders has been greatly oversimplified.

Recent studies of the relationships of plasma HVA to brain DA metabolism and function provide compelling arguments that this measure may be of significant clinical utility in monitoring CNS DA function in psychiatric disorders. To the best of our knowledge this area of clinical investigation has been seriously neglected. However, caution is indicated in view of what has been learned with regard to MHPG. Clearly, further investigation of the significance of plasma HVA is indicated. Ultimately, the utility of peripheral monoamine metabolites to assess CNS monoamine function will require a better understanding not only of the sources of these metabolites, but also of the multiplicity of factors governing their formation and disposition both in neuronal and nonneuronal pools.

REFERENCES

Ader, J.P., Muskiet, F.A.J., Jeuring, H.J. and Korf, J. (1978). *J. Neurochem., 30,* 1213
Aghajanian, G.K. (1978). In *Essays in Neurochemistry and Neuropharmacology,* ed. M.B.H. Youdim, W. Lovenberg, D.R. Sharman and J.R. Lagnado (Wiley, New York) p. 1
Aizenstein, M.L. and Korf, J. (1979). *J. Neurochem., 32,* 1227
Bacopoulos, N.G., Heninger, G.R. and Roth, R.H. (1978). *Life Sci., 23,* 1805
Bacopoulos, N.G., Hattox, S.E. and Roth, R.H. (1979). *Eur. J. Pharmacol., 56,* 225
Beckmann, H. and Goodwin, F.K. (1975). *Arch. Gen. Psychiat., 32,* 17
Blombery, P., Kopin, I.J., Gordon, E.K. and Ebert, M.H. (1979). In *Catecholamines: Basic and Clinical Frontiers,* ed. E. Usdin, I.J. Kopin and J. Barchas (Pergamon Press, New York) p. 681
Bond, P.A., Jenner, F.A. and Sampson, G.A. (1972). *Psychol. Med., 2,* 81
Braestrup, C., Nielsen, M. and Kruger, J.S. (1974). *J. Neurochem., 23,* 569
Breese, G.R., Prange, A.J. Jr., Howard, J.L., Lipton, M.A., McKinney, W.T. Bowman, R.E. and Bushnell, P. (1972). *Nature New Biol., 240,* 286
Cheifetz, S. and Warsh, J.J. (1980). *J. Neurochem., 34,* 1093
Collard, K.J. (1978). *Br. J. Pharmacol., 62,* 137
Coppen, A., RamaRao, V.A., Ruthven, C.R.J., Goodwin, B.L. and Sandler, M. (1979). *Psychopharmacology 64,* 95
Crawley, J.N., Maas, J.W. and Roth, R.H. (1979a). *Psychopharmacol. Bull., 15,* 27
Crawley, J.N., Roth, R.H. and Maas, J.W. (1979b). *Brain Res., 166,* 180
Deakin, J.F.W., Baker, H.F., Frith, C.D., Joseph, M.H. and Johnstone, E.C. (1979). *J. Psychiat. Res., 15,* 57
Dedek, J., Baumes, R., Tien-Duc, N., Gomeni, R. and Korf, J. (1979). *J. Neurochem., 33,* 687
DeLeon-Jones, F., Maas, J.W. and Dekirmenjian, H. (1975). *Am. J. Psychiat., 132,* 1141
DeLeon-Jones, F.A., Steinberg, J. Dekirmenjian, H. and Garver, D. (1978). *Commun. Psychopharmacol., 2,* 267
DeMet, E.M. and Halaris, A.E. (1979). *Biochem. Pharmacol., 28,* 3043
Duncan, R.J.S. and Sourkes, T.L. (1974). *J. Neurochem., 22,* 663
Ebert, M.H. and Kopin, I.J. (1975). *Trans. Ass. Am. Phys., 88,* 256
Ebert, M.H., Post, R.M. and Goodwin, F.K. (1972). *Lancet ii,* 766
Elchisak, M.A., Polinsky, R.J., Ebert, M.H., Powers, K.J. and Kopin, I.J. (1978). *Life Sci., 23,* 2339
Farah, M.B., Adler-Graschinsky, E. and Langer, S.Z. (1977). *Naunyn-Schmiedeberg's Arch. Pharmacol., 297,* 119
Foldes, A. and Meek, J.L. (1974). *J. Neurochem., 23,* 303

Gale, St. W. and Maas, J.W. (1977). *J. Neural Transm., 41*, 59
Garfinkel, P.E., Warsh, J.J., Stancer, H.C. and Godse, D.D. (1979). *Am. J. Psychiat., 136*, 535
Goode, D.J., Dekirmenjian, H., Meltzer, H.Y. and Maas, J.W. (1973). *Arch. Gen. Psychiat., 29*, 391
Goodwin, F.K. and Potter, W.Z. (1978). In *Depression*, ed. J.O. Cole (Plenum Press, New York) p. 41
Greenspan, K., Schildkraut, J.J., Gordon, E.K., Baer, L., Arnoff, M.S. and Durell, J. (1970). *J. Psychiat. Res., 7*, 171
Halmi, K.A. (1978). *Ann. Rev. Med., 29*, 137
Helmeste, D.M., Stancer, H.C., Coscina, D.V., Takahashi, S. and Warsh, J.J. (1979). *Life Sci., 25*, 601
Heninger, G., Bacopoulos, N., Roth, R., Bowers, M., Jr. and Sweeney, D. (1979). In *Catecholamine: Basic and Clinical Frontiers*, Vol. 2, ed. E. Usdin, I.J. Kopin and J. Barchas (Pergamon Press, New York) p. 1887
Hoeldtke, R., Rogawski, M. and Wurtman, R.J. (1974). *Br. J. Pharmacol., 50*, 265
Hollister, L.E., Davies, K.L., Overall, J.E. and Anderson, T. (1978). *Arch. Gen. Psychiat., 35*, 1410
Howlett, D.R., Jenner, F.A. and Naharski, S.R. (1975). *J. Pharm. Pharmacol., 27*, 447
Joseph, M.H., Baker, H.F., Johnstone, E.C. and Crow, T.J. (1976). *Psychopharmacology 51*, 47
Karasawa, T., Furukawa, K. and Shimizu, M. (1978). *J. Neurochem., 30*, 1525
Karoum, F., Wyatt, R.J. and Costa, E. (1974). *Neuropharmacology 13*, 165
Karoum, F., Neff, N.H. and Wyatt, R.J. (1976). *J. Neurochem., 27*, 33
Karoum, F., Moyer-Schwing, J., Potkin, S.G. and Wyatt, R.J. (1977a). *Brain Res., 125*, 333
Karoum, F., Neff, N.H. and Wyatt, R.J. (1977b). *Eur. J. Pharmacol., 44*, 311
Karoum, F., Ranscher, F. and Wyatt, R.J. (1978). *Trans. Am. Soc. Neurochem., 9*, 196
Kessler, J.A., Fenstermacher, J.D. and Patlak, C.S. (1976). *Brain. Res., 102*, 131
Kopin, I.J. (1978). In *Psychopharmacology: A Generation of Progress*, ed. M.A. Lipton, A. Dimascio and K. Killam (Raven Press, New York) p. 933
Korf, J., Aghajanian, G.K. and Roth, R.H. (1973). *Eur. J. Pharmacol., 21*, 305
Korf, J., Grasdijk, L. and Westerink, B.H.C. (1976). *J. Neurochem., 26*, 579
Lorens, S.A. and Guldberg, H.C. (1974). *Brain Res., 78*, 45
Lovenberg, W., Weissbach, H. and Udenfriend, S. (1962). *J. Biol. Chem., 237*, 89
Maas, J.W. (1975). *Arch. Gen. Psychiat., 32*, 1357
Maas, J.W. and Landis, D.H. (1968). *J. Pharmacol. Exp. Ther., 163*, 147
Maas, J.W., Fawcett, J.A. and Dekirmenjian, H. (1968). *Arch. Gen. Psychiat., 19*, 129
Maas, J.W., Dekirmenjian, H., Garver, D., Redmond, D.E., Jr and Landis, D.H. (1972). *Brain Res., 41*, 507
Maas, J.W., Landis, D.H. and Dekirmenjian, H. (1976). *Comm. Psychopharmacol., 2*, 403
Maas, J.W., Hattox, S.E., Landis, D.H. and Roth, R.H. (1977). *Eur. J. Pharmacol., 46*, 221
Maas, J.W., Hattox, S.E., Greene, N.M. and Landis, D.H. (1979a). *Science 205*, 1025
Maas, J.W., Hattox, S.E. and Landis, D.H. (1979b). *Biochem. Pharmacol., 28*, 3153
Maas, J.W., Hattox, S.E., Martin, D.M. and Landis, D.H. (1979c). *J. Neurochem., 32*, 839
Maas, J.W., Green, N.M., Hattox, S.E. and Landis, D.H. (1979d). In *Catecholamine: Basic and Clinical Frontiers*, Vol. 2, ed. E. Usdin, I.J. Kopin and J. Barchas (Pergamon Press, New York) p. 1878
Mannarino, E., Kirshener, N. and Nashold, B.S. Jr. (1963). *J. Neurochem., 10*, 373
Meek, J.L. and Neff, N.H. (1972a). *J. Pharmacol. Exp. Ther., 181*, 457
Meek, J.L. and Neff, N.H. (1972b). *Brit. J. Pharmacol., 45*, 435
Meek, J.L. and Neff, N.H. (1973). *J. Pharmacol. Exp. Ther., 184*, 570
Muscettola, G., Wehr, T. and Goodwin, F.K. (1977). *Am. J. Psychiat., 134*, 914
Muskiet, F.A.J., Jeuring, H.J., Ader, J.P. and Wolthers, B.G. (1978). *J. Neurochem., 30*, 1495
Nielsen, M. and Braestrup, C. (1976). *J. Neurochem., 27*, 1211
Nielsen, M. and Braestrup, C. (1977). *Naunyn-Schmiedeberg's Arch. Pharmacol., 300*, 93
Peyrin, L., Cottet-Emard, J.M., Javoy, F., Agid, Y., Herbert, A. and Glowinski, J. (1978). *Brain Res., 143*, 567
Roth, R.H., Murrin, L.C. and Walters, J.R. (1976). *Eur. J. Pharmacol., 36*, 163
Sangdee, C. and Franz, D.N. (1979). *Psychopharmacology 62*, 9

Schildkraut, J. (1973). *Ann. Rev. Pharmacol., 13,* 427

Schildkraut, J.J., Oysulak, P.J., Schatzberg, A.F., Gudeman, J.E., Cole, J.O., Rohde, W.A. and LaBrie, R.A. (1978). *Arch. Gen. Psychiat., 35,* 1427

Sedvall, G. (1975). In *Handbook of Psychopharmacology,* Vol. 6, ed. L.L. Iversen, S.D. Iversen and S.H. Snyder (Plenum Press, New York) p. 127

Sharman, D.F. (1969). *Brit. J. Pharmacol., 36,* 523

Sheard, M.H. and Aghajanian, G.K. (1968). *J. Pharmacol. Exp. Ther., 163,* 425

Silva, P., Landsberg, L. and Besarab, A. (1979). *J. Clin. Invest., 64,* 850

Smith, R.C. Narasimhachari, N. and Davis, J.M. (1978). *J. Neural Transm., 42,* 159

Stancer, H.C., Warsh, J.J., Tang, S.W., Takahashi, S. and Sheppard, R.J. (1980). In *Enzymes and Neurotransmitters and Mental Disease* ed. M.B.H. Youdim, E. Usdin and T.L. Sourkes (Wiley, Chichester, England) p. 221

Subrahmanyam, S. (1975). *Brain Res., 87,* 355

Swahn, C.G. and Wiesel, F.A. (1976). *J. Neural Transm., 39,* 281

Swahn, C.G., Sangaide, B., Wiesel, F.A. and Sedvall, G. (1976). *Psychopharmacology 48,* 147

Sweeney, D., Maas, J.W. and Heninger, G.R. (1978a). *Arch. Gen. Psychiat., 35,* 1418

Sweeney, D., Pickar, D., Redmond, D.E. and Maas, J.W. (1978b). *Lancet 1,* 872

Takahashi, S., Godse, D.D., Naqvi, A., Warsh, J.J. and Stancer, H.C. (1978). *Clin. Chim. Acta 84,* 55

Tang, S.W., Helmeste, D.M. and Stancer, H.C. (1978a). *Naunyn-Schmiedeberg's Arch. Pharmacol., 305,* 207

Tang, S.W., Stancer, H.C. and Warsh, J.J. (1978b). *Brain Res. Bull., 3,* 669

Taube, H.D., Starke, K. and Borowski, E. (1977). *Naunyn-Schmiedeberg's Arch. Pharmacol., 299,* 123

Tipton, K.F., Houslay, M.D. and Turner, A.J. (1977). In *Essays in Neurochemistry and Neuropharmacology,* Vol. 1, ed. M.B.H. Youdim, W. Lovenberg, D.F. Sharman and J.R. Lagnado (Wiley, New York) p. 103

Tyce, G.M. (1977). *Res. Commun. Chem. Pathol. Pharmacol., 16,* 669

Van Praag, H.M. (1978) In *Handbook of Psychopharmacology,* Vol. 13, ed. L.L. Iversen, S.D. Iversen and S.H. Snyder (Plenum Press, New York) p. 187

Warsh, J.J. and Stancer, H.C. (1975). *Eur. J. Pharmacol., 32,* 128

Westerink, B.H.C. (1979). *Eur. J. Pharmacol., 56,* 313

Wilk, S. and Stanley, M. (1978). *Psychopharmacology 57,* 77

Wurtman, R.J. and Watkins, C.J. (1977). *Nature 265,* 79

Zis, A.P., Cowdry, R.W., Wehr, T.A., Muscettola, G. and Goodwin, F.K. (1979). *Psychiat. Res., 1,* 93

CONTRIBUTORS

N. Barden
Laboratory of Molecular Endocrinology
Le Centre Hospitalier de l'Université Laval
2705 Laurier Blvd
Quebec G1V 4G2
Canada

C. Beyer
Division de Neurociencias
Instituto Mexicano del Seguro Social
Apartado Postal 73–032
Mexico 73, DF
Mexico

M.S. Briley
Department of Biology
Synthélabo LERS
58, rue de la Glacière
75013 Paris
France

G.M. Brown
Department of Neurosciences
Faculty of Health Sciences
McMaster University
1200 Main St W
Hamilton, Ontario L8S 4J9
Canada

P. Brown
Department of Psychiatry
McMaster University
Hamilton, Ontario L8S 4J9
Canada

F. Camanni
Department of Endocrinology
University of Turin
Turin, Italy

F. Casanueva
Department of Pharmacology
University of Milan
Milan, Italy

J.M. Cleghorn
Department of Psychiatry
Faculty of Health Sciences
McMaster University

1200 Main St W
Hamilton, Ontario L8S 4J9
Canada

D. Cocchi
Department of Pharmacology
University of Milan
Milan, Italy

D.H. Coy
Department of Psychology
University of New Orleans
New Orleans, LA 70122
USA

L. Cusan
Laboratory of Molecular Endocrinology
Le Centre Hospitalier de l'Université Laval
2705 Laurier Blvd
Quebec G1V 4G2
Canada

P.L. Darby
Psychosomatic Medicine Unit
Clarke Institute of Psychiatry
Toronto, Ontario M5T 1R8
Canada

A. Dupont
Laboratory of Molecular Endocrinology
Le Centre Hospitalier de l'Université Laval
2705 Laurier Blvd
Quebec G1V 4G2
Canada

P.G. Ettigi
Department of Psychiatry
Medical College of Virginia
Virginia Commonwealth Station
Richmond, Virginia
USA

L. Ferland
Laboratory of Molecular Endocrinology
Le Centre Hospitalier de l'Université Laval
2705 Laurier Blvd
Quebec G1V 4G2
Canada

P.E. Garfinkel
Psychosomatic Medicine Unit
Clarke Institute of Psychiatry
Toronto, Ontario M5T 1R8
Canada

A.R. Genazzani
Department of Obstetrics and Pathology
University of Siena
Siena, Italy

P.E. Harrison-Read
Department of Pharmacology
The Medical College of
St Bartholomew's Hospital
Charterhouse Square
London EC1M 6BQ
England

N. Hatotani
Department of Psychiatry
Mie University
School of Medicine
Tsu, Mie
Japan 514

M.E. Henry
Department of Pharmacology
School of Medicine
Faculty of Health Sciences
University of Ottawa
275 Nicholas St
Ottawa, Ontario K1N 9A9
Canada

P.D. Hrdina
Department of Pharmacology
School of Medicine
Faculty of Health Sciences
University of Ottawa
275 Nicholas St
Ottawa, Ontario K1N 9A9
Canada

A. Jakubovic
Department of Neurological Sciences
University of British Columbia
2255 Wesbrook Mall
Vancouver, BC V6T 1W5
Canada

A.J. Kastin
Tulane University and
Veterans Administration
New Orleans, LA
USA

I. Kitayama
Department of Psychiatry
Mie University
School of Medicine
Tsu, Mie
Japan 514

V. Knott
Department of Psychiatry
Royal Ottawa Hospital
1145 Carling Ave
Ottawa, Ontario K1Z 7K4
Canada

F. Labrie
Laboratory of Molecular Endocrinology
Le Centre Hospitalier de l'Université Laval
2705 Laurier Blvd
Quebec G1V 4G2
Canada

S. Lal
Department of Psychiatry
Montreal General Hospital
and Douglas Hospital Center
Montreal, Quebec
Canada

Y.D. Lapierre
Department of Psychiatry
University of Ottawa
Royal Ottawa Hospital
1145 Carling Ave
Ottawa, Ontario
Canada

K. Larsson
Division de Neurociencias
Instituto Mexicano del Seguro Social
Apartado Postal 73-032
Mexico 73, DF
Mexico

P.P. Li
Department of Psychiatry
Clarke Institute of Psychiatry
250 College St
Toronto, Ontario M5T 1R8
Canada

V. Locatelli
Department of Pharmacology
University of Milan
20129 Milan
Italy

P. Mantegazza
Department of Pharmacology
University of Milan
20129 Milan
Italy

A. Martinez-Campos
Department of Pharmacology
University of Milan
20129 Milan
Italy

F. Massara
Department of Endocrinology
University of Turin
Turin
Italy

E.G. McGeer
Department of Neurological Sciences
University of British Columbia
2255 Wesbrook Mall
Vancouver, BC V6T 1W5
Canada

P.L. McGeer
Department of Neurological Sciences
University of British Columbia
2255 Wesbrook Mall
Vancouver, BC V6T 1W5
Canada

E.E. Müller
Department of Pharmacology
University of Milan
20129 Milan
Italy

J. Nomura
Department of Psychiatry
Mie University
School of Medicine
Tsu, Mie
Japan 514

G.A. Olson
Department of Psychology
University of New Orleans
New Orleans, LA 70122
USA

R.D. Olson
Department of Psychology
University of New Orleans
New Orleans, LA 70122
USA

R.E. Poland
Division of Biological Psychiatry
Department of Psychiatry
Harbor/UCLA Medical Center
1000 W Carson St
Torrance, CA 90509
USA

R.B. Rastogi
Psychopharmacology Research Labs
Douglas Hospital Research Center
6875 Lasalle Blvd
Verdun, Quebec H4H 1R3
Canada

R.T. Rubin
Division of Biological Psychiatry
Department of Psychiatry
Harbor/UCLA Medical Center
Torrance, CA 90509
USA

C.A. Sandman
Department of Psychiatry and Human
  Behaviour
University of California
Medical Center
101 City Dr S
Orange, CA 92668

A.V. Schally
Tulane University and
Veterans Administration Hospital
New Orleans, LA

R.L. Singhal
Department of Pharmacology
School of Medicine
Faculty of Health Sciences
University of Ottawa
275 Nicholas St
Ottawa, Ontario K1N 9A9
Canada

H.C. Stancer
Section of Biochemical Psychiatry
Clarke Institute of Psychiatry
250 College St
Toronto, Ontario M5T 1R8
Canada

J.J. Warsh
Section of Biochemical Psychiatry
Clarke Institute of Psychiatry
250 College St
Toronto, Ontario M5T 1R8
Canada

D. de Wied
Rudolf Magnus Institute of Pharmacology
Medical Faculty
University of Utrecht
Vondellaan 6, 3521 GD Utrecht
The Netherlands

V.M. Wiegant
Rudolf Magnus Institute of Pharmacology
Medical Faculty
University of Utrecht
Vondellaan 6, 3521 GD Utrecht
The Netherlands

INDEX

Abnormal Thyroid Function 212;
anxiety 212; mania 212;
neurotransmitter systems 208;
schizophrenia 212
L-α-Acetylmethadol 161, 162, 165,
166
ACTH & ACTH-Like Neurpeptides
5, 8, 40, 352; behavioural effects
of 30; brain calcium 32, 44; brain
cyclic AMP 42; brain phospholipids
44; brain phosphoproteins 43;
effect of neuroleptics 179;
excessive grooming behaviour 32;
mechanism of action 40
Androgens 104, 105, 108, 109, 111
Animal Model, Depression 189, 265,
270; adrenal responses to
separation 270; antidepressant
drugs 201; biological rhythm 198;
food and water intake 194; growth
hormone 271; imipramine 195;
learned helplessness model 265;
LH 273; monoamines 201; NA
systems 196; plasma cortisol 270;
in primates 267; prolactin 272;
psychoendocrine model 193;
rectal temperature 194; reserpine
model 265; social isolation 266;
spontaneous running activity 194;
stress-induced model 193;
testosterone 273; TRH 272;
turnover rate of catecholamines
197; uncontrollable stress 198;
vaginal smear 194
Anorexia Nervosa 279, 289; ACTH
stimulation 287; adrenals 286;
body image 279; carbohydrate
metabolism 289; cortisol 286;
fertility 280; free plasma
tryptophan 290; FSH 282;
gonadotropins 281; growth
hormone 284; HVA & 5-HIAA
290; hypothalamic function 281,
282; insulin resistance 289;
ketosteroids 287; luteinizing
hormone (LH) 282; melatonin
288; MHPG excretion 290;

monoamine metabolism 290;
prolactin 287; testosterone 289;
thermoregulation 289; thyroid 285
Antidepressant Drugs 307; binding
studies 303; tricyclics, action 217
Anxiety 327; electrodermal activity
328; physiological accompaniments
328; skin conductance 328;
vasomotor responses 329
Apomorphine, effects 210, 257;
behavioural activity in
hyperthyroid rats 210; Li on
motor activity changes 257, 258
Avoidance Behaviour 32; ACTH-like
neuropeptides 32; fragments of
β-LPH 31

Benzodiazepines 213, 214
Bromocriptine 85, 88

cAMP in brain 22, 32, 42; ACTH-like
neuropeptides 32, 42
Cannabinoids: in animals 154, 156,
157, 165; DNA synthesis 158;
gonadal function 153, 154, 157,
160, 161, 166; in humans 154,
155, 165; hypothalamus 160;
Leydig cells 156, 160, 161; LH
155-7; nucleic acid 154, 158;
nucleic acid synthesis 158, 161,
166; ovary 157; pituitary 160,
166; progesterone 157; prolactin
155, 156; protein 158, 161, 166;
sexual activity 165;
spermatogenesis 156, 161; testes
156; testosterone 155, 159, 161;
Δ^9-THC 156; uterus 157, 161
Catecholamines (CA) 53; dopamine
53, 394; epinephrine 53; in Li
action 228-30; norepinephrine 53
Codeine 161, 165
Cortisol 286, 350; in anorexia
nervosa 286; cirrcadian variation
369; dexamethasone suppression
test 375; levels in saliva 367;
perturbation tests 366
Cushing's Disease 88; ACTH 88;

bromocriptine 88

Depression 21, 189, 310, 330, 349;
adrenocorticotrophic hormone
351; biogenic amines 192;
cardiovascular studies 332;
catecholamines 353; classification
330; cortisol 350, 375;
dexamethasone suppression test
352, 374; electrodermal activity
331; endocrine changes 374;
growth hormone 191, 353;
hypothalamo-pituitary-adrenal
axis 189; hypothalamo-pituitary-
gonadal axis 192; hypothalamo-
pituitary-thyroid axis 190;
indoleamine hypothesis of 300;
luteinizing hormone 356; prolactin
191, 354; skin conductance 331;
subtypes 357; thyroid stimulating
hormone 355; TRH 355
DHPG Formation 387
L-DOPA, effects on 229; motor
activity in rodents 229
Dopamine 53, 129, 137, 341;
apomorphine 210, 394; DA
antagonist drugs 69; direct DA
agonists 60; extrahypothalamic
structure 59; indirect DA agonists
60; mesocortical 54; mesolimbic
342, 394; metabolites 383, 341,
393, 395; neurohormonal
interaction 59; neuroleptics,
effects of 394; nigrostriatal 54,
342, 394; pathways, action of
Li 257; presynaptic receptors 258;
receptor blockade 394; receptor
supersensitivity 210; spontaneous
locomotor activity 210;
stereotyped behaviour 210; L-
triiodothyronine 210; tuberoinfundibular 342

Endorphins 11, 31, 123, 124, 132,
137, 164, 347; (D-ala^2)-analogs
145; behavioural effects 31;
deficiency 21; α-endorphin 12, 31,
146, 147; β-endorphin 12, 31, 37,
144-7; γ-endorphin 12, 31, 147;
(D-ala^2)-β-endorphin 144; (des-
try^1)-γ-endorphin 147, 148;
(leu^5)-β-endorphin 148; gonadal
function 164; intraventricular
injection 123; naloxone 164;
neuroleptic-like effects 37;

schizophrenia 347; sexual activity
164
Enkephalins 11; leu-enkephalin 123,
146; met-enkephalin 123, 132,
144, 146; (D-ala^2)-met-enkephalin
144; (D-ala^2)-met-enkephalin-
NH$_2$ 146; (D-ala^2,F$_5$phe^4)-met-
enkephalin-NH$_2$ 144, 145, 147;
(D-phe^4)-met-enkephalin 145,
146
Estrogens, 104, 111, 137, 161

Fertility 280; in anorexia nervosa 280
FSH 282; in anorexia nervosa 282;
effect of neuroleptics 178

Gonadal Functions 153, 161, 163;
effect of cannabinoids 153;
effect of narcotics 161
Gonadotropins 281; in anorexia
nervosa 281
Growth Hormone 77, 123, 284, 343,
353; in anorexia nervosa 284;
apomorphine 343; DA agonist
& DA antagonists drugs 77;
L-DOPA 344; effect of
neuroleptics 176-8; insulin
hypoglycemia 353; schizophrenia
343; tardive dyskinesia 344;
TRH 353

Hamilton Depression Rating Scale 312
Heroin 161, 162, 163, 165, 166
^{3}H-Imipramine: high afinity binding
in brain 299, 305; high affinity
binding in depressed patients
310; high affinity binding in
human platelets 309, 311;
inhibition of binding 306; regional
distribution of binding 306
Hyperprolactinemic State 69, 78;
DA agonist drugs, treatment 78;
DA antagonist drugs, diagnosis
69; indirectly acting DA agonist
69; PRL secreting adenoma 70;
PRL secretion 69
Hypothalamic Function 281; in
anorexia nervosa 281

Infertility 81; DA 81; ergot drugs 81;
pituitary hormones 81

Lactation 79; bromocriptine 79;
lisuride 80; metergoline 79; PRL
levels 79

LH 282, 356; in anorexia nervosa 282; effect of neuroleptics 178; and LHRH 356
LH-Releasing Hormone 282; in anorexia nervosa 282
Lithium 215, 225; antimanic action 215; behavioural effects in rodents 224-59; effect on 5-HT induced serotonin syndrome 252; effects on NE & DA metabolism 215; plasma concentrations 225; use in manic depressive disorders 225
Locomotor Activity: effect of amphetamine 228-9; effects of Li in rodents 227-8; in rats 239
Locus Coeruleus (LC) 387, 389; clonidine 387; phenoxybenzamine 387

Mania 226; animal models based on drug-induced hyperactivity 226, 228-9
Manic Depressive Disorders 259; dopamine and 259; model of Li's action in 259; serotonin and 259
Melatonin 288; in anorexia nervosa 288
Memory 33; effects of ACTH-like neuropeptides 32
Menstrual Function 280; in anorexia nervosa 280
MHPG 290; in anorexia nervosa 290; in plasma 389
Morphine 161-6; hypersensitivity to 14; induced behavioural arousal 24
Motivational Processes 32; effects of ACTH-like neuropeptides 32
MSH 6, 8, 31; behavioural effects of 31; developmental studies 9; inhibiting factor 5

Naloxone 11, 125, 145
Narcotics 153, 161, 162-6; in animals 163, 165; DNA 165; female 164; FSH 162, 163, 164; gonadal function 153, 164-6; in humans 161, 164, 165; hypothalamus 166; *in vitro* 164, 165, 166; Leydig cells 160; LH 162, 163, 164; naloxone 164; nucleic acid synthesis 166; pituitary 164, 166; prolactin 162;

protein 165, 166; RNA 160, 165; semen 164; sexual activity 163, 164, 165; testes 165; testosterone 162, 164
Neuroleptics, effects on: growth hormone secretion 176-8; LH and FSH secretion 178; posterior pituitary secretion 179; prolactin secretion 174, 176, 375; TSH and ACTH secretion 179
Noradrenaline Metabolites 384; brain phenolsulphotransferase 385; conjugation 385; 3, 4-didydroxy-phenylethylene glycol 384; free MHPG 387; glucuronide conjugates 386; MHPG 383, 384, 388; normetanephrine 385; probenecid-sensitive active efflux 385; sulphate and glucuronide conjugates 387; sulphate esters 386; total MHPG in rat, mouse & human 385; vanillylmandelic acid (VMA) 384

Opiates, effects: after i.c. administration 144; after i.p. administration 144, 145, 146; after i.v. administration 144, 145; i.m. administration 148; i.v. injection 144, 145, 147
Opioid Peptides 11; analgesic effect 23
Orienting Reflex (OR) 317, 321

Phospholipids in Brain 44; ACTH-like neuropeptides 44
Phosphorylation of Brain Proteins 43; effects of ACTH-like peptides 32, 43
Pituitary Hormones 6, 30, 154, 155; behavioural effects of 30; effect of neuroleptics 174-9; pituitary gland 5; response to perturbation tests 366
Polydipsia 233; due to Li treatment 233
Pregnancy 83; bromocriptine 85; ergot drugs 83
Primate Models of Depression 267; behavioural changes 268; desmethylimipramine 269; imipramine 270; α-methylpara-tyrosine 267; mother-infant separation 267; peer-peer separation 268; reserpine-induced 267

Prolactin 6, 123, 128, 137, 287, 344, 354; in anorexia nervosa 287; apomorphine 345; depression 354; effect of neuroleptics 174-6; effect of stress 366; neural regulation 344; neuroleptics 345; response to haloperidol 375; schizophrenia 344; secretion tests 71; serial blood sampling 367; TRH 354
Propranolol, effect on 252; serotonin syndrome 252-7
Protein Kinase in Brain 43; ACTH-sensitive 43
Pseudomale Behaviour 98, 102
Psychophysiological Assessment 332; limitations 332; patient medication 332; secondary aspects of disease 333; statistical analysis 334; subdiagnosis 332
Psychophysiological Methods in Psychiatry 318; attention, information processing 320; autonomic response stereotype 320; defensive response 321; electrodermal activity 318; heart rate 319; orienting response 321; phasic heart rate response 319; skin conductance (SC) 319; skin potential (SP) 318; stimulus-induced phasic response 319; stimulus/task specificity 320; tonic resting levels 319; unitary concept of arousal 320; vasomotor activity 319

Rearing Activity in Rats 239
Receptor Changes 302; following antidepressant treatment 302
REM Sleep 309
Reserpine-Like Drugs 230; interaction with Lithium 230

Schizophrenia 21, 323, 341; attention & arousal processes 326; blood pressure levels 326; cardiovascular activity 326; D- & DL-propranolol 325; defensive response 326; endogenous opiates 346; endorphins 346; enkephalins 346; fast habituators 324; FSH 346; growth hormone 343; heterogeneity 349; high risk research 325;

17-hydroxycorticosteroids 342; laterality differences 324; LH 346; limbic forebrain pathology 324; medication 325; neural regulation 343; overarousal 323; physiological types 323; prolactin 344; pupillometric measures 325; skin conductance 324; TRH 346; underarousal 323
Self-Stimulation 20
Serotonin (5-HT) 129, 230; behavioural syndrome 231; desensitization, effect of Li 252; increased rate of synthesis due to Li 231; intraneuronal deamination affected by Li 230; intraneuronal storage affected by Li 230
Serotonin Metabolites 396; CSF 5-HIAA 397; dorsal & median raphe nuclei 396; electrical stimulation 396; 5-HIAA 396; 5-hydroxyindolealdehyde 396; 5-hydroxytryptophol 396; in mode of antianxiety action 214; rat brain 396
Sex Steroids 111, 112; species variation 112
Sexual Behaviour 97-112; brain substrates 100; in females 97; hormonal control of 108; hormones 101; in males 98; sexual activity 157, 161, 165; species variation 112
Sleep Deprivation 309
Statistical Analysis: repeated measure data 370; statistical power analysis 372
Stress 350; cortisol 350, 366; effect on circulating hormone levels 367; prolactin 366
Supersensitivity: in serotonin systems 257

Tardive Dyskinesia 344, 348
Testosterone 155, 156, 157, 160, 162, 163, 164, 165, 289; in anorexia nervosa 289; FSH 164; Leydig cells 158, 161; LH 164; progesterone synthesis 160; sperm 160
Δ^9-Tetrahydrocannabinol 153, 155, 156, 157, 158, 159, 160; FSH 158; hypothalamus 158; *in vitro* effects 158; LH 158

Thyroid Hormone 207, 208; in
   anorexia nervosa 285;
   hyperthyroidism 208;
   hypothyroidism 208
Thyrotropin 6, 285, 355; in anorexia
   nervosa 285; TRH 355
Thyrotropin Releasing Hormone 217;
   antidepressant action 217; TRH &
   imipramine 218
L-Triiodothyronine 217, 285; in
   anorexia nervosa 285; depressed
   patients 217
Tryptophan 231, 248; behavioural
   effect in combination with MAO
   231; behavioural effect of short-
   term Li 248; Neuronal uptake
   increased by Li 231
Tryptophan Hydroxylase 232;
   decrease in activity due to Li 232

Urinary HVA & DOPAC 395; CNS
   contribution to peripheral HVA &
   DOPAC 396; stable isotope
   infusion 395; V-A difference
   technique 396
Urinary MHPG 388, 389, 390, 392;
   brain contribution to 390;
   carbidopa 390; debrisoquine 390;
   exercise 392; peripheral
   decarboxylase inhibitor 390;
   peripheral monoamine oxidase
   inhibitor 390; peripheral
   sympathetic activity 389; response
   to antidepressant medications
   392; schizophrenic patients 391;
   variability in excretion 392